Chromosomes Today

VOLUME 12

Chromosomes Today

VOLUME 12

Edited by

N. Henriques-Gil
Department of Genetics
University of San Pablo
Madrid, Spain

J.S. Parker
Professor of Plant Cytogenetics
The University of Cambridge, UK

and

M.J. Puertas
Department of Genetics
University of Complutense
Madrid, Spain

CHAPMAN & HALL
London · Weinheim · New York · Tokyo · Melbourne · Madras

Published by Chapman & Hall, 2–6 Boundary Row, London SE1 8HN, UK

Chapman & Hall, 2–6 Boundary Row, London SE1 8HN, UK

Chapman & Hall GmbH, Pappelallee 3, 69469 Weinheim, Germany

Chapman & Hall USA, 115 Fifth Avenue, New York, NY 10003, USA

Chapman & Hall Japan, ITP-Japan, Kyowa Building, 3F, 2-2-1 Hirakawacho, Chiyoda-ku, Tokyo 102, Japan

Chapman & Hall Australia, 102 Dodds Street, South Melbourne, Victoria 3205, Australia

Chapman & Hall India, R. Seshadri, 32 Second Main Road, CIT East, Madras 600 035, India

First edition 1997

Printed in Great Britain by TJ Press, Padstow, Cornwall

ISBN 0 412 75240 9

A Catalogue record for this book is available from the British Library

Library of Congress Catalog Card Number: 96-71013

Printed on permanent acid-free text paper, manufactured in accordance with ANSI/NISO Z39.48-1992 (Permanence of Paper).

Contents

Colour plates appear between pages 32–3, 64–5, 128–9, 192–3, 224–5 and 352–3

List of contributors ix

Preface xv

PROLOGUE

1 **Cytogenetics: yesterday, today and forever. A conceptual and historical view** 3
J.R.Lacadena

PART ONE: SEQUENCE ORGANIZATION IN CHROMOSOMES

2 **Chromosome differentiation using nucleases: an overview** 23
J. Gosálvez, C. López-Fernández, V. Goyanes and R. Mezzanotte

3 **Y-linked genes and spermatogenesis in primates** 50
W. Schempp

4 **Subchromosomal painting libraries (SCPLs) from somatic cell hybrids** 56
M. Rocchi, R. Antonacci, R. Marzella, P. Finelli, C. Cassano, A. Lonoce, C. Cino, A. Forabosco and N. Archidiacono

5 **The relationship between gene density and chromosome banding patterns in mammalian nuclei** 65
J.M. Craig and W.A. Bickmore

PART TWO: MAINTENANCE AND BREAKDOWN OF STABILITY

6 **Genes controlling chromosome structure in *Drosophila melanogaster*** 87
F. Verni', M.P. Somma, R. Gandhi, M.L. Goldberg and M. Gatti

7 **Retrotransposons at *Drosophila* telomeres and terminal chromosome deficiencies** 104
H. Biessmann, M.F. Walter, E. Kurenova and J.M. Mason

8 **Interspecific micronucleus model for the study of induced chromosome aberrations in human male germ cells** 122
J. Egozcue, L. Tusell, R. Álvarez, A. Genescà, M.R. Caballín, L. Barrios and R. Miró

9 **Characterization of amplified DNA sequences in human cancers** 132
M. Muleris, A. Almeida, M. Gerbault-Seureau, A.M. Dutrillaux, B. Malfoy and B. Dutrillaux

PART THREE: CONTROL MECHANISMS

10 **Transcription factories and chromosome structure** 147
A. Pombo, J. McManus, T. A. Hughes, F.J. Iborra, D.A. Jackson and P.R. Cook

11 **The replication of ribosomal RNA genes in eukaryotes** 161
C. López-Estraño, J.B. Schvartzman and P. Hernández

12 **Genomic interactions: Gene expression, DNA methylation and nuclear architecture** 182
N. Neves, A. Castilho, M. Silva, J.S. Heslop-Harrison and W. Viegas

13 **Gametic imprinting at the mouse and human IGF2R/MPR300 locus** 202
O.W. Smrzka and D.P. Barlow.

PART FOUR: MEIOTIC MOBILITY AND RECOMBINATION

14 Chromosome behaviour in earliest meiotic prophase 217
H. Scherthan

15 Univalent orientation in living meiocytes 249
E. Rebollo and P. Arana

16 Proteins Controlling Sister-Chromatid Cohesion 270
T.L. Orr-Weaver, D.P. Moore, A.W. Kerrebrock, S.E. Bickel and D.W. Wyman

17 Expanding genetic maps: reevaluation of the relationship between chiasmata and crossovers 283
B.S. Gill, K.S. Gill, B. Friebe and T.R. Endo

PART FIVE: EVOLUTIONARY DYNAMICS

18 Evolution of a near–neutral B chromosome 301
J.P.M. Camacho, J. Cabrero, M.D. López–León, and M.W. Shaw

19 Chromosome painting in wheat 320
J.M. Vega, S. Abbo, M. Feldman and A.A. Levy

20 New insights into chromosome evolution in plants from molecular cytogenetics 333
I.J. Leitch, A.S. Parokonny and M.D. Bennett

21 A saltation model of karyotypic evolution in the Octodontoidea (Mammalia, Rodentia) 347
M.H. Gallardo

EPILOGUE

22 The who is who of chromosomes 369
P.B. Moens

Index 375

List of contributors

Abbo, S.
Department of Plant Genetics, Weizmann Institute of Science, Rehovot, ISRAEL.

Almeida, A.
Laboratoire de Cytogénétique Moléculaire et Oncologie, UMR 147 CNRS, Institut Curie, Paris, FRANCE.

Álvarez, R.
Unitat de Biologia, Departament de Biologia Cellular i Fisiologia, Facultat de Medicina. Universitat Autònoma de Barcelona. 08193 Bellaterra, SPAIN.

Antonacci, R.
Institute of Genetics, University of Bari, Bari, ITALY.

Arana, P.
Departamento de Genética, Facultad de Biología, Universidad Complutense. E-28040 Madrid, SPAIN.

Archidiacono, N.
Institute of Genetics, University of Bari, Bari, ITALY.

Barlow, D.P.
Institute of Molecular Pathology (I.M.P.), Dr. Bohr-Gasse 7, A-1030 Vienna, AUSTRIA. Telephone: +431 797300; Fax: +431 7987153.

Barrios, L.
Unitat de Biologia Cellular, Departament de Biologia Cellular i Fisiologia, Facultat de Ciències. Universitat Autònoma de Barcelona. 08193 Bellaterra, SPAIN.

Bennett, M.D.
Jodrell Laboratory, Royal Botanic Gardens, Kew, Richmond, Surrey TW9 3DS, UK.

Bickel, S.E.
Whitehead Institute for Biomedical Research and the Dept. of Biology, Massachusetts Institute of Technology, Cambridge, MA 02142, USA.

Bickmore, W.A.
Medical Research Council, Human Genetics Unit, Western General Hospital, Crewe Road, Edinburgh, EM4 2XU. U.K.

Biessmann, H.
Developmental Biology Center, University of California, Irvine, CA 92717. USA.

Caballín, M.R.
Unitat d'Antropologia, Departament de Biologia Animal, Biologia Vegetal i Ecologia, Facultat de Ciències. Universitat Autònoma de Barcelona. 08193 Bellaterra, SPAIN.

Cabrero, J.
Departamento de Genética, Facultad de Ciencias, Universidad de Granada, E-18071 Granada, SPAIN.

Camacho, J.P.M.
Departamento de Genética, Facultad de Ciencias, Universidad de Granada, E-18071 Granada, SPAIN.

Cassano, C.
Institute of Genetics, University of Bari, Bari, ITALY.

Castilho, A.
Departamento de Botânica e Eng. Biológica, Instituto Superior de Agronomía. 1399 Lisboa. PORTUGAL. Phone +3511 3638161. FAX +3511 3635031.

Cino, C.
Istituto Anatomia Umana Normale, Modena, ITALY

Cook, P.R.
CRC Nuclear Structure and Function Research Group, Sir William Dunn School of Pathology, University of Oxford, South Parks Road, Oxford, OX1 3RE, U.K.

Craig, J.M.
Medical Research Council, Human Genetics Unit, Western General Hospital, Crewe Road, Edinburgh EH4 2XU, U.K.

Dutrillaux, A.M.
Laboratoire de Cytogénétique Moléculaire et Oncologie, UMR 147 CNRS, Institut Curie, Paris, FRANCE.

Dutrillaux, B.
Laboratoire de Cytogénétique Moléculaire et Oncologie, UMR 147 CNRS, Institut Curie, Paris, FRANCE.

Egozcue, J.
Unitat de Biologia Cellular, Departament de Biologia Cellular i Fisiologia, Facultat de Ciències. Universitat Autònoma de Barcelona, 08193 Bellaterra, SPAIN.

Endo, T.R.
Laboratory of Genetics, Kyoto University, 631 Kyoto, JAPAN.

Feldman, M.
Department of Plant Genetics, Weizmann Institute of Science, Rehovot, ISRAEL.

Finelli, P.
Institute of Genetics, University of Bari, Bari, ITALY.

Forabosco, A.
Institute of Genetics, University of Bari, Bari, ITALY.

Friebe, B.
Wheat Genetics Resource Center, Department of Plant Pathology, Kansas State University, Manhattan, KS 66506-5502, USA.

Gallardo, M.H.
Instituto de Ecología y Evolución, Universidad Austral de Chile, Casilla 567, Valdivia, CHILE.

Gandhi, R.
Section of Genetics and Development, Cornell University, Ithaca, NY 14853. USA.

Gatti, M.
Dipartimento di Genetica e Biologia Molecolare, Universita' di Roma 'La Sapienza', P. A. Moro 5, 00185 Roma. ITALY.

Genescà, A.
Unitat de Biologia, Departament de Biologia Cellular i Fisiologia, Facultat de Veterinària. Universitat Autònoma de Barcelona. 08193 Bellaterra, SPAIN.

Gerbault-Seureau, M.
Laboratoire de Cytogénétique Moléculaire et Oncologie, UMR 147 CNRS, Institut Curie, Paris, FRANCE.

Gill, B.S.
Wheat Genetics Resource Center, Department of Plant Pathology, Kansas State University, Manhattan, KS 66506-5502, USA.

Gill, K.S.
Wheat Genetics Resource Center, Department of Plant Pathology, Kansas State University, Manhattan, KS 66506-5502, USA.

Goldberg, M.L.
Section of Genetics and Development, Cornell University, Ithaca, NY 14853. USA.

Gosálvez, J.
Departamento de Biología, Unidad de Genética. A-201. Universidad Autónoma de Madrid. 28049 Madrid. SPAIN. TEL: 341-3978210; FAX: 341-3978344; E-mail: gosalvez@vm1.sdi.uam.es

Goyanes, V.
Departamento de Genética, Instituto Universitario de Ciencias de la Salud. Hospital Juan Canalejo. La Coruña. SPAIN.

Hernández, P.
Departamento de Biología Celular y del Desarrollo, Centro de Investigaciones Biológicas (CSIC). Velázquez 144, 28006 Madrid, SPAIN. Tel.: +341-5611800 Ext.4232, Fax: +341-5648749, E-mail address: cibph67@cc.csic.es

Heslop-Harrison, J.S.
Karyobiology Group, John Innes Institute, Colney Lane, Norwich NR4 7UH, U.K. Phone +44/01603 452571. FAX +44/01603 456844. Email: hharrison@bbsrc.ac.uk

Hughes, T.A.
CRC Nuclear Structure and Function Research Group, Sir William Dunn School of Pathology, University of Oxford, South Parks Road, Oxford, OX1 3RE, UK.

Iborra, F.J.
CRC Nuclear Structure and Function Research Group, Sir William Dunn School of Pathology, University of Oxford, South Parks Road, Oxford, OX1 3RE, U.K.

Jackson, D.A.
CRC Nuclear Structure and Function Research Group, Sir William Dunn School of Pathology, University of Oxford, South Parks Road, Oxford, OX1 3RE, U.K.

Kerrebrock, A.W.
Whitehead Institute for Biomedical Research and the Dept. of Biology, Massachusetts Institute of Technology, Cambridge, MA 02142, USA.

Kurenova, E.
Laboratory of Molecular Genetics. National Institute of Environmental Health Sciences. Research Triangle Park, NC 27709. USA.

Lacadena, J.R.
Departamento de Genética, Facultad de Biología. Universidad Complutense, Madrid, SPAIN.

Leitch, I.J.
Jodrell Laboratory, Royal Botanic Gardens, Kew, Richmond, Surrey TW9 3DS, U.K.

Levy, A.A.
Department of Plant Genetics, Weizmann Institute of Science, Rehovot, ISRAEL.

Lonoce, A.
Institute of Genetics, University of Bari, Bari, ITALY.

López-Estraño, C.
Departamento de Biología Celular y del Desarrollo, Centro de Investigaciones Biológicas (CSIC). Madrid, SPAIN.

López-Fernández, C.
Departamento de Biología, Unidad de Genética. A-201. Universidad Autónoma de Madrid. 28049 Madrid. SPAIN. TEL: 341-3978210; FAX: 341-3978344.

López-León, M.D.
Departamento de Genética, Facultad de Ciencias, Universidad de Granada, E-18071 Granada, SPAIN.

Malfoy, B.
Laboratoire de Cytogénétique Moléculaire et Oncologie, UMR 147 CNRS, Institut Curie, Paris, FRANCE.

Marzella, R.
Institute of Genetics, University of Bari, Bari, ITALY.

Mason, J.M.
Laboratory of Molecular Genetics. National Institute of Environmental Health Sciences. Research Triangle Park, NC 27709. USA.

McManus, J.
CRC Nuclear Structure and Function Research Group, Sir William Dunn School of Pathology, University of Oxford, South Parks Road, Oxford, OX1 3RE, U.K.

Mezzanotte, R.
Istituto di Biologia Generale, Facoltá di Medicina e Chirurgia. Universitá di Cagliari. Cagliari. ITALY.

Miró, R.
Unitat de Biologia, Departament de Biologia Cellular i Fisiologia, Facultat de Medicina. Universitat Autònoma de Barcelona. 08193 Bellaterra, SPAIN.

Moens, P.
Department of Biology, York University, 4700 Keele Street. Downsview Ontario. M3J 1P3, CANADA. Phone: 416 736-5358; fax: 416 736-5731; email: FS300025@sol.yorku.ca.

Moore, D.P.
Whitehead Institute for Biomedical Research and the Dept. of Biology, Massachusetts Institute of Technology, Cambridge, MA 02142, USA.

Muleris, M.
Laboratoire de Cytogénétique Moléculaire et Oncologie, UMR 147 CNRS, Institut Curie, Paris, FRANCE.

Neves, N.
Departamento de Botânica e Eng. Biológica, Instituto Superior de Agronomía. 1399 Lisboa. PORTUGAL. Phone +3511 3638161. FAX +3511 3635031.

Orr-Weaver, T.L.
Whitehead Institute for Biomedical Research and the Dept. of Biology, Massachusetts Institute of Technology, Cambridge, MA 02142, USA.

Parokonny, A.S.
Jodrell Laboratory, Royal Botanic Gardens, Kew, Richmond, Surrey TW9 3DS, U.K.

Pombo, A.
CRC Nuclear Structure and Function Research Group, Sir William Dunn School of Pathology, University of Oxford, South Parks Road, Oxford, OX1 3RE, UK.

Rebollo, E.
Departamento de Genética, Facultad de Biología, Universidad Complutense. E-28040 Madrid, SPAIN.

Rocchi, M.
Institute of Genetics, University of Bari, Bari, ITALY.

Schempp, W.
Institute of Human Genetics and Anthropology. University of Freiburg. 79106 Freiburg, GERMANY.

Scherthan, H.
Abt. Humanbiologie und Humangenetik der Universität, Postf.: 3049, D-67653 Kaiserslautern, GERMANY.

Schvartzman, J.B.
Departamento de Biología Celular y del Desarrollo, Centro de Investigaciones Biológicas (CSIC). Madrid, SPAIN.

Shaw, M.W.
Department of Agricultural Botany, School of Plant Sciences, University of Reading, Whiteknights, Reading RG6 6AU, UK.

Silva, M.
Departamento de Botânica e Eng. Biológica, Instituto Superior de Agronomía. 1399 Lisboa. PORTUGAL.

Smrzka, O.W.
Institute of Molecular Pathology (I.M.P.), Dr. Bohr-Gasse 7, A-1030 Vienna, AUSTRIA. Telephone: +431 797300; Fax: + 431 7987153.

Somma, M.P.
Dipartimento di Genetica e Biologia Molecolare, Universita' di Roma 'La Sapienza', P. A. Moro 5, 00185 Roma. ITALY.

Tusell, L.
Unitat de Biologia, Departament de Biologia Cellular i Fisiologia, Facultat de Medicina. Universitat Autònoma de Barcelona. 08193 Bellaterra, SPAIN.

Vega, J.M.
Department of Plant Genetics, Weizmann Institute of Science, Rehovot, ISRAEL.

Verni', F.
Dipartimento di Genetica e Biologia Molecolare, Universita' di Roma 'La Sapienza', P. A. Moro 5, 00185 Roma. ITALY.

Viegas, W.
Departamento de Botânica e Eng. Biológica, Instituto Superior de Agronomía. 1399 Lisboa. PORTUGAL. Phone +3511 3638161. FAX +3511 3635031.

Walter, M.F.
Developmental Biology Center, University of California, Irvine, CA 92717. USA.

Wyman, D.W.
Whitehead Institute for Biomedical Research and the Dept. of Biology, Massachusetts Institute of Technology, Cambridge, MA 02142, USA.

Preface

The *12th International Chromosome Conference* (11—15 September, 1995) was held in San Lorenzo de El Escorial to the north west of Madrid on the slopes of the Sierra de Guadarrama. Both the natural beauty of the mountains and the austere granite blocks of the Monastery buildings combined to give an intense sense of community. From this emerged a strong feeling of collaboration and friendship amongst the Conference participants.

The present volume of the series *Chromosomes Today* contains all the invited lectures given at the *12th ICC*. The programme, the poster presentations, and the poster abstracts, as well as the full list of organisers, the Advisory Committee, and the name and addresses of all contributors to the Conference were published in *Chromosome Research* (Vol. 3, suppl. 1, September, 1995).

In the preceding volume of the series molecular cytogenetics was regarded as an 'ill-defined discipline'. Three years later we have witnessed the complete incorporation of molecular biology into cytogenetics. Initially, the only relationship between cytogenetics and molecular biology concerned the DNA sequence organisation of chromosomes but now we see molecular techniques applied to such diverse cytogenetic problems as the chromosome stability, meiotic mobility and evolution. Volume 12 of *Chromosomes Today* illustrates this integration with sections on Sequence Organisation, Maintenance and Breakdown of Stability, Control Mechanisms, Meiotic Mobility and Recombination, and Evolutionary Dynamics. Indeed, most contributions to this volume take advantage of molecular techniques, although some others reveal cytogenetic novelty using more traditional methods. We are sure that C.D. Darlington would have appreciated the addition of colour to the classical beauty of our chromosomes.

The peripatetic Conference now moves on to Italy at the University of Ancona under the guidance of Ettore Olmo. The enormous advances in cytogenetics over the years makes it difficult for us to predict the state of our science even three years ahead and we look forward to the next *Chromosomes Today* with excitement. The thought of C.D. Darlington written in the preface of *Chromosomes Today*, vol. 2 (1967) 'our past progress in unravelling these results does not seem disappointing; but it has left enough loose ends to occupy the attention of this and of many more Chromosome Conferences' is equally encouraging and true today.

The editors wish to express their gratitude to the International Advisory Committee, the Local Organising Committee and the volunteer helpers for their splendid support before and during the 12th ICC.

PROLOGUE

Juan-Ramón Lacadena

1

Cytogenetics: yesterday, today and forever. A conceptual and historical view

J.R. Lacadena

1.1 FROM THE CONCEPT OF GENETICS TO THE CONCEPT OF CYTOGENETICS

The birth of Genetics, a new science able to explain biological hereditary phenomena, should have taken place when it was able to answer the following two fundamental questions: firstly, what are the laws by which biological characters are transmitted from parents to offspring? and secondly, what is the physical basis by which such characteristics are conserved and transmitted? or, in other words, what is the molecular basis of inheritance? The answer to the first question was given by Mendel in 1865. The answer to the second question was given in 1944 when Avery, McLeod and McCarty identified the deoxyribonucleic acid (DNA) as the hereditary material; that is to say, Mendel's abstract 'hereditary factors' — named 'genes' by Johannsen (1909) — are DNA. Consequently, one should accept that Genetics was not born in 1900 when de Vries, Correns and Tschermak rediscovered Mendel's laws, but that the birth of Genetics lasted for eighty years: it started in 1865(66) when Mendel published his experimental results and finished in 1944 when Avery, McLeod and McCarty identified Griffith's transforming principle.

After Bateson defined Genetics in 1906 as the 'science that studies inheritance and variation in living beings', its growth and diversification was very fast and wide-ranging; perhaps not equated by any other science. The specialization of Genetics at the levels of organism (Viral

Genetics, Bacterial Genetics,... Human Genetics), organization (Molecular Genetics, Cytogenetics,... Population Genetics) or process (Developmental Genetics, Evolutionary Genetics) led occasionally to a real difficulty of understanding among scientists from different fields. However, there is something that maintains a unitary and universal concept of the basic principles of Genetics as a science. It is the hereditary material, the common denominator to all the above mentioned diversification (Rubio, 1973). It led me to propose a new concept of Genetics as the 'science that studies hereditary material at any level (molecular, cellular, individual, populational) or dimension (analytic—structural, dynamic and space—temporal)' (Lacadena, 1981).

The chromosome can be defined as

> 'the organized hereditary material which acquires an increasing complexity in evolution, passing from simple naked molecules of nucleic acids in prokaryotes to complex interactions with histone and non histone proteins as main chemical components in eukaryotes. Chromosomes have a triple function to accomplish: to conserve, to transmit and to express the genetic information that they carry' (Lacadena, 1981).

According to the outlined concepts, and making an analogy with Genetics, one could define Cytogenetics as 'the science that studies the organized hereditary material (the chromosome) at any level or dimension'. This definition justifies fully the content of Cytogenetics.

Historically, Cytogenetics arose as a hybrid science: it is the F_1 of Cytology and Genetics, inheriting from the former its qualitative, physical and descriptive aspects, while its quantitative and physiological approaches come from the latter (Swanson, 1960). The subsequent development of Cytogenetics looks like an F_2 transgressive segregation since its field is between Cell Biology and Genetics, but their frontiers are not always well defined and very often researchers and teachers of Cytogenetics can go beyond the frontiers and find themselves doing or teaching Cell Biology or Genetics, but not Cytogenetics in the strict sense.

1.2 FROM THE CHROMOSOME THEORY OF HEREDITY TO THE MODERN CYTOGENETICS

When in 1900 de Vries, Correns and Tschermak rediscovered Mendel's work, the cytological knowledge of chromosome behaviour in mitosis (Flemming, 1879, 1880) and meiosis (Boveri, 1892) were sufficiently advanced to understand that chromosomes could be the vehicles of the abstract Mendelian hereditary factors.

Although *a posteriori* one can assume that the link between the cytological facts and the Mendelian laws were immediately established, the story was not so simple. Of a great interest is the critical and historical study made by Baxter and Farley (1979). As these authors point out, the assumption that everything had fallen into place by 1900, and that the Mendelian results immediately clarified the cytological findings, springs in part from a rather naive misconception of the nature of scientific enterprise.

In fact, Correns (1900) explained Mendel's results by assuming that before the definitive formation of the sexual nuclei in the hybrids a separation of both *anlage* always occurs in such a manner that half the sexual nuclei contain the *anlage* for the recessive character and half the *anlage* for the dominant. The separation takes place at the earliest in the formation of sperm and pollen *anlage*.

The 1:1 ratio very much supports nuclear division taking place by Weismann's reduction division. As Baxter and Farley (*l.c.*) indicate, this statement gave credence to the widely held belief that by 1900 the processes of maturation division, or meiosis, had been fully worked out. However, the story was not so simple; thus one can remember for instance, that great cytologists of that era as Strassburger denied the existence of the reductional division.

As a matter of fact, in the nineteenth century, cytologists were unable to agree on what they saw under the microscope because they lacked a common theory within which their observations could be interpreted. That is to say, the chromosome behaviour could not simply be observed but it had to be interpreted according to scientist's assumptions about the nature of the hereditary process and the hereditary material. In this sense, Baxter and Farley conclude that 'it was only after the rediscovery of Mendel's laws that cytologists came to share a set of common assumptions which led them to agree on what they saw under the microscope and eventually to accept the link between chromosomes and Mendelian factors'.

After many vicissitudes, it was Sutton who in 1902 and 1903 showed the significance of the reductional division and pointed out the parallelism between chromosome behaviour and the Mendelian segregation. Thus, Sutton stated in 1902: 'I may finally call attention to the probability that the association of paternal and maternal chromosomes in pairs and their subsequent separation during the reducing division as indicated above may constitute the physical basis of the Mendelian laws of heredity'. His own investigations together with those of Boveri (1902) constitute the cytological basis of the genetic theory, giving rise to the Sutton—Boveri hypothesis or Chromosome Theory of Heredity.

From this moment, many experimental data confirming the theory were obtained mainly by Morgan and coworkers with *Drosophila melanogaster*. Thus, Morgan (1910) inferred the relation between a gene (the mutation *white*) and a definite chromosome (the X chromosome) from the genetic analysis of the eye colour character of the fly. Later, Bridges (1916) used the nondisjunction phenomenon as a proof of the chromosome theory of heredity by relating the vermilion mutation to X chromosome behaviour. On the other hand, the linkage maps obtained by Sturtevant (1913) demonstrated that genes are linearly ordered on the chromosomes. New evidence was obtained using the polytene chromosomes of the salivary glands of *Drosophila* (Painter, 1933). Finally, the relation between interchange of homologous chromosomes and genetic recombination was simultaneously demonstrated in *Drosophila* (Stern, 1931) and in maize (Creighton and McClintock, 1931).

To sum up, the chromosome theory of heredity was defined by the following three points:

1. Genes are located on chromosomes;
2. Genes are linearly ordered; and
3. To the genetic recombination phenomenon corresponds a cytological phenomenon of interchange of homologous chromosome segments produced by crossing-over (Morgan and Cattell, 1912).

Two periods can be considered in the historical development of Cytogenetics. In the first period (Classical or Conventional Cytogenetics), which extends from 1902 (Sutton—Bovery hypothesis) to the middle sixties, cytogenetic investigations were mainly descriptive and based on light microscopy and from 1956 also on electron microscopy (Ris, 1956; Moses, 1956; Fawcett, 1956). Summarizing this period, one should note the works of synthesis made, among others, by E.B. Wilson (*The Cell in Development and Heredity*, 3rd ed., 1925), C.D. Darlington (*Recent Advances in Cytology*, 1932, revised in 1965), C.P. Swanson (*Cytology and Cytogenetics*, 1960), M.J.D. White (*The Chromosomes*, 6th ed., 1973; *Animal Cytology and Evolution*, 3rd. ed., 1973), G.L. Stebbins (*Variation and Evolution in Plants*, 5th ed., 1963), etc. Their books have been guides for generations of cytogeneticists. Among these authors, I would like to do justice to Cyril Dean Darlington as the father of Cytogenetics, not only for his work of synthesis but also his own original investigation. In relation to criticisms which arose some time ago against Darlington's work, it seems to me appropriate to repeat here what Macgregor said on making reference to a certain problem concerning the lampbrush chromosomes: 'a hypothesis has not to be necessarily correct to be useful'. In fact, although some Darlingtonian theories were not right, they served as a stimulus for others to ratify or reject and, in this way, Cytogenetics progressed.

As a discipline included in Genetics, Cytogenetics changed as a consequence of the introduction of new techniques and methodologies, giving rise to a new period (Molecular Cytogenetics) which began in 1957 when Taylor and coworkers used tritiated thymidine to demonstrate the semiconservative model of chromatid reproduction. The new Cytogenetics expanded very fast with the application of the *in situ* hybridization techniques of nucleic acids (Ritossa and Spiegelman, 1965; Gall and Pardue, 1969; Pardue and Gall, 1969), the visualization of gene expression by the spreading technique in electron microscopy (Miller and Beatty, 1969), the chromosome banding techniques (Caspersson *et al.*, 1968; Sumner *et al.*, 1971), the biochemical analysis of the cell cycle (see Prescott, 1976) and of meiosis (Hotta *et al.*, 1966; Stern and Hotta, 1973), the analysis of chromosome structure both in the nucleosomal organization of chromatin (Olins and Olins, 1974; Kornberg and Thomas, 1974; Kornberg, 1974) and the structural role of non histone proteins (Paulson and Laemmli, 1977; Adolph *et al.*, 1977*a,b;* Laemmli *et al.*, 1977). The progress achieved in the first part of the second period of Cytogenetics was synthesized in books by Lima-de-Faria (1969, 1983), Bostock and Sumner (1978), Lewin (1980), Swanson *et al.* (1981), among others.

The second part of modern Cytogenetics begins in the eighties with the molecular and functional analysis of chromosome structures such as centromeres (Clarke and Carbon, 1980, 1985) and telomeres (Blackburn and Szostak, 1984; Greider and Blackburn, 1985; Blackburn, 1991, 1992) as well as the genetic analysis of the cell cycle (Nurse and Bisset, 1981; reviews by Nurse, 1990, 1994). Another growing field of research is the molecular analysis of chromosome movements during mitosis (see Hyams and Brinkley, 1989).

Of great importance was the introduction of the fluorescence *in situ* hybridization technique (FISH) when a non isotopic label substituted for an isotopic one. Probes were synthesized using biotin (Langer *et al.*, 1981), digoxigenin (Heiles *et al.*, 1988), and so on (see reviews by Lichter and Ward, 1990; Lichter and Cremer, 1991). When probes are obtained with total DNA from a chromosome isolated by flux cytometry (Lichter *et al.*, 1988) a chromosome painting technique is obtained (Pinkel *et al.*, 1988). Chromosome painting can be extended in interspecific hybrids of plant species and derivatives to the complete set of chromosomes belonging to one of the parental species, giving rise to the genome *in situ* hybridization technique (GISH) (Heslop-Harrison *et al.* 1988; Schwarzacher *et al.*, 1989).

The application of new techniques permits us to approach again old cytogenetic problems not yet solved. In fact, the methodology of modern Cytogenetics is giving very good results when the biochemical and physiological approaches of cell biologists are complemented with a genetic experimental design. The research on the cell cycle is a paradigmatic example (Nurse, 1990, 1994).

This is the reason why I will not make an argument in defence of Cytogenetics in a molecular wonderland (John, 1983) against such radical opinions as those expressed by the Nobel laureate F.H.C. Crick (1977) who, in the epilogue of the 6th International Chromosome Conference held in Helsinki, put in doubt cytogenetic investigations made with metaphase chromosomes

> 'the dullest form of chromosome: an inert package needed to make orderly mitosis possible. The meiotic chromosome, in its less active phases is not much better... It is not enough, in order to understand the Book of Nature, to turn over the pages looking at the pictures and not reading the text. Painful though it may be, it will also be necessary to learn to read the text. Only with the assistance of molecular biology will this be possible'.

As I wrote on a previous occasion (Lacadena, 1984), this is a suggestive simile: in order to explain the Book of Nature, molecular geneticists write the text while cytogeneticists put in the pictures. Nevertheless, everybody will agree that pictures and drawings often help the understanding of content. In summary, in my opinion, there is no denying the convenience of applying molecular techniques in cytogenetical research but avoiding a radical reductionism that gave us a distorted view of chromosome structure and function. I agree with Ferguson-Smith (1993) who stated that 'molecular biologists who ignore Cytogenetics do so at their peril as, without it, they are likely to have an incomplete understanding of the fundamentals of Genetics'.

Thus, the methodology of modern Cytogenetics allows us to know the reason why concerning many cytogenetical phenomena described during the first or classic period. Making an analogy to Crick's simile, one could say that modern Cytogenetics is projecting the talkie film corresponding to an old silent film. The molecular analysis of chromosome behaviour in mitosis, the most basic process of chromosome life (see reviews by Hyams and Brinkley, 1989; Koshland, 1994; Fuller, 1995), seems to be a paradigmatic example. The ideal situation would be that all the cytogenetical phenomena described by classical Cytogenetics could be analyzed and explained in molecular and genetical terms.

1.3 CYTOGENETICS AS A MODEL EXPERIMENTAL SCIENCE

In a conceptual context, it is interesting to relate some of the facts which happened in Cytogenetics life which can be taken as examples of what should not occur in an experimental science, namely 1) the axioms, 2) to consider as obsolete old experimental data because of the advent of sophisticated modern technology, and 3) to consider a scientific assertion as immutable.

1. The phenomenon of chiasma terminalization can be taken as an example of axiom. Chiasma terminalization was defined in 1929 by Darlington as a progressive shift between diplotene and metaphase of the first meiotic division in the distribution of chiasmata along the arms from their points of origin to more distal positions. Its occurrence was axiomatically accepted for many years with no new

experimental evidence. Almost fifty years later, experimental evidence was accumulated firstly questioning (Fox, 1973; Hultén, 1974) and after demonstrating that chiasma terminalization (in the Darlingtonian sense) does not occur (Jones, 1977; Tease, 1978; Tease and Jones, 1978; Kanda and Kato, 1980; Santos *et al.*, 1987, 1989). Consequently, it would be more correct to call the phenomenon chiasma resolution.

2. Modern scientists can commit an error when they reject *a priori* experimental data obtained in the classical period of Cytogenetics. A good example is the one band—one gene hypothesis proposed in the thirties (Painter, 1934; Bridges, 1935) to explain the genetic significance of the bands of polytene chromosomes of *Drosophila melanogaster.* The hypothesis was rejected by molecular geneticists (how could it be possible to see the genes with the light microscope!). However, forty years later, Judd and coworkers (1972) ratified the theory by means of a fine cytogenetic experimental design. The theory was reformulated as one band—one genetic function.
3. In Cytogenetics, as in any other experimental science, conclusions cannot be considered as immutable: it is the Popperian falsifiability of the hypothesis.

   Although many examples can be found throughout the history of Cytogenetics, one of the most representative is the investigation of the genetic determination of sexual differentiation. Thus, it will be remembered how initially the H-Y antigen was considered responsible for testes differentiation (see reviews by Ohno, 1979; Wachtel, 1983), to be followed by the testicular determination factors TDF and Tdy in humans and mice respectively, to conclude after spectacular molecular cytogenetic research with the identification, isolation and characterization of the Y chromosome region responsible for sex determination, SRY and Sry in humans and mice, respectively (Sinclair *et al.*, 1990; Koopman *et al.*, 1991; see review by Goodfellow and Lovell-Badge, 1993). Nevertheless, as Wiberg (1989) pointed out, whatever the function of the H-Y antigen may be, the falsifiability of the hypothesis and not its corroboration is the force which drives science.

1.4 THE CONTENT OF CYTOGENETICS

The accomplishment of the structure—function binomial is a constant in Biology. After defining Cytogenetics as the science that studies chromosomes at any level and dimension, the aim of Cytogenetics is to analyze how the structural organization and behaviour of chromosomes guarantee the conservation, transmission and expression of the genetic information that they carry.

From an academic (teaching and research) point of view, Cytogenetics must explain how chromosomes conserve the genetic information from cell to cell in a multicellular organism, how they transmit it from parents to offspring and how they release it orderly in fashion to control cellular functions and the development of the organism. Likewise, it must analyze the genetic control of the structure and behaviour of chromosomes, their variation and genetic and evolutionary consequences as well as the applications to Plant and Animal Breeding and to Medicine.

Many good books on 'Cytogenetics' or 'Chromosomes' have been published; however, I ought to say that they are not complete because usually lack some important topics or these are not treated in depth. In my opinion, the contents of a textbook on general Cytogenetics should be (Lacadena, 1996):

1. Introduction: Concept and history of Cytogenetics
2. Structure of the eukaryotic chromosome
3. Chromosomes as carriers of genetic information: Conservation, transmission and expression
   (a) Mitosis: Cell division and variations of the cell division cycle
   (b) Meiosis: The Chromosome Theory of Heredity and fundamental processes
   (c): Chromosomes and genetic function
      - Polytene chromosomes
      - Lampbrush chromosomes
      - Sex chromosomes
      - Accessory chromosomes
4. Causes that modify chromosome behaviour

5. Structural chromosome variation
   (a) Deletions
   (b) Duplications
   (c) Inversions
   (d) Translocations
6. Numerical chromosome changes
   (a) Polyploidy
   (b) Haploidy
   (c) Aneuploidy
7. Applications of Cytogenetics to Plant and Animal Breeding
8. Human Cytogenetics

1.5 EPILOGUE

In the opening address of the 2nd Oxford Chromosome Conference held in 1967 (which corresponds to the 2nd International Chromosome Conference), Darlington speculated on 'what we do not know about chromosomes'. Ever since, the progress of Cytogenetics has been fantastic, especially during the two last decades. However, our knowledge of chromosome structure and behaviour is not yet complete. When shall we display of a complete structural model of the eukaryotic chromosome? When shall we show our students a wire and plastic scale model of the chromosome similar to the human scale models used to teach anatomy to medicine students? When shall we be able to draw a chart representing diagrammatically the intricate network of genetic signals which control the cell cycle and mitotic and meiotic processes in a similar way as biochemists show all the steps in the regulation of metabolism? When shall we produce a cartoon showing the dynamic aspects of chromatin in the processes of replication, transcription, recombination and repair? These are the challenges to Cytogenetics for the near future.

REFERENCES

Adolph,K.W.; Cheng,S.M. and Laemmli,U.K. (1977*a*) Role of non histone proteins in metaphase chromosome structure. *Cell,* **12**, 805.

Adolph,K.W., Cheng,S.M., Paulson,J.R. and Laemmli,U.K. (1977*b*) Isolation of a protein scaffold from mitotic HeLa cell chromosomes. *Proceedings of the National Academy of Science*, **74**, 4937.

Avery,O.T., McLeod,C.M. and Mc Carty,M. (1944) Studies on the chemical nature of the substance inducing transformation of pneumococcal types. Induction to transformation by a deoxyribonucleic acid fraction isolated from *Pneumococcus* type III. *Journal of Experimental Medicine*, **79**, 137—58.

Baxter, A. and Farley, J. (1979) Mendel and meiosis. *Journal Hist. Biol.*, **12**, 137—73.

Blackburn, E.H. (1991) Structure and function of telomeres. *Nature*, **350**, 569—73.

Blackburn, E.H. (1992) Telomerases. *Annual Review of Biochemistry*, **61**, 113—29.

Blackburn, E.H. and Szostak, J.W. (1984) The molecular structure of centromeres and telomeres. *Annual Review of Biochemistry*, **53**, 163—94.

Bostock, C.J. and Sumner, A.T. (1978) *The eukaryotic chromosome.* Nort-Holland Publishing Co., Amsterdam.

Boveri, T. (1892) Über die Entstehung des Vegensatzes zwischen den Geschlechtszellen und den somatichen Zellen bei *Ascaris megalocephala* nebst Bemerkungen zur Entstechungsgeschichte der Nematoden. *Sitzungsber. Ges. Morph. Phys.*, **8**, 114—25.

Boveri, T. (1902) Über mehrpolige Mitosen als Mittel zur Analyse des Zellkerns. *Verh. phys.-med. Gessellsc.*, **35**, 67—90.

Bridges, C.B. (1916) Non disjunction as a proof of the chromosome theory of heredity. *Genetics*, **1**, 1—52, 107—63.

Bridges, C.B. (1935) Salivary chromosome maps. *Journal of Heredity*, **26**, 60—4.

Caspersson, T., Farber, S., Foley, G.E. *et al.* (1968) Chemical differentiation along metaphase chromosomes. *Experimental Cell Research*, **49**, 219—22.

Clarke, L. and Carbon, J. (1980) Isolation of a yeast centromere and construction of functional small circular chromosomes. *Nature*, **287**, 504—09.

Clarke, L. and Carbon, J. (1985) The structure and function of yeast centromeres. *Annual Review of Genetics*, **19**, 29—56.

Correns, C. (1900) G. Mendel's Regel über das Verhalten der Nachkommenschaft der Rassenbastarde. *Deut. Bot. Ges. Berlin,* **18**, 158—69.

Creighton, H.B. and McClintock, B. (1931) A correlation of cytological and genetical crossing-over in *Zea mays*. *Proceedings of the National Academy of Science*, **17**, 492—97.

Crick, F.H. (1977) Postcript. *Chromosomes Today*, **6**, 403—6.

Darlington, C.D. (1929) Chromosome behaviour and structural hybridity in the *Tradescantiae. Journal of Genetics*, **21**, 207—86.

Fawcett, D. (1956) The fine structure of chromosomes in the meiotic proph~se of vertebrate spermatocytes. *Journal of Biophysical and Biochemical Cytology*, **2**, 403.

Ferguson-Smith, M.A. (1993) From chromosome number to chromosome map: the contribution of human cytogenetics to genome mapping, in *Chromosomes Today,* **11**, (eds. A.T.Sumner and A.C.Chandley), Chapman & Hall, London, pp. 3—19.

Flemming, W. (1879) Contributions to the knowledge of the cell and its life phenomena. *Arch. Mikr. Anat.*, **16**, 302—406.

Flemming, W. (1880) Beitrage zur Kenntniss der Zelle und ihrer Lebenserscheinungen. Teil II. *Arch. Mikr. Anat.*, **18**, 151—259.

Fox, D.P. (1973) The control of chiasma distribution in the locust *Schistocerca gregaria* (Forksal). *Chromosoma*, **43**, 289—328.

Fuller, M.T. (1995) Riding the polar winds: Chromosomes motor down East. *Cell*, **81**, 5—8.

Gall, J.G, and Pardue, M.L. (1969) Formation and detection of RNA—DNA hybrid molecules in cytological preparations. *Proceedings of the National Academy of Science*, **63**, 378—83.

Goodfellow, P.N. and Lovell-Badge, R. (1993) *SRY* and sex determination in mammals. *Annual Review of Genetics*, **27**, 71—92.

Greider, C.W. and Blackburn, E.H. (1985) Identification of a specific telomere terminal transferase activity in *Tetrahymena* extracts. *Cell*, **43**, 405—13.

Heiles,H.B.J. *et al.* (1988) *In situ* hybridisation with digoxigenin-labelled DNA of papilloma viruses (HPV 16/18) in HeLa and Sitla cells. *Biotechniques*, **6**, 978—81.

Heslop-Harrison, J.S., Schwarzacher, T., Leicht, A.R. and Anamthawat-Jónsson, K. (1988) A method of identifying DNA sequences in chromosomes of plants. *European Patent Application Number 8828130. 8. December 2.*

Hotta, Y., Ito, M. and Stern, H. (1966) Synthesis of DNA during meiosis. *Proceedings of the National Academy of Science*, **56**, 1184—91.

Hultén, M. (1974) Chiasma distribution at diakinesis in the normal human male. *Hereditas*, **76**, 55—78.

Hyams, J.S. and Brinkley, B.R.(eds)(1989). *Mitosis. Molecules and mechanisms.* Academic Press, London.

Johannsen, W. (1909) *Elemente der exakten Erblichkeitslehre*. Fischer, Jena.

John, B. (1983) Through the looking glass: a sceptical cytologist in a molecular wonderland, in *Kew Chromosome Conference II* (eds, P.E.Brandham and M.D.Bennett), *George Allen & Unwin*, pp. 305—10.

Jones, G.H. (1977) A test for early terminalisation of chiasmata in diplotene spermatocytes of *Schistocerca gregaria. Chromosoma*, **63**, 287—94.

Judd, B.H., Shen, M.W. and Kaufman, T.C. (1972) The anatomy and function of a segment of the X chromosome of *Drosophila melanogaster. Genetics*, **71**, 139—56.

Kanda, N. and Kato, H. (1980) Analysis of crossing-over in mouse meiotic cells by BrdU labelling technique. *Chromosoma*, **78**, 113—21.

Koopman, P., Gubbay, J., Vivian, N. *et al.* (1991) Male development of chromosomally female mice transgenic for *Sry. Nature*, **351**, 117—21.

Kornberg, R.D. (1974) Chromatin structure: a repeating unit of histone and DNA. *Science*, **184**, 868—71.

Kornberg, R.D. and Thomas, J.O. (1974) Chromatin structure: oligomers of the histones. *Science*, **184**, 865—8.

Koshland, D. (1994) Mitosis: Back to the basics. *Cell*, **77**, 951—4.

Lacadena, J.R. (1981) *Genética*, 3rd edn, A.G.E.S.A., Madrid.

Lacadena, J.R. (1984) De la teoría cromosómica de la herencia a la Citogenética moderna como modelo de ciencia experimental, in *En el centenario de Mendel: la Genética ayer y hoy* (coor. J.R.Lacadena), Editorial Alhambra S.A., Madrid, pp. 167—88.

Lacadena, J.R. (1996) *Citogenética*. Editorial Complutense, Madrid, (in press).

Laemmli, U.K., Cheng, S.M., Adolph, K.W. *et al.* (1977) Metaphase chromosome structure: the role of non histone proteins. *Cold Spring Harbor Symposia on Quantitative Biology*, **42**, 351—60.

Langer, P.R., Waldrop, A.A. and Ward, D.C. (1981) Enzymatic synthesis of biotin labelled polynucleotides: novel nuclei acid affinity probes. *Proceedings of the National Academy of Science*, **78**, 6633—7.

Lewin, B. (1980) *Gene expression 2. Eucaryotic chromosome,* 2nd edn, John Wiley & Sons, New York.

Lichter, P. and Cremer, T. (1991) Chromosome analysis by nonisotopic in situ hybridization, in *Human Cytogenetics: A practical approach* (eds. D.E.Rooney and B.H.Czepulkowski), IRL, Oxford, pp. 157—92.

Lichter, P., Cremer, T., Borden, J. *et al.* (1988) Delination of individual chromosomes in metaphase and interphase cells by *in situ* suppression hybridization using recombinant DNA libraries. *Human Genetics*, **80**, 224—34.

Lichter, P. and Ward, D.C. (1990) Is nonisotopic in situ hybridization finally coming of age? *Nature*, **345**, 93—5.

Lima-de-Faria, A. (ed)(1969) *Handbook of Molecular Cytology.* North-Holland Publishing Co., Amsterdam.

Lima-de-Faria, A. (1983) *Molecular evolution and organization of the chromosome.* Elsevier Science Publishers, Amsterdam.

Mendel, G. (1866) Versuche über Pflanzenhybriden. *Verh. des Naturf. Vereines in Brünn (Abhandlungen)*, **4**, 3—47.

Miller, O.L. and Beatty, B.R. (1969) Visualization of nucleolar genes. *Science*, **164**, 955—7.

Morgan, T.H. (1910) Sex limited inheritance in *Drosophila. Science*, **32**, 120—2.

Morgan, T.H. and Cattell, E. (1912) Data for the study of sexlinked inheritance in *Drosophila. Experimental Zoology*, **13**, 79.

Moses, M.J. (1956) Chromosomal structures in Crayfish spermatocytes. *Journal of Biophysical and Biochemical Cytology*, **2**, 215.

Nurse, P. (1990) Universal control mechanisms regulating onset of M-phase. *Nature*, **344**, 503—8.

Nurse, P. (1994) Ordering S phase and M phase in the cell cycle. *Cell*, **79**, 547—50.

Nurse, P. and Bissett, Y. (1981) Gene required in G_1 for commitment to cell cycle and in G_2 for control of mitosis in fission yeast. *Nature*, **292**, 558—60.

Ohno, S. (1979) *Major sex-determining genes.* Monographs on endocrinology, vol. 11, Springer-Verlag, Berlin.

Olins, D.E. and Olins, A.L. (1974) Spheroid chromatin units (nu bodies). *Science*, **183**, 330—2.

Painter, T.S. (1933) A new method for the study of chromosome rearrangements and the plotting of chromosome maps. *Science*, **78**, 585—6.

Painter, T.S. (1934) Salivary chromosomes and the attack on the gene. *Journal of Heredity*, **25**, 465—76.

Pardue, M.L. and Gall, J.G. (1969) Molecular hybridization of radioactive DNA to the DNA of cytological preparations. *Proceedings of the National Academy of Science*, **64**, 600—4.

Paulson, J.R. and Laemmli, U.K. (1977) The structure of histone-depleted metaphase chromosomes. *Cell*, **12**, 817—28.

Pinkel, D., Landegent, J., Collins, C. *et al.* (1988) Fluorescence *in situ* hybridization with human chromosome-specific libraries: detection of trisomy 21 and translocations of chromosome 4. *Proceedings of the National Academy of Science*, **85**, 9138—42.

Prescott, D.M. (1976) The cell cycle and the control of cellular reproduction. *Advances in Genetics*, **18**, 99—177.

Ris, H. (1956) A study of chromosomes with the electron microscope. *Journal of Biochemical and Biophysical Cytology*, Suppl. **2**, 385—92.

Ritossa, F.M. and Spiegelman, S. (1965) Localization of DNA complementary to RNA in the nucleolus organizer region of *D. melanogaster. Proceedings of the National Academy of Science*, **53**, 737—45.

Rubio, J. (1973) Genética. Su posición entre las ciencias biológicas. *Boletín Estación Experimental Aula Dei*, **12**, 80 pp.

Santos, J.L., Ciprés, G. and Lacadena, J.R. (1987) Metaphase I chiasmata in silver stained cores of bivalents in grasshopper spermatocytes. *Genome*, **29**, 235—8.

Santos, J.L., Ciprés, G. and Lacadena, J.R. (1989) A quantitative study on chiasma terminalization in the grasshopper *Chorthippus jucundus*. *Heredity*, **62**, 51—7.

Schwarzacher, T., Leicht, A.R., Bennett, M.D. and Heslop-Harrison, J.S. (1989) *In situ* localization of parental genomes in a wide hybrid. *Annals of Botany*, **64**, 315—24.

Sinclair, A.H., Berta, P., Palmer, M.S. *et al.* (1990) A gene for the human sex-determining region encoding a protein with homology to a conserved DNA-binding motif. *Nature*, **346**, 240—4.

Stern, C. (1931) Zytologisch-genetische Untersuchungen als Beweise für die Morgansche Theorie des Faktoresaustanch. *Biol. Zentralbl.*, **51**, 547—87.

Stern, H. and Hotta, Y. (1973) Biochemical control of meiosis. *Annual Review of Genetics*, **7**, 37—66.

Sturtevant, A.H. (1913) The linear arrangement of six sex-linked factors in *Drosophila*, as shown by their mode of association. *Journal Experimental Zoology*, **14**, 43—59.

Sumner, A.T., Evans, H.J. and Buckland, R.A. (1971) New technique for distinguishing between human chromosomes. *Nature New Biology*, **232**, 31—2.

Sutton, W.S. (1902) On the morphology of the chromosome group in *Brachystola magna*. *Biological Bulletin*, **4**, 24—39.

Sutton, W.S. (1903) The chromosomes in heredity. *Biological Bulletin*, **4**, 231—51.

Swanson, C.P. (1960) *Cytology and Cytogenetics*. Macmillan & Co., Ltd., London.

Swanson, C.P., Merz, T. and Young, W.J. (1981) *Cytogenetics. The chromosomes in division, inheritance and evolution*. Prentice—Hall, Inc., Englewood Cliffs, New Jersey.

Taylor, J.H., Woods, P.S. and Hughes, W.L. (1957) The organization and duplication of chromosomes as revealed by autoradiographic studies using tritium-labeled thymidine. *Proceedings National Academy of Science*, **43**, 122—8.

Tease, C. (1978) Cytological detection of crossing-over in BUdR substituted meiotic chromosomes using the fluorescent plus Giemsa technique. *Nature*, **272**, 823—4.

Tease, C. and Jones, G.H. (1978) Analysis of exchanges in differentially stained meiotic chromosomes of *Locusta migratoria* after BrdU-substitution and FPG staining. I. Crossover exchanges in monochiasmate bivalents. *Chromosoma*, **69**, 163—78.

Wachtel, S.S. (1983) *H-Y antigen and the biology of sex determination.* Grune & Stratton, New York.

Weismann, A. (1889) The continuity of the germ—plasm as the foundation of a theory of heredity, in *Essays upon heredity and kindred biological problems. I, Oxford*, pp. 162—248.

Wiberg,U.H. (1989) H-Y antigen: Dogmas and enigmas, in *Evolutionary mechanisms in sex determination* (ed S.S. Wachtel), CRC Press, Inc., Boca Raton, Florida, pp. 111—21.

PART ONE

SEQUENCE ORGANISATION IN CHROMOSOMES

Jaime Gosálvez

Jeffrey Craig

Mariano Rocchi

Werner Schempp

2

Chromosome differentiation using nucleases: an overview

J. Gosálvez, C. López-Fernández, V. Goyanes and R. Mezzanotte

2.1 THE FINE BRUSHSTROKES ON THE HISTORICAL CANVAS OF CHROMOSOMAL DIFFERENTIATION

The era of the longitudinal differentiation of metaphase chromosomes (chromosome banding) dawned in 1968 when Caspersson and co-workers described the use of quinacrine mustard as a fluorescent compound which heterogeneously stained fixed human chromosomes, thus producing the so-called Q-bands. Around the same time, Pardue and Gall (1970) were experimenting with the *in situ* hybridization of nucleic acids to localize satellite DNAs in mouse chromosomes. Twenty five years later this technique is an indispensable tool in the approach to basic problems in various fields of Genetics and Biology. It was also in the early 1970's that Sumner *et al.* (1971) and Sumner (1972) unveiled techniques of easy application to produce G- and C-bands by means of chemical pretreatment of metaphase chromosomes. The basis of chromosomal differentiation was established. Thus, the use of fluorochromes which are able to recognize GC- or AT-enriched DNA motifs, physicochemical modifications of the chromatin, and the incorporation of modern molecular techniques into cytogenetics would prove to be the first steps along new pathways in the production of longitudinal differentiation of metaphase chromosomes or interphase nuclei.

Nevertheless, before Caspersson's studies, non-uniform staining of the chromatin, dependent on specific chromosome organization, was known to exist. Polytene chromosomes in Diptera and the chromomeric organization of pachytene chromosomes in maize are good examples of such longitudinal differentiation. Delving further back into cytogenetic history, differences along the chromosome arms after silver staining (kinetochores and nucleoli) were reported around a century ago (reviewed in Sumner, 1990). Undoubtedly, the progress of chromosome banding shows itself to be a compromise between resolving power and specificity for localization of chromosomal markers. In this sense, *in situ* hybridization of nucleic acids and its modifications using fluorescence (FISH) is one of the most powerful methods to map and analyze the distribution of DNA sequences in the chromosome. Additionally, the functional role of the different chromosome domains and their evolutionary implications has been the second motor which has driven the development of this technology. Although the information about chromosome proteins is very limited compared with that of DNA, the influence of these chromosomal components, especially the non-histone proteins, cannot be ignored. Certain techniques assume the DNA molecule to be free of interactions with other chromosome components. However, proteins (or proteins modified after fixation) have frequently been shown to modulate the capacity of a particular protocol to operate in a similar fashion in different biological systems. In other words, proteins may modify the expected specificity of a cytogenetic technique. Since specificity is one of the main goals of the new chromosome banding era, most of the non-specific or difficult to explain methods of chromosome differentiation have been replaced by techniques derived from molecular technology. Sumner (1990) opined that 'the subject of chromosome banding is thus a very varied one, ranging from the highly practical to the esoteric'. In fact, the esoteric is present in most of our experimental approaches. For example, when we discovered that the DNA molecule, immersed in a proteinaceous matrix, is so stable that after resisting alcohol/acid fixation it is still susceptible to selective recognition by certain enzymes and that these enzymes operate, as predicted, on spread chromosomes, we let our imagination run wild in an attempt to reproduce in our molecular technological protocols the modes of enzyme action that operate on naked DNA. The esoteric magic of molecular technology harnessed to modern cytogenetics is broad in its scope: from the selective labelling of the DNA molecule within a metaphase chromosome by *in situ* nick translation or *in situ* random priming, to the hybridization of specific primers which can be extended by polymerases in a similar fashion to PCR experiments.

The search for greater specificity in the production of chromosome bands prompted the first approaches to chromosome differentiation using nucleases. It was also at the beginning of the 1970's that Alfi *et al.* (1973) and Kato *et al.* (1974) produced C-like banding patterns on mammalian chromosomes using non-specific nucleases such as DNase I. However, Gall (1963) and Nash and Plaut (1964) may be considered to be the pioneers in the use of DNase I and RNase to analyze the breakage kinetics and denatured states of DNA in lampbrush and polytene chromosomes. Ten years after the discovery of the first Restriction Endonucleases (REs), these enzymes were used to produce *in situ* cleavage of the DNA molecule housed in the chromosome. Gould *et al.* (1976) and Jones (1977) were the first to digest fixed chromosomes, although the real potential of these techniques was only properly realized after 1983 when Mezzanotte *et al.*, (1983a) and Miller *et al.*, (1983) published a set of parallel results showing the capacity of these molecular 'scalpels' to produce longitudinal differentiation on fixed human chromosomes.

2.2 RESTRICTION ENDONUCLEASES: MOLECULAR SCULPTORS OF THE CHROMOSOME

As was predicted, REs recognize relatively short DNA base sequences, cleave at specific sites and permit DNA fragments to be removed from conventional methanol—acetic acid fixed nuclei. The result is a differential DNA removal that is revealed by Giemsa or fluorochrome staining. Regions of the chromosome that are resistant to RE attack appear dark or highly fluorescent with these respective stains, while the lightly stained or the dully fluorescent areas are susceptible to DNA removal. One would expect as many banding patterns as there are REs available, since the DNA cleavage should be different for each RE. However, in reality, most of the banding patterns obtained by treatment with different REs overlap. Moreover, only 4-, 5- or 6-bp cutters induce enough chromosome differentiation to permit analysis under the light microscope. In general, in mammalian chromosomes, REs produce banding patterns which resemble those obtained by other methods, i.e. C-, G- or R-like bands or an absence of chromosome differentiation, but which exhibit their own characteristics. Interestingly, some REs greatly reduce or even abolish the staining ability of some chromosome regions, especially those containing repetitive DNA sequences (see Table 2.1). For example, after digestion

Table 2.1 Banding patterns obtained in different species after *in situ* digestion of metaphase chromosomes with different REs.

Species	*C-Like*			*G-Like*	
Humans(*)	Alu I[4,5]	Msp I[9]	Taq I[9]	Hae III[4,5,7]	Stu I[3]
	Dde I[4,7]	Ple I[1]	Hind III[5]	Eco RI[5]	Ava I[3]
	Mbo I[4,7]	Bst NI[6]	Eco RII[4]	Msp I[8]	Bst NI[6]
	Rsa I[7]			Hpa II[8]	
Chimpanzee	Alu I[10,37]	Rsa I[10]	Mbo I[10]	Hae III[10]	
	Dde I[10]	Hae III[10]			
Gorilla	Alu I[10,37]	Rsa I[10]	Mbo I[10]	Hae III[10]	
	Dde I[10]	Hae III[10]			
Orangutan	Alu I[10,11]	Rsa I[10]	Mbo I[10]	Hae III[10]	
	Dde I[10]	Hae III[10]			
Mouse	Alu I[6,14,15]	Hind III[14]	Hae III[6,16]	Hae III[6]	Eco RII[6]
	Dde I[6]	Hinf I[6]	Eco R II[6]	Ava II[6]	Hind III[14]
	Mbo I[6,15]			Bst NI[6]	Bam HI[14]
Muntiacus muntjac	Alu I[13]	Hinf I[13]		Hae III[13,17]	
	Mbo I[13]	Sau 3A[17]			
	Hae III[13]				
Gallus domesticus	Hae III[18]			Hae III[18]	
Rana erythrea	Hae III[18]				
Bufo marinus	Hae III[18]				
Bufo americanus	Hae III[18]				
Gastrotheca riobambae	Hae III[18]				
Salmo gairdneri	Alu I[19]	Pvu II[19]	Hae III[19]		
	Mbo I[19]	Hinf I[19]			
Murena helena	Hae III[20]	Dde I[20]	Mbo I[20]		
Anguilla anguilla	Alu I[32]	Dde I[32]	Hae III[32]		
	Mbo I[32]				
Chorthippus parallelus	Msp I[42]*	Hpa II[42]*			
Oedipoda germanica	Alu I[21]	Dde I[21]	Hae III[21]		
	Hinf I[21]	Mbo I[21]			
Eyprepocnemis plorans	Alu I[41]*	Hinf I[41]*	Msp I[41]*		
	Mbo I[41]*	Hae III[41]*	Rsa I[41]*		
	Dde I[41]*	Bamb HI[41]*			
Pyrgomorpha conica	Hae III[22]	Mbo I[22]			
Arcyptera tornosi	Alu I[23]	Hae III[23]	Hinf I[23]		
Dociostaurus genei	Alu I[40]*				
D. maroccanus	Taq I[40]*				
Baetica ustullata	Hae III[24]	Hha I[24]	Msp I[24]		
Tenebrio molitor	Alu I[27]				
Tenebrio obscurus	Alu I[28]	Dde I[28]	Hae III[28]		
	Hind III[28]	Hinf I[28]	Cla I[28]		
	Acc I[28]	Eco RI[28]	Dra I[28]		
	Rsa I[28]				

(*) *R-Like*: Mse I[29] Tfi I[2]

Table 2.1 (continued)

Species	*C-Like*			*G-Like*
Culiseta longiareolata	Alu I[25]	Hha I[25]	Msp I[25]	
	Hae III[25]	Hinf I[25]	Hpa II[25]	
	Bamb HI[25]			
Aedes albopictus	Hpa II[26]			
Sarcophaga bullata	Alu I[31]			
Drosophila melanogaster	Alu I[12]	Hae III[12]		
Drosophila virilis	Alu I[15]			
Drosophila nasuta	Alu I[33]	Sau 3A[33]	Eco RI[33]	
	Sin I[33]	Hae III[33]		
Vicia faba	Alu I[30,34]	Ava I[34]	Eco RV[34]	
	Sph I[34]	Hae III[34]	Rsa I[34]	
	Mbo I[30,34]	Cla I[34]	Eco RI[34]	
Hordeum vulgare	Alu I[34]	Cla I[34]		
Secale cereale	Dra I[35]	Hpa II[35]	Msp I[35]	
Scilla sibirica	Hae III[38]			
Scilla autumnalis	Alu I[39]*	Rsa I[39]*	Dra I[39]*	
Muscari comosum	Alu I[39]*	Rsa I[39]*	Hae III[39]*	
	Dra I[39]*			
Dasypyrum villosum	Alu I[36]*	Dde I[36]*	Hae III[36]*	
	Taq I[36]*			

1 Tagarro, I. *et al.* (1992) *Cytogenetics and Cell Genetics* 60, 102. **2** Tagarro, I. *et al.* (1992) *Cytogenetics and Cell Genetics* 60, 154. **3** Tagarro, I. (1993) *Tesis Doctoral*. Universidad Autónoma de Madrid. **4** Miller, D.A. *et al.* (1983) *Science* 219, 395. **5** Mezzanotte, R. *et al.* (1983) *Journal of Histochemistry and Cytochemistry* 29, 115. **6** Kaelbling, M. *et al.* (1984) *Chromosoma* 90, 128. **7** Bianchi, M. *et al.* (1985) *Journal of Molecular and Evolution* 22, 323. **8** Fernández-Peralta, A.M. *et al.* (1994) *Genome* 37, 770. **9** Bianchi, N.O. and Bianchi, M. (1987) In: *Cytogenetics*, Obe & Basler. Springer, Berlin. pp.280. **10** Babu, A. and Verma, R.S. (1986) *Cytogenetics and Cell Genetics* 41, 96. **11** De Stefano, G.F. *et al.*(1986) *Human Genetics* 72, 268. **12** Mezzanotte, R. (1986) *Chromosoma* 93, 249. **13** Babu, A. and Verma, R.S. (1985c, 1986b) *American Journal of Human Genetics* 37, A43; *Cytogenetics and Cell Genetics* 41, 96. **14** Mezzanotte, R. and Ferrucci, L. (1984) *Genetica* 64, 123. **15** Mezzanotte, R. *et al.* (1986) *Genetica* 70, 107. **16** Mezzanotte, R. *et al.* (1985) *Basic and Applied Histochemistry* 29, 395. **17** Ueda, T. *et al.* (1987) *Chromosoma* 95, 251. **18** Schmid, M. and Almeida, C.G. (1988). *Chromosoma* 96, 283. **19** Lloyd, M. and Thorgaard, G.H. (1988) *Chromosoma* 96, 171. **20** Cau, A. *et al.* (1988) *Cytogenetics and Cell Genetics* 47, 223. **21** López-Fernández, C. *et al.* (1989) *Heredity* 62, 269. **22** López-Fernández, C. *et al.* (1988) *Genome* 30, 621. **23** Gosálvez, J. *et al.* (1987) *Heredity* 59, 173. **24** Sentís, C. *et al.* (1989) *Genome* 32, 208. **25** Marchi, A. and Mezzanotte, R. (1988) *Heredity* 60, 21. **26** Bianchi, N. *et al.*(1986) *Chromosoma* 94, 362. **27** Juan, C. *et al.* (1990) *Heredity* 65, 157. **28** Ugarkovic, D. *et al.* (1994) *Chromosome Research* 2, 217. **29** Ludeña, P. *et al.* (1991) *Cytogenetics and Cell Genetics* 57, 82. **30** Frediani, M. *et al.* (1987) *Theoretical and Applied Genetics* 75, 46. **31** Bultmann, H. and Mezzanotte, R. (1987) *Journal of Cell Science* 88, 327. **32** Cau, A. *et al.* (1992) *Genome* 35, 838. **33** Tiwari, P.K. and Lakhotia, S.C. (1991) *Journal of Bioscience* 16, 187. **34** Schubert, I. (1990) *Caryologia* 43, 117. **35** Stober, T. *et al.* (1993) *Genome* 36, 998. **36** Pignone, D. *et al.* (1995) *Chromosome Research* 3, 109. **37** Ferrucci, L. *et al.* (1987) *Cytogenetics and Cell Genetics* 44, 13. **38** Lozano, R. *et al.* (1991) *Chromosoma* 100, 439. **39** Lozano, R. *et al.* (1990) *Heredity* 64, 185. **40** Rodríguez-Iñigo (1994) *Tesis Doctoral*, Universidad Autónoma de Madrid. **41** López-Fernández, C. *et al.* (1992) *Heredity* 68, 365. **42** Rubio, J.M. (1992) *Tesis Doctoral*. Universidad Autónoma de Madrid.

with Alu I human chromosomes are only densely stained in the centromeric regions of chromosomes 1, 9, 16, and on the distal regions of chromosome Y (a partial C-like banding pattern; Mezzanotte *et al.*, 1983a). On the other hand, mouse chromosomes show an orthodox C-banding pattern after digestion with the same RE and the centromeric region of all the chromosomes stains intensely (Kaelbling *et al.*, 1984).

2.2.1 RE-banding as a result of DNA cleavage and extraction

There is now enough evidence from different sources to demonstrate that, as predicted from the first experiments, the banding pattern obtained after *in situ* digestion with REs is due to differences in enzyme digestibility of the DNA. Miller *et al.*, (1984) used *in situ* hybridization to demonstrate a reduction of mouse satellite DNA following digestion with Bst NI, a RE which cleaves mouse satellite DNA into 234 bp monomers (Southern, 1975). Likewise, Lloyd and Thorgaard (1988) observed a notable reduction in the hybridization signal after digestion with Alu I of salmonid chromosomes. Other approaches have demonstrated that this DNA loss results in a staining reduction when stoichiometric methods for DNA detection are used. For example, digestion of human metaphase chromosomes with Alu I and staining with Feulgen gives rise to C-like bands along some chromosomes and shows a loss of more than half the DNA content (Mezzanotte *et al.*, 1985). Likewise, using X-ray microanalysis, where the phosphorus X-ray signal is a marker for DNA, it was demonstrated that the action of a RE such as Hae III on mouse chromosomes induces loss of phosphorous from both the centromeres and the chromosome arms although the loss in the arms is very much greater (Gosálvez *et al.* 1991). In this case, the structure of the centromeric chromatin, which the strong staining with Giemsa or fluorochromes would lead us to believe is resistant to RE attack, is affected by the enzyme after all. An intuitive representation of the partial resistance of a chromosome region to RE cleavage is given in Figure 2.1. Undercondensed mouse chromosomes (cell line L 929) directly stained with Giemsa show centromeric regions with lower staining intensities than that achieved in the chromosome arms (Fig. 2.1a, c). When these chromosomes are digested with Alu I and stained with Giemsa, a C-like banding pattern is produced (Fig. 2.1b, d). However, densitometric quantification of staining intensities along control (Fig. 2.1e) and Alu I-treated chromosomes (Fig. 2.1f) clearly shows that the centromere has been attacked by the RE, thereby increasing the capacity of Giemsa to bind to this chromosome region. In other words, we must be cautious in any conclusion that a chromosome region is resistant to RE digestion.

Theoretically, DNA digestion should produce polynucleotide fragments whose length depends on the frequency of the target present within the DNA. Therefore, it has been suggested that REs producing fragments shorter than 350 bp should produce massive DNA loss (Miller *et al.*, 1983; Bianchi *et al.*, 1985), while digestion producing longer fragments would not allow DNA extraction from fixed chromatin. Burkholder (1989) showed that DNA fragments longer than 4 kb can be removed from unfixed mouse chromosomes. Gel electrophoresis of DNA fragments produced after *in situ* digestion of fixed chromatin of *Tenebrio molitor* (Coleoptera) chromosomes suggested that the size of DNA fragments is not after all a limiting factor in their extraction (Juan *et al.*, 1991). In this case a 142 bp repeated unit (and multiples thereof) produced with Eco RI was targeted and these fragments were able to be recovered from solubilized DNA and from DNA remaining on the slide. This has also been observed in mouse chromosomes, using the same method of analysis (Mezzanotte *et al.*, 1992).

2.2.2 Specificity of action

The genome of *T. molitor* has large heterochromatin blocks distributed in every autosome. Petitpierre *et al.* (1988) pointed out that there is a large amount of satellite DNA in this species (close to 60% of the whole genome). Eco RI cuts the satellite DNA localized in the centromeric heterochromatin into monomeric subunits of 142 bp in naked DNA and acts on fixed chromosomes cleaving and extracting these tandemly repeated sequences in a similar fashion. Subsequently, a dramatic decrease in the staining intensity with Giemsa or fluorochromes and in the capacity to hybridize the 142 bp probe was detected (Juan *et al.*, 1991). Additionally, the series of DNA fragments produced from solubilized DNA after Eco RI *in situ* cleavage is similar to that obtained on naked total DNA. In the same species, Alu I is unable to attack the satellite DNA sequences but does cleave the main band DNA both in naked DNA and in fixed chromosomes. This remarkable coincidence of Eco RI and Alu I action on naked DNA and fixed chromosomes suggests that the DNA in fixed chromatin

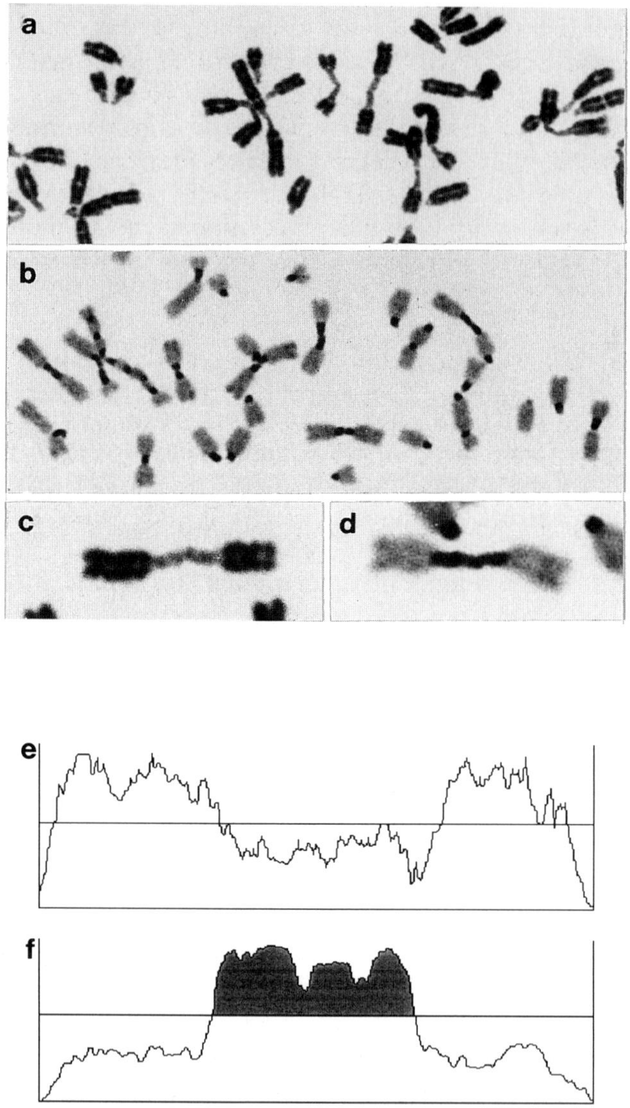

Fig. 2.1 a) Hoechst 33258 undercondensed mouse L929 chromosomes after Giemsa staining. b) Same type of chromosomes after digestion with the RE Alu I and stained under the same conditions as chromosomes in (a). e) Densitometric profile of a control selected chromosome (c) and a chromosome digested with Alu I (d). Area in black shows increase of staining achieved after Alu I digestion in the centromere region.

might be solubilized independently of the interactions of this molecule with chromosome proteins, DNA modifications after fixation or with the tridimensional structure of the RE itself. The decrease of staining intensity in the centromeric regions of mouse chromosomes obtained after RE digestion with Bst NI or Ava II (Miller *et al.*, 1984) — REs which are known to cut the DNA fraction in monomers of low molecular weight — might have been considered as a simple example of the irrelevance of chromatin structure to active *in situ* DNA extraction. However, evidence from different experimental approaches now suggests a much greater complexity whereby the structure of the chromatin in each species probably determines the capacity of an RE to cleave and remove the DNA from a fixed chromosome. Alu I produces C-like bands on fixed and unfixed mouse chromosomes (Kaelbling *et al.*, 1984; Burkholder, 1989). It is incapable of attacking major satellite DNA, unlike Bst NI that is known to cut this DNA fraction into monomers of 234 bp. The analysis of DNA fragments produced after *in situ* digestion with these enzymes showed that the major mouse satellite DNA is not extracted from the centromeres of fixed chromosomes after Alu I digestion and is only removed to a limited extent by Bst NI digestion. Thus, Bst NI acts *in situ* with a different efficiency than on naked DNA (Mezzanotte *et al.*, 1992). There are similar results reported in *Tenebrio obscurus*, where a 344 bp monomer localized in the constitutive heterochromatin is resistant to *in situ* cleavage with different REs. However, the DNA could be digested by chromatin structure modification using proteinase K (Ugarkovic *et al.*, 1994).

In addition to cleavage frequency, RE cytological effects and, in particular, G-band production have been interpreted as a consequence of the different accessibility of the RE to conformational changes of the chromatin (Mezzanotte *et al.*, 1985). Differences in DNA base composition and organization of chromatin between G- and R-banded regions are now well established (Korenberg and Rikoski, 1988) and might affect DNA accessibility and/or extraction produced by endonuclease digestion. Differential accessibility of chromosome regions to macromolecules has been demonstrated by using non-specific nucleases (Sahasrabudhe *et al.*, 1978) and isoschizomers (Gosálvez *et al.*, 1989). Additionally, the fact that only a few REs are known to induce R-bands (see Table 2.1 for details) could be related to the differential accessibility of REs in different chromosome regions (G- or R-bands in mammalian chromosomes). Moreover, chromatin structure is expected to be different in suspensions and spreads of chromosomes. In fact, chromatin fibre digestion analyzed by electron microscopy shows clear differences (compare Gosálvez and Goyanes, 1988 and Gosálvez *et al.*, 1990). DNA fixation is an important characteristic that cannot be ignored when trying to understand DNA removal. It has been demonstrated that digestion of isolated and unfixed mammalian chromosomes removes DNA fragments of variable size (0.2—4 kb; Burkholder, 1989). The maximum length of DNA fragment that can optimally be removed from fixed chromatin is about 4 kb. This implies that fixation, though being an important factor in determining RE-induced banding (e.g. formaldehyde and glutaraldehyde prevent band formation with any RE (Mezzanotte unpublished data), does not play any role in determining the fragment length of DNA removed (Mezzanotte *et al.*, 1992). Other factors such as ageing the DNA on the slides are technical variables which influence cleavage and subsequent extraction of DNA from eukaryotic chromosomes.

Taking into account these possibilities for modulating DNA removal we must assume that the genome of each species is a singular case and that no general rule governs the DNA recognition, cleavage and extraction mediated by an RE. However, as a practical rule of thumb for *in situ* digestion experiments, we recommend the use of known restriction sites within the DNA sequence as the most effective DNA cutters for those species where DNA sequence distribution data exist. In those species with unknown DNA sequence distribution, assays of random cutting are more effective using 4—5 bp cutters, since the possibility of finding a restriction site within the DNA sequence decreases as the number of base pairs per target increases.

2.2.3 Chromosome bands produced

In situ digestion of metaphase chromosomes with REs to produce chromosome differentiation has been systematically carried out since 1983 in an attempt to understand the evolutionary relationships between related taxa as well as many structural and functional aspects of their genomes. The species studied range from plants to the most complex mammalian organisms (Table 2.1). However, although a large number of REs have been assayed and the distribution of their restrictions sites is expected to differ among unrelated taxa, the banding patterns produced are restricted to variations of C-, G- or R-bands (although the latter two types occur almost exclusively in

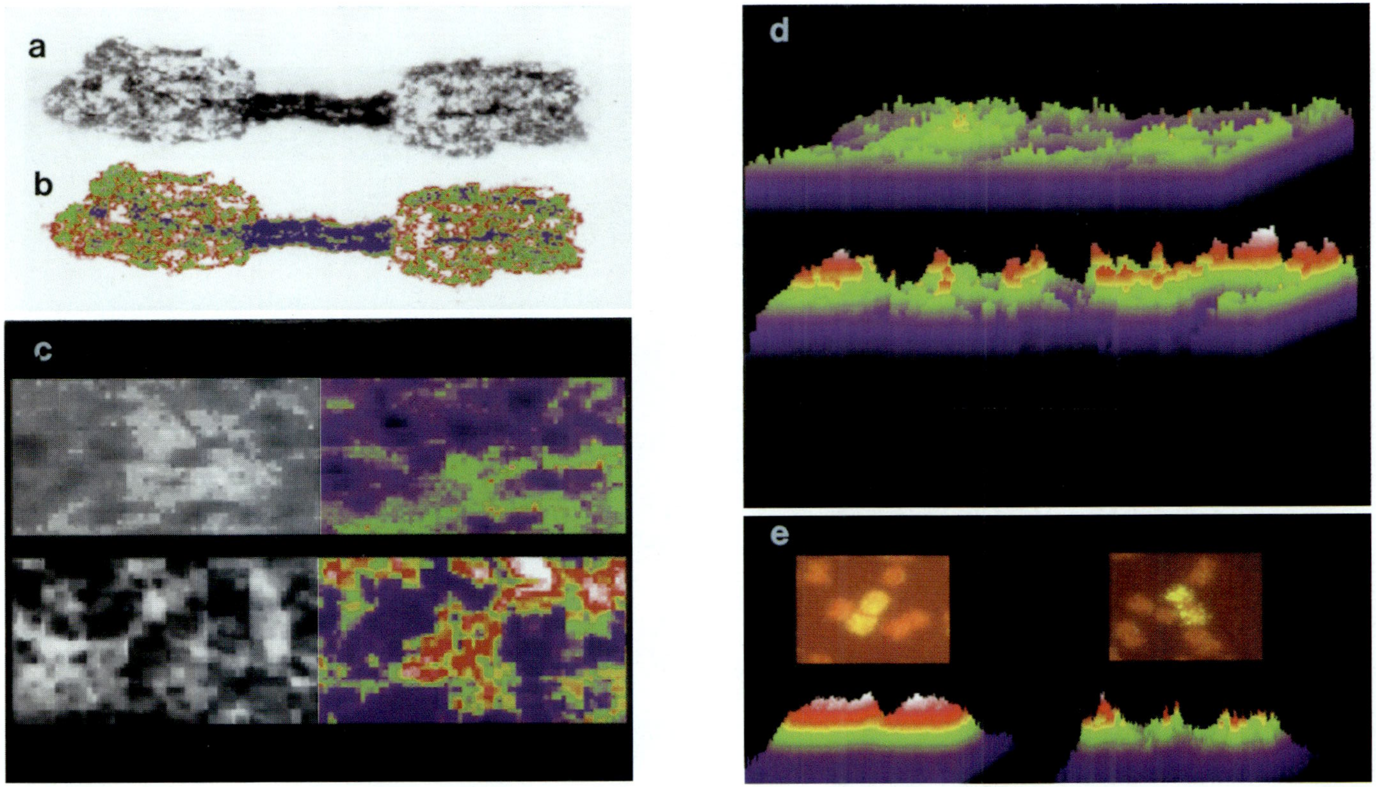

Plate 1 **Fig. 2.2** a–e (see p.35)

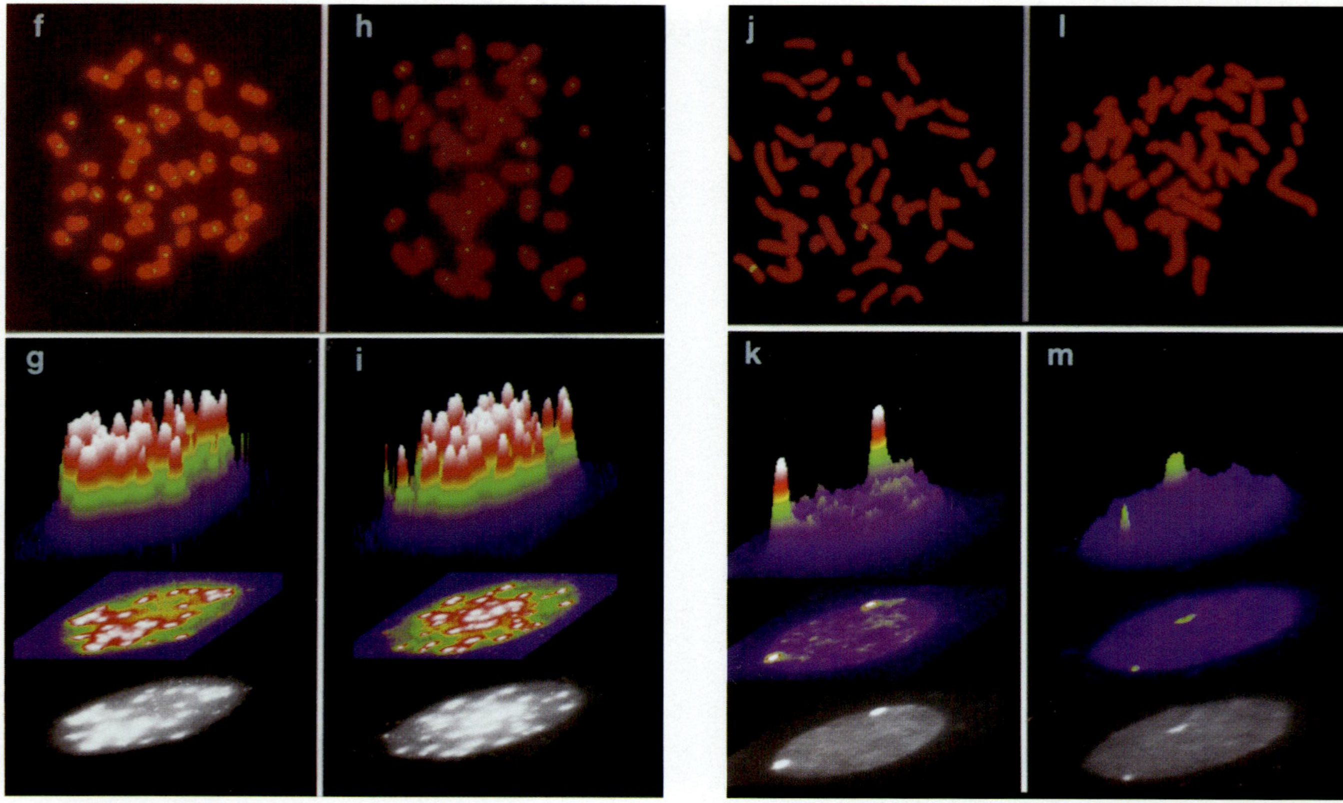

Plate 1 **Fig. 2.2** f–m (see p.35)

vertebrate chromosomes). Massive DNA extractions produce chromosome gaps (negative bands) which sometimes coincide, and other times not, with the location of constitutive heterochromatin (Herrero *et al.*, 1993). Cryptic bands — those not localized by other techniques of longitudinal chromosome differentiation — have also been reported after RE digestion in mouse and in grasshopper chromosomes (López-Fernández *et al.*, 1991).

As may be appreciated from Table 2.1, the most common effect of REs on fixed chromosomes is the production of a C-like banding pattern. It should be stressed that true C-banding (complete identification of constitutive heterochromatic regions) is not reproduced. Thus, for example, Ple I, which yields C-like bands in humans, fails to produce centromeric bands on chromosomes 6, 10, 11 and diminishes the size of the heterochromatic bands on chromosome 15 and the distal block on the Y chromosome (Tagarro, 1993). This behaviour could be mimicked by other REs and on other chromosomes. For example, Alu I, Mbo I, and Dde I permit only the partial detection of C-bands in human chromosomes, although some of these REs are able to detect polymorphisms generated by differential homologous DNA removal within homologous identical C-banded regions (Babu, 1988). The variability detected in C-banded regions raises the possibility of the existence of specific subclasses of repetitive DNAs localized within apparently homogeneous blocks of constitutive heterochromatin (see López-Fernández *et al.*, 1991 for a review).

G- and R-bands have regularly been found on chromosomes of higher vertebrates (mammals, birds and reptiles) but only sporadically in other animals and in plants (see Sumner, 1990 for a review). Some REs can produce a G-like banding pattern on mammalian chromosomes (see Table 2.1), although this has usually been a bone of contention. For example, the different response of human and mouse chromosomes after Hae III digestion is dependent upon the stain used and has been interpreted as a consequence of the different accessibility of the RE to different chromosome regions and to chromatin conformational changes (Mezzanotte *et al.*, 1983b; 1985). Mse I and Tfi I are the only REs known to produce R bands on human chromosomes. In both cases this pattern is produced at low RE concentrations, whereas higher concentrations yield less intense R-banding, favouring the production of a C-banding pattern. It has been suggested that R-bands are very sensitive to endonucleases (Ludeña *et al.*, 1991). However, in this case REs appear to be capable of cutting and extracting DNA from G-bands according to the base sequence composition, while the conformational differences between R- and G-bands represent a secondary level of modulation of their action. In any case, some of the G- and R-like bands produced by REs give rise to simultaneous C- and G-like banding patterns when longer periods of digestion are used.

As restriction sites do indeed play a role *in situ* RE-chromosome digestion, although the same banding pattern is obtained with different REs, the chromatin fibre would be expected to be affected to a different extent. For example, *in situ* digestion with Alu I, Mbo I or Dde I allows a similar level of discrimination between the pericentromeric constitutive heterochromatin and the euchromatic chromosome arms of most human chromosomes (Babu, 1988), even though the distribution of restriction sites for each RE on the chromatin is expected to be very different. The possible differential chromatin alterations are not easy to compare and quantify under light microscopy since the structure of the chromatin fibre cannot be resolved. However, this may be achieved with electron microscopy (Goyanes, 1985) especially if assisted by digital image analysis. The latter technique permits differences obtained from various digested chromosomes to be analyzed by:

1. the generation of different coloured images by application of a Look Up Table (LUT) which assigns a different colour to a grey image (Fig. 2.2a, b);
2. the analysis of similarly coloured pixels and production of pseudo-3D representations of the colour values for each pixel (Fig. 2.2c, d). This generates a distinct chromatin topography for each digestion (Fig. 2.2d). These images therefore yield information which permits electron density in a chromosome region to be correlated with the pattern and intensity of RE-cleavage;
3. transformations of these or similar images may be mathematically compared in order to avoid the subjectivity inherent in comparisons with the naked eye. Digital image analysis allows the estimation of various parameter values for the visual differences established in i) and ii). There exists a series of routines within image analysis software that permits the modification of digitalized images. Using the VISILOG program (Noesis, 1992), we have optimized a sequence of operations for obtaining a simple structure susceptible to analysis (Fig. 2.3).

Fig 2.2 ***Colour Plate 1*** a—d) L929 mouse chromosomes. a) Whole-mounted metaphase chromosome digested with the RE Hae III. b) The same chromosome after application of a specific LUT to assign a determined color to a grey image. c—d) Selected digitalized chromosome area for a comparison of the chromatin texture after digestion with two different REs. Alu I (above) and Hae III (below). Original digitalized images of a selected region of a chromosome arm (c, left) give rise to coloured images (c, right) after application of the same LUT as in (b). d) Pseudo 3D topographic representation of the chromatin resulting from RE digestion shown in (c). e—m) Fluorescence *in situ* hybridization in human chromosomes. e) Chromosome painting and pseudo-3D representation of the hybridization signals obtained after thermal denaturation (above) and Exonuclease III treatment (below). f—i) *In situ* hybridization of alphoid DNA sequences after thermal denaturation (f) and Exonuclease III treatment (h); topographic representation of signal intensities in an interphase nucleus after thermal denaturation (g) and Exonuclease III treatment (i). j—m) *In situ* hybridization of classical satellite DNA D9Z1 probe after thermal denaturation (j) and Exonuclease III treatment (l); topographic representation of the signal intensities in an interphase nucleus after thermal denaturation (k) and Exonuclease III treatment (m).

After digitization (Fig. 2.3a1—4) each image is sharpened using a 7×7 kernel high-pass filter (Fig. 2.3b1—4) to maximize the contrast between the chromatin and the background. Binary images (BIs) of the whole chromatin fibre are derived from the filtered images by employing a fixed level of segmentation of $\delta=1$ (Fig. 2.3c1—4). Finally, BIs are iteratively eroded to produce skeletons (Fig. 2.3d1—4). The area, perimeter, shape, Euler number of BIs and the number of triple points of skeletons may then be derived and statistically compared from samples of images, thereby allowing the patterns of chromatin digestion of different REs to be quantified (Gosálvez, in preparation). This system of chromatin textural measurements could provide the basis for the analysis and comparison of chromatin fibre phenotypes produced under other experimental conditions such as the organization of G- and R-bands within the mammalian genome, the arrested chromatin assembly in BrdU or 5-azacytidine substituted DNA, or the modifications of chromatin after interaction with specific DNA binding molecules such as Hoescht 33258 or Dystamycin A.

Finally, we want to highlight the application of RE digestion polymorphisms in the medical sphere. Sau 3A used on metaphase chromosomes and interphase lymphocyte nuclei and on bone marrow cells produces a differential removal of DNA from different chromosomes. All chromosome regions are cleaved and removed, except for the constitutive heterochromatin of both chromosomes 9 and the pericentromeric heterochromatin of chromosome 3 which is polymorphic for its removal. Thus, homozygous uncleaved (3++), homozygous cleaved (3− −) and heterozygous (3+−) chromosome pairs may be identified when the cells are stained with DAPI (Gosálvez *et al.*, in press). There exists the possibility of exploiting this type of variation in the treatment of leukaemia by bone marrow transplant. In cases where the polymorphism on chromosome 3 can discriminate a bone marrow recipient and their donor the evolution of haemopoiesis, as measured by the frequency of donor versus recipient cells, can be determined at intervals following transplantation (Gosálvez, in preparation).

2.3 ALTERNATIVE METHODS TO VISUALIZE *IN SITU* RE-DIGESTION

2.3.1 Restriction endonucleases combined with *in situ* nick translation

There is an alternative method to visualize the sites of RE attack on fixed chromosomes: the restriction endonuclease/nick translation technique (RE/NT). This selectively labels those chromosome regions with a high frequency of restriction sites and whose DNA has not been removed. The method requires short *in situ* cleavage of preparations with appropriate REs to produce enough DNA nicking to allow polymerization of new DNA motifs in the presence of labelled nucleotides. The cuts generated by the REs are used as starting points for DNA extension and the newly synthesized DNA motifs are *in situ* catalyzed by polymerase. Finally, those incorporated labelled nucleotides are detected by different immunocytochemical methods. It is widely believed that because the RE/NT banding method is not based on DNA extraction, it is more sensitive than direct DNA removal using massive RE cleavage and staining. However, the problems of accessibility of REs to their targets in the fixed chromatin still remain. Thus, the action of some REs, which induces a clear cutting pattern on naked DNA, could be prohibited by a specific chromatin structure in a chromosome region. Interestingly, there are cases where REs cannot gain access to their targets unless the conformation of the chromatin is altered. For example, the restriction enzyme Bst NI cuts the mouse major DNA satellite, localized in each centromeric region, into fragments of 234 bp. However, this enzyme does not seem to be very effective to *in situ* digest the mouse centromeric heterochromatin unless it is undercondensed by treatment in the presence of 5-azacytidine (de la Torre *et al.*, 1991). This demonstrates an important limitation of the RE/NT technique and indicates that negative results (the absence of nick translated chromosome regions) do not necessarily imply the absence of restriction sites.

Given the resolution of light microscopy and the proven sensitivity of this technique, one of the most easily understood applications of the RE/NT procedure is for detecting highly or moderately repetitive DNA sequences on chromosomes. Adolph and Hameister (1985) and Bullerdiek *et al.* (1985) first reported results using this method on human chromosomes, although firmer conclusions about the banding patterns obtained came from subsequent experiments with mouse chromosomes (Adolph, 1988; de la Torre *et al.*, 1991). In the latter case, although the response of the DNA to RE cleavage is time-dependent, enzymes such as Sau 96I produce C-like bands which are complementary to those produced after Eco RI digestion, where all centromeric regions stain only palely (de la Torre *et al.*, 1991). Adolph (1988) applied the RE/Giemsa and RE/NT procedures to mouse chromosomes using the restriction enzymes Ava II and Eco RI. Both enzymes gave complementary results with nick translation and Giemsa staining. However, this complementarity, as a consequence of the distribution of restriction sites, does not always occur. Sumner *et al.* (1990), in an extensive study of human chromosomes, found that the patterns obtained after RE/NT were not necessarily the reverse of those seen after RE/NT/Giemsa staining. They concluded that the patterns obtained after RE digestion and Giemsa staining or nick translation depend not only on the distribution of the enzyme recognition sites on the chromosomes, but also on the accessibility of the REs and extraction of the chromosomal DNA.

Recently, this method has also been shown to operate in the genetic systems of amphibians and insects and has provided new insights into the *in situ* analysis of the structure of repetitive DNA sequences. Thus,

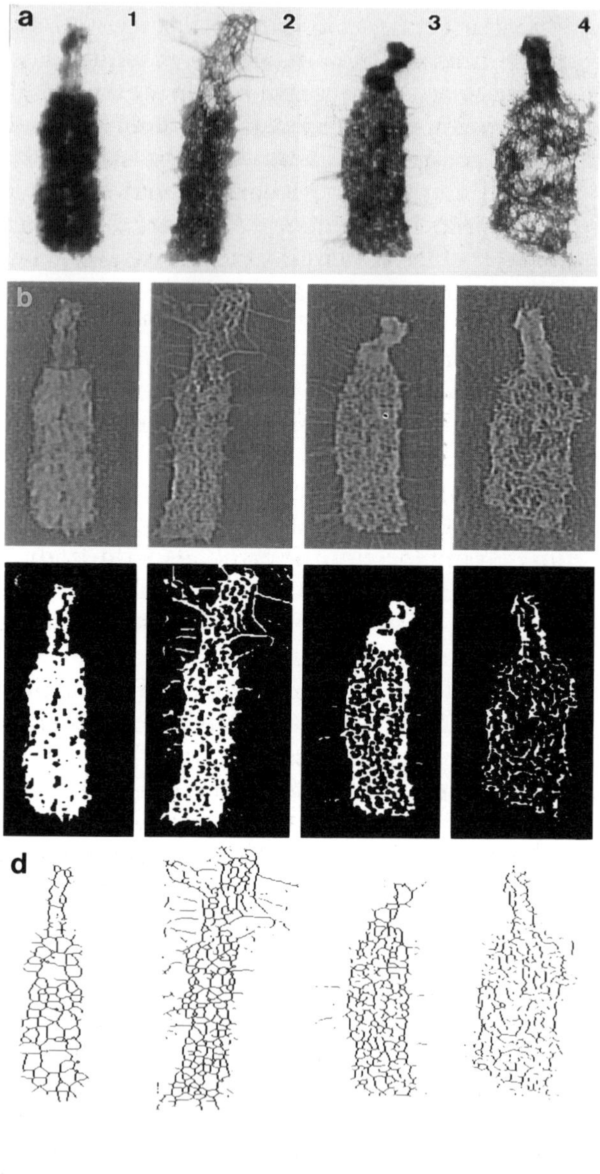

Fig. 2.3 Whole-mounted L929 mouse chromosomes digested with 3 different REs and processed with the same image analysis routine to show the differences obtained in the chromatin structure. a—d) Sequence of image proccessing. a1) control chromosomes. a2) chromosome digested with Alu I. a3) chromosome digested with Msp I. a4) chromosome digested with Hae III. The skeletons derived from a1—4 provide a series of functions that can be mathematically compared.

RE/NT has been of use in species where RE/Giemsa provides poor chromosome differentiation. *Triturus marmoratus* is a good example, since fixed chromosomes of this species do not yield chromosome bands after digestion with various restriction enzymes and Giemsa or ethidium bromide staining. However, when the same REs were used for *in situ* nick translation, specific banding patterns were obtained (Herrero *et al.*, 1993). This study demonstrates the presence of specific subsets of repetitive DNAs within the apparently homogeneous C-heterochromatin of this species, as revealed by standard C-banding. The RE/NT method has also been effective in demonstrating homologous heterochromatin heterogeneity possibly due to a difference in the presence of restriction sites within a tandemly repeated DNA sequence. In the grasshopper *Arcyptera fusca*, it was found in one individual that only one of the two homologous constitutive heterochromatic blocks of one autosome pair was differentially labelled after Sau 3A/NT. Interestingly, further studies demonstrated that frequencies of individuals with both homologous centromeres labelled and those with neither labelled co-existed in a population in a Hardy—Weinberg equilibrium (de la Torre *et al.*, 1993).

Some REs have the dinucleotide CpG within their recognition sequence and of these, some are affected by methylation. Several pairs of isoschizomers have been reported which can be affected by methylation of cytosines adjacent to guanines in such a way that while one of them is inhibited by methylation, the other is not (McClelland, 1981). The isoschizomer pair Hpa II (CCGG, sensitive to methylation of the internal cytosine) and Msp I (same restriction site, insensitive to methylation of the internal cytosine) has been used both in direct RE/Giemsa and in RE/NT experiments to distinguish relatively unmethylated regions from methylated ones in an attempt to establish a difference between active gene-enriched chromosome regions and those where inactive sequences are resident. It is known that a high level of methylation occurs in most eukaryotic genomes and that CpG sequences are those commonly affected. Most genes in mammals appear to be associated with high concentrations of the dinucleotide CpG, which is otherwise rare in other parts of the genome (Bird, 1986, 1987). These concentrations of CpG dinucleotides, which have also been found in higher vertebrates and in some plants, are generally unmethylated and can be digested selectively by appropriate REs that include this dinucleotide in their recognition sequence. Their frequency in invertebrates is not currently known, but the available evidence from a variety of eukaryotes suggests that CpG islands may be restricted to certain taxonomic groups. Use of the RE/NT technique to localize CpG islands has been carried out largely in humans where the cytological evidence of the distribution of active gene regions, inferred from the results obtained from DNase I hypersensitivity, indicates that they tend to occupy distal positions on the chromosomes (see below). The isoschizomers Hpa II and Msp I as well as the REs Cfo I and Npa II (both containing the CpG doublet) produce a banding pattern similar to the conventional R-banding pattern (Sumner *et al.*, 1990; Adolph and Hameister, 1990; de la Torre *et al.*, 1992). At least in human chromosomes, therefore, there is a good correlation between distribution of active genes and chromosome regions with higher concentrations of CpG.

2.4 THE EFFECT OF NON-SPECIFIC NUCLEASES ON FIXED CHROMOSOMES

2.4.1 Action of DNase I

As we have mentioned above, potentially active genes are organized in a conformation that renders them hypersensitive to digestion with DNase I (Weintraub and Groudine, 1976), a property which is maintained even in fixed metaphase chromosomes (Gazit *et al.*, 1982). Taking advantage of this, several workers (reviewed in Sumner *et al.*, 1993) have mapped the distribution of genes on chromosomes by *in situ* nick translation using those nicks produced by the DNase I at low concentrations i.e. before massive DNA removal. The pattern of DNAse I hypersensitivity on mammalian chromosomes was found to resemble that of R-banding but was more variable in general. Recently, studies of human metaphase chromosomes have shown that the H3 isochore family, which contains the highest gene concentrations, is found mainly in distal regions of chromosomes and on those bands that are positive to staining with chromomycin A3 (Saccone *et al.* 1992). These results establish a clear correspondence between the locations of DNase I hypersensitive sites and potentially active genes. Since mammalian chromosomes (compartmentalized in G/R bands) were the subject of the first experiments, it was not known for some time whether this technique would produce results with the genomes of other taxa where no such bands had been detected. We now know that in *Rana perezi* (Amphibia) and in *Chorthippus parallelus* (Orthoptera) the chromosomes sites for preferential DNase-nicking are concentrated at the distal chromosome regions where a preferential nick translation is detected (Herrero *et al.*, 1995; de la Torre *et al.*, 1996). In *R. perezi* the clusters of rDNA (where there is a particularly high concentration of gene activity) are hypersensitive to DNase I, a pattern similar to that found in human and mouse chromosomes. Overall, DNase I hypersensitive chromosome regions detected with DNase/NT seem to be related to the distribution of active or potentially active genes, and so inactive genes — probably those that are tissue-specific — will not be detected. However, in meiotic cells, DNase I hypersensitivity need not be indicative of the potential for transcription, but may instead signal sites of pairing and crossing-over (Chandley and McBeath, 1987).

2.4.2 Action of Exonuclease III

Single-stranded DNA, as required for *in situ* hybridization purposes, is generally obtained by DNA denaturation using either high temperatures or reagents such as NaOH and formamide. Enzymatic degradation of one of the DNA strands is possible in chromosomes prepared for light microscopy by nicking with a RE, followed by digestion with Exonuclease III (Exo-III) permitting the localization of alphoid DNA sequences (van Deken *et al.*, 1988). This enzyme releases 5' mononucleotides from the 3' ends of nicked double-stranded DNA and amplifies the breaks to form long single-stranded stretches of DNA. Exo-III can be directly used on methanol—acetic acid-fixed and spread air-dried chromosomes without previous RE-nicking. In such circumstances, the enzyme removes nucleotides at spontaneously occurring nicks and breaks in the DNA. In both cases extensive single-stranded DNA motifs are produced that can act as targets for *in situ* hybridization of labelled DNA probes. Consequently, these single-stranded DNA areas can also hybridize with oligonucleotides of random sequences, which can then be extended by the Klenow fragment of *E. coli* DNA polymerase I. Incorporation of labelled nucleotides by the polymerase can then be detected under fluorescence microscopy using a fluorochrome-tagged reporter molecule.

The RE/Exo-III method could be adapted to whole-mounted chromosomes prepared for electron microscopy using gold-tagged reporter molecules, permitting a detailed visualization of the chromosome labelling produced. In undercondensed mouse chromosomes *in situ* random priming and refilling of the single-stranded DNA motifs produced with Exo-III shows a differential distribution of gold particles along the chromosomes, depending on the RE used. After nicking with Eco RI the gold particles are homogeneously distributed along the chromosome arms but not within the centromeric heterochromatin. In fact, a short chromatin area within juxtacentromeric undercondensed heterochromatin is not labelled by this method (Gosálvez *et al.*, 1993a). Only those heterochromatic regions close to the chromosome arms appear positively labelled. According to the heterochromatin distribution obtained after C-banding, the unlabelled region is included within the pericentric constitutive heterochromatin. This intra-heterochromatic unlabelled region specifically obtained after Eco RI cleavage would thus correspond to the chromosome domain where the minor satellite is located. This interpretation coincides with that obtained after immunolocalizing a 120 bp genomic consensus sequence of the minor satellite which occupies the central domain within the centromeric regions (Wong and Rattner 1988). These sequences strictly co-localize with the kinetochore.

Exo-III has also been used to test the effect of nicking on chromatin of several agents that can induce DNA damage such as ionizing radiation, H_20_2 treatment and photosensitization of DNA that has incorporated BrdU. The results were compared with those from control untreated chromatin and an increase in chromosome damage was always observed (Gosálvez *et al.,* 1993b). Exo-III was also used to produce sister chromatid differentiation (SCD) in chromosomes after differential incorporation of 5-Bromodeoxyuridine (BrdU) in sister chromatids. In fact, short UV-treatments, capable of inducing nicks in chromosomal DNA corresponding to BrdU residues, made subsequent Exo-III digestion able to produce SCD. Under these conditions, the bifilarly BrdU-substituted chromatids are able to be attacked to a greater extent than are the unifilarly BrdU-substituted chromatids (Mezzanotte *et al.*, 1989).

Perhaps the most promising use of Exo-III *in situ* treatment of chromosomes is to reveal structural differences in chromatin organization. Such is its efficiency that it is even able to unveil structural differences between chromosomes of different tissues. Thus, when Exo-III is directly applied to recently spread mitoses from human amniotic fluid cells, the chromosomes show a faint R-banding pattern after random priming. This banding pattern coexists with highly labelled pericentromeric constitutive heterochromatin of chromosome 9. Constitutive heterochromatin of chromosomes 1, 16, and Yq was also highlighted, though less intensely than in chromosome 9. Nevertheless, when chromosomes from human peripheral blood lymphocytes were similarly processed, only a highly contrasted R-banding pattern was obtained (Fernández *et al.*, 1995). These R-bands are similar to those obtained from thermally denatured human lymphocyte chromosomes that have been *in situ* random primed and extended (García de la Vega *et al.*, 1990). The euchromatic bands appear to be more sensitive to Exo-III digestion and so generate more single-stranded DNA. The comparative study of the effects of thermal DNA denaturation on the distribution of chromosomal painting probes and the distribution of the same probes after Exo-III digestion also indicates that the action of Exo-III is not homogeneously distributed throughout the chromosomal arms. In fact, painting probes do not hybridize homogeneously along the chromosome as they do when thermal DNA denaturation is used (Fig. 2.2e). The differential sensitivity between R and G bands to Exo-III is probably also dependent on the factors which determine the DNase I sensitivity commented above.

Finally, Exo-III has also revealed differences in the chromosome organization of repetitive DNA sequences of human chromosomes (Fernández *et al.*, 1995). Only alphoid sequences are susceptible to hybridization after Exo-III digestion to the same extent as on thermally denatured chromosomes and only a decrease of 10% in the hybridized signal is noted in Exo-III-treated chromosomes (Fig. 2.2f, g; h, i). However, classical satellite DNA probes specific for chromosome 9 and Y (D9Z1 and DYZ1) show a drastic reduction of approximately 50% in their capacities to hybridize the respective probes after Exo-III digestion. Also, the intensity of their hybridization signals is clearly weaker than in thermally denatured chromosomes (Fig. 2.2j, k; l, m). The effectiveness of Exo-III in demonstrating differences in the level of organization of the chromatin offers us the possibility of establishing regional differences that would otherwise remain cryptic to other techniques.

ACKNOWLEDGEMENTS

The authors thank the selfless contributions of our colleagues P.L. Mason, J. de la Torre, P. Herrero, J.L. Bella, C. García de la Vega, I. Buño, J.L. Fernández and J.L. Díez-Martín. Probably, without their help, this manuscript would have taken forever to produce. Recent results of workers in our groups are the fruits of the generous financial support of FIS (94/0212) and SAF (93—0093) (Spain).

REFERENCES

Adolph, S. (1988) 'In situ' nick translation distinguishes between C-band positive regions on mouse chromosomes. *Chromosoma,* **96**, 102—06.

Adolph, S. and Hameister, S. (1985) 'In situ' nick translation of metaphase chromosomes with biotin-labeled dUTP. *Human Genetics,* **69**, 117—21.

Adolph, S. and Hameister, S. (1990) 'In situ' nick translation of human metaphase chromosomes with the restriction enzymes Msp I and Hpa II reveals an R-band pattern. *Cytogenetics and Cell Genetics*, **54**, 132—36.

Alfi, O.S., Donnell, G.N. and Derencsenyi, A. (1973) C-banding of human chromosomes produced by DNase. *Lancet*, **2**, 505.

Babu, A. (1988) Heterogeneity of heterochromatin of human chromosomes as demonstrated by restriction endonuclease treatment, in *Heterochromatin. Molecular and structural aspects.* (ed R.S. Verma), Cambridge University Press, Cambridge, pp. 250—75.

Bianchi, M.S., Bianchi, N.O., Pantelias, G.E. and Wolff, S. (1985) The mechanism and pattern of banding induced by restriction endonucleases in human chromosomes. *Chromosoma* **91**, 131—6.

Bird, A.P. (1986) CpG islands and the function of DNA methylation. *Nature*, **321**, 209—13.

Bird, A.P. (1987) CpG islands as gene markers in the vertebrate nucleus. *Trends in Genetics*, **3**, 342—47.

Bullerdiek, J., Dittmer, J., Faehre, A. *et al* (1985) A new banding pattern of human chromosomes by 'in situ' nick translation using Eco RI and biotin-dUTP. *Clinical Genetics,* **28**, 173—76.

Burkholder, G.D. (1989) Morphological and biochemical effects of endonucleases on isolated mammalian chromosomes 'in vitro'. *Chromosoma*, **97**, 347—55.

Caspersson, T., Farbel, S., Folley, G.E. *et al* (1968) Chemical differentiation along metaphase chromosomes. *Experimental Cell Research*, **49**, 219—22.

Chandley, A.C. and McBeath, S. (1987) DNase I hypersensitive sites along the XY bivalent at meiosis in man include the XpYp pairing region. *Cytogenetics and Cell Genetics*, **44**, 22—31.

Fernández J.L., Campos A., López-Fernández *et al* (1995) Difference in constitutive heterochromatin behaviour between human amniocytes and lymphocytes detected by a sequential *in situ* exonuclease III digestion-random primer extension procedure. *Journal of Medical Genetics*, **32**, 32—35.

Gall, J.G. (1963) Kinetics of deoxyribonuclease action on chromosomes. *Nature*, **198**, 36—38.

García de la Vega, C., Martínez-Zapater, J.M., López-Fernández, C. *et al* (1990) 'In situ' random primer extension of metaphase chromosomes. *Cytogenetics and Cell Genetics,* **53**, 221—12.

Gazit, B., Cedar, H., Lerer, I. *et al* (1982) Active genes are sensitive to deoxyribonuclease I during metaphase. *Science*, **217**, 648—50.

Gosálvez, J. and Goyanes, V. (1988) Selective digestion of mouse chromosomes with restriction endonucleases.I Scaffold like structures and bands by electron microscopy. *Cytogenetics and Cell Genetics*, **48**, 198-00.

Gosálvez, J., López-Fernández, C., Buño, I., *et al* (in press) Restriction endonucleases in situ digestion (REISD) and fluorescence in situ hybridization (FISH) as complementary methods to analyze chimaerism and residual disease after bone marrow transplantation. *Cancer Genetics and Cytogenetics*.

Gosálvez, J., López-Fernández, C., Fernández, J.L. *et al* (1993b) 'In situ' oligonucleotide priming to detect induced DNA breaks under electron microscopy. *Trends in Genetics,* **9**, 156—57.

Gosálvez, J., López-Fernández, C., Ferrucci, L. *et al* (1989) DNA base sequence is not the only factor accounting for restriction endonuclease activity on metaphase chromosomes: evidence using isoschizomeres. *Cytogenetics and Cell Genetics*, **50**, 142—44.

Gosálvez, J., López-Fernández, C., García de la Vega, C. *et al* (1993a) Selective digestion of mouse chromosomes with restriction endonucleases. Oligonucleotide priming of single-stranded DNA produced with exonuclease III. *Genome*, **36**, 230—34.

Gosálvez, J., Mezzanotte, R., López-Fernández, C. *et al* (1991) Selective digestion of mouse chromosomes with restriction endonucleases. II. X-ray microanalysis of Hae III-treated chromosomes. *Cytogenetics and Cell Genetics*, **56**, 82—86.

Gosálvez, J., Sumner, A.T., López-Fernández, C. *et al* (1990) Electron microscopy and biochemical analysis of mouse metaphase chromosomes after digestion with restriction endonucleases. *Chromosoma*, **99**, 155-57.

Gould, D.C., Callan, H.G. and Thomas, C.A. (1976) The action of restriction endonucleases on lampbrush chromosomes. *Journal of Cell Science*, **21**, 303—13.

Goyanes, V. (1985) Electron microscopy of chromosomes: toward an ultrastructural cytogenetics?. *Cancer Genetics and Cytogenetics*, **15**, 349—67.

Herrero, P., Torre, J. de la., Arano, B. *et al* (1995) Patterns of DNase sensitivity in the chromosomes of *Rana perezi* (Amphibia: Anura). *Genome*, **38**, 339—43.

Herrero, P., Torre, J. de la., Gosálvez, J. *et al* (1993) Heterochromatin heterogeneity in *Triturus marmoratus* (Urodela: Salamandridae) demonstrated with specific DNA-binding fluorochromes and 'in situ' Restriction endonuclease/Nick translation. *Caryologia,* **4**, 343—53.

Jones, K.W. (1977) Repetitive DNA and primate evolution, in *Molecular structure of human chromosomes*.(ed J.J.Yunis), Academic Press, New York, pp. 295—10.

Juan, C., Gosálvez, J., Mezzanotte, R. *et al* (1991) Cytological and biochemical characterization of the *in situ* endonuclease digestion of fixed *Tenebrio molitor* chromosomes. *Chromosoma,* **100**, 432—38.

Kaelbling, M., Miller, D.A. and Miller, O.J. (1984) Restriction enzyme banding of mouse metaphase chromosomes. *Chromosoma*, **90**, 128—32.

Kato, H., Tsuchiya, K. and Yosida, T.H. (1974) Constitutive heterochromatin of indian muntjack chromosomes revealed by DNase treatment and a C-banding technique. *Canadian Journal of Genetics and Cytology*, **16**, 273—80.

Korenberg, J.R. and Rykowski, M.C. (1988) Human genome organization: Alu, LINES and the molecular structure of metaphase chromosome bands. *Cell,* **53**, 391—00.

Lloyd, M.A. and Thorgaard, G.H. (1988) Restriction endonuclease banding of rainbow trout chromosomes. *Chromosoma*, **96**, 171—77.

López-Fernández, C., Gosálvez, J., Ferrucci, L, *et al* (1991) Restriction endonucleases in the study of eukaryotic chromosomes. *Genetica*, **83**, 257—74.

Ludeña, P., Sentís, C., De Cabo, S. F., *et al* (1991) Visualization of R-bands in human metaphase chromosomes by the restriction endonuclease Mse I. *Cytogenetics and Cell Genetics*, **57**, 82—86.

McClelland, M. (1981) The effect of sequence specific DNA methylation on restriction endonuclease cleavage. *Nucleic Acids Research*, **9**, 1859—66.

Mezzanotte, R., Bianchi, U., Vanni, R. *et al* (1983b) Chromatin organization and restriction nuclease activity on human metaphase chromosomes. *Cytogenetics and Cell Genetics*, **36**, 562—66.

Mezzanotte, R., Ferrucci, L., Vanni, R. *et al* (1983a) Selective digestion of human metaphase chromosomes by Alu I restriction endonuclease. *Journal of Histochemistry and Cytochemistry,* **31**, 553—56.

Mezzanotte, R., Ferrucci, L., Vanni, R. *et al* (1985) Some factors affecting the action of restriction endonucleases on metaphase chromosomes. *Experimental Cell Research*, **161**, 247—53.

Mezzanotte, R., Rossino, R., Nieddu, M. *et al* (1992) The DNA fragments produced by Alu I and Bst NI digestion of fixed mouse chromosomes. *Chromosoma*, **101**, 641—44.

Mezzanotte, R. Rossino, R., Orru, S. *et al* (1989) Nuclease activity in human metaphase chromosomes substituted with 5-bromodeoxyuridine. *Chromosoma*, **97**, 334—38.

Miller, D.A., Choi, Y.C. and Miller, O.J. (1983) Chromosome localization of highly repetitive human DNA's and amplified ribosomal DNA with restriction enzymes. *Science*, **219**, 395—97.

Miller, D.A., Gosden, J.R., Hastie, N.D. *et al* (1984) Mechanism of endonuclease banding of chromosomes. *Experimental and Cell Research,* **155**, 294—98.

Nash, D. and Plaut, W. (1964) On the denaturation of chromosomal DNA 'in situ'. *Proceedings of National Academy of Science,* **51**, 731—35.

Noesis (1992) Visilog 4. Reference Guide. Velizy, France.

Pardue, M.L. and Gall, J.G. (1970) Chromosomal localization of mouse satellite DNA. *Science*, **186**, 1356—58.

Petitpierre, E., Gatewood, J.M. and Schmid, C.W. (1988) Satellite DNA from the beetle *Tenebrio molitor*. *Experientia*, **44**, 498—99.

Saccone, S., de Sario, A., della Valle, G. *et al* (1992) The highest gene concentrations in the human genome are in telomeric bands on metaphase chromosomes. *Proceedings of the National Academy of Science*, **89**, 4913—17.

Sahasrabudhe, C.G., Pathak, S. and Hsu, T.C. (1978) Response of mammalian metaphase chromosomes to endonuclease digestion. *Chromosoma*, **69**, 331—38.

Southern E.M. (1975) Long range periodicity in mouse satellite DNA. *Journal of Molecular Biology*, **94**, 51—69.

Sumner, A.T. (1972) A simple technique for demonstrating centromeric heterochromatin. *Experimental Cell Research,* **75**, 304—06.

Sumner, A.T. (1990) *Chromosome banding.* Unwin Hyman Inc. Ltd. London.

Sumner, A.T., Evans, H.J. and Buckland, R.A. (1971) New technique for distinguishing between human chromosomes. *Nature New Biology*, **232**, 31—32.

Sumner, A.T., Taggart, M., Mezzanotte, R. *et al* (1990) Patterns of digestion on human chromosomes by restriction endonucleases demonstrated by 'in situ' nick translation. *Histochemistry Journal,* **22**, 639—52.

Sumner A.T., de la Torre J. and Stupia L. (1993) The distribution of genes on chromosomes: A cytological approach. *Journal of Molecular Evolution*, **37**, 117—22.

Tagarro, I. (1993) *Caracterización citogenético-molecular del ADN satélite humano de secuencia simple.* Tesis Doctoral. Universidad Autónoma de Madrid. Madrid.

Torre, J. de la, Herrero P., García de la Vega, *et al.* (1996) Patterns of DNase I sensivity in the chromosomes of the grasshopper *Chorthippus parallelus* (Orthoptera). *Chromosome Research*, **4**, 56—60.

Torre, J. de la, López-Fernández, C., Herrero, P. *et al* (1993) 'In situ' nick translation of meiotic chromosomes to demonstrate homologous heterochromatin heterogeneity. *Genome*, **36**, 268—70.

Torre, J. de la, Mitchell, A.R. and Sumner, A.T. (1991) Restriction endonuclease/nick translation of fixed mouse chromosomes: a study of factor affecting digestion of chromosomal DNA 'in situ' *Chromosoma,* **100**, 203—11.

Torre, J. de la, Sumner, A.T., Gosálvez, J. *et al* (1992) The distribution of genes on human chromosomes as studied by 'in situ' nick translation. *Genome*, **35**, 890—94.

Ugarkovic, D., Plohl, M., Petitpierre, E. *et al* (1994) *Tenebrio obscurus* satellite DNA is resistant to cleavage by restriction endonucleases *in situ*. *Chromosome Research*, **2**, 217—23.

van Dekken, H., Pinkel, D., Mulikin, J. *et al* (1988) Enzymatic production of single-stranded DNA as target for fluorescence 'in situ' hybridization. *Chromosoma*, **97**, 1—15.

Weintraub, H. and Groudine, M. (1976) Chromosomal subunits in active genes have an altered conformation. *Science*, **93**, 848—56.

Wong, A.K.C. and Rattner, J.B. (1988) Sequence organization and cytological localization of the minor satellite of mouse. *Nucleic Acid Research*, **16**, 1645—61.

3

Y-linked genes and spermatogenesis in primates

W. Schempp

3.1 INTRODUCTION

Phylogenetically, the human Y is of recent origin, having reached its present configuration during the divergent evolution of the branches leading to great apes and to human.

Owing to its evolutionary origin and fate the Y chromosome is characterized by its bipartite structure. There is the still X—Y homologous pairing segment, the pseudoautosomal region (PAR) with several pseudoautosomal genes on it, as well as genes or pseudogenes that have counterpart genes on the X chromosome. Within the PAR pairing and recombination occurs during spermatogenesis and this may be important for the correct segregation of the X and Y chromosomes. On the other hand, the Y-specific part including *SRY*, the primary sex-determining locus, must avoid recombination as otherwise the sex-determining pathway would break down. In this context it is interesting to note the hazardous proximity of the *SRY* gene to the PAR on the human Y chromosome and furthermore that this proximity of the *SRY* gene to the Y chromosomal PAR seems to be maintained in all simian primate species. Despite numerous molecular genetic studies of the human Y chromosome during the past decade, only a few Y-specific functional genes have been identified, and this paucity of genes may be characteristic for this chromosome. In addition to *SRY* only three further Y-specific expressed genes have been described so far on the human Y chromosome. All three genes are expressed specifically in the testis and therefore might be functionally included in spermatogenesis.

Cytogenetic investigations of sterile men suggested the location of a gene controlling spermatogenesis, the azoospermia factor (AZF), in band q11.23 of the Y chromosome (Tiepolo and Zuffardi, 1976; Chandley *et al.*, 1989). AZF has been mapped to a deletion in Y chromosome interval 6 (Ma *et al.*, 1992) and, using probes known to map within this Y chromosome deletion, a novel gene family, the Y chromosome RNA recognition motif (*YRRM*) family, has been isolated and characterized (Ma *et al.*, 1993). Members of this *YRRM* family may consitute candidates for the *AZF* locus. Most recently, a further testis-specific gene, the *DAZ* (deleted in azoospermia) gene in Yq11.23, could be identified by Reijo *et al.* (1995). The *DAZ* gene is transcribed in the adult testis and appears to encode an RNA binding protein. As the *DAZ* gene is deleted in 12 of 89 men with azoospermia it serves as the new candidate gene for AZF. Another testis-specific gene, *TSPY* (testis specific protein Y encoded), was originally identified in an attempt to detect transcribed DNA sequences specific for the Y chromosome. Analysis of the cDNA clone and subsequent comparison with genomic DNA led to the initial speculation of an intronless gene (Arnemann *et al.*, 1987; 1991). Work by Zhang *et al.* (1992) and Manz *et al.* (1993), however, provided evidence that *TSPY* is present in several copies and that at least three differently sized *TSPY* transcripts exist, generated by different patterns of splicing. The exact number of transcibed genes or pseudogenes is not yet known. The description of eight cosmids representing 15 *TSPY* elements (Manz *et al.*, 1993) provided evidence that all *TSPY* sequences are organized as constitutive parts of the DYZ5 repeat unit, and, as was already shown for the DYZ5 repeat unit (Tyler-Smith *et al.*, 1988), interindividual variability with respect to the *TSPY* copy number occurs.

The *TSPY* sequence family remains Y-specifically conserved and multicopy in various primates, the cow and the pig (Arnemann *et al.*, 1987; Zhang *et al.*, 1992; Jakubiczka *et al.*, 1993), while no signals have been detected in small mammals such as rat, mouse and hamster (Jakubiczka *et al.*, 1993). Interestingly, the *YRRM* sequence family also remains male specific and multicopy in various higher and lower primate species tested. Among several non-primates analysed, only the shrew and the mouse show male-specific conservation of *YRRM* sequences, while in most other species additional signals are shared by both sexes, and no Y specificity at all is seen in the marsupial potoroo (Ma *et al.*, 1993; Schempp *et al.*, 1995).

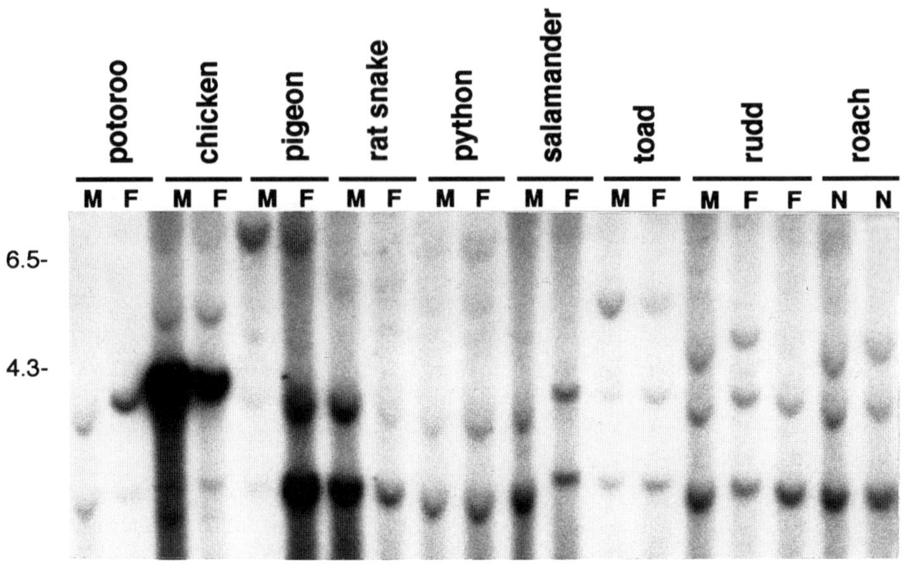

Fig. 3.1 Conservation of *YRRM* sequences in vertebrates. Southern blot analysis of *Eco*RI-digested male (M) and female (F) DNAs of the marsupial potoroo and several species of birds, reptiles and fishes. N: sex not determined. The signals are predominantly shared by both sexes and no male specific signals could be detected in any species tested. Fragment lengths are shown on the left in kilobases.

3.2 CONSERVATION AND ORGANIZATION OF Y-LINKED GENES

Here we present our hitherto unpublished Southern blot data demonstrating the conservation of *YRRM* sequences in several species of birds, reptiles and fishes, respectively (Fig. 3.1). As was already shown for the marsupial potoroo, no sex-specificity is seen in these species.

Using chromosomal *in situ* hybridization (FISH) we have recently demonstrated that the members of the *YRRM* and *TSPY* gene families are Y chromosome specific in hominoids (Schempp *et al.*, 1995). After hybridization with a *YRRM* cosmid the main signals are detected in human Yq11.23 as well as weaker secondary signals in Yp11.2. The reverse situation is found after hybridization with *TSPY* cosmids, the

Fig. 3.2 ***Colour Plate 2*** Two-colour FISH with the *YRRM* cosmid A5F (red) and the *TSPY* cosmid 2.2133 (green) on mitotic Y chromosomes of the human (a), the pygmy chimpanzee (b) and the pig-tail macaque (c). The yellow signal spots are due to overlapping *YRRM* (red) and *TSPY* (green) signals.

Fig. 3.3 ***Colour Plate 2*** Two-colour FISH with the *YRRM* cosmid A5F (red) and the *TSPY* cosmid 2.2133 (green) on released Y chromatin in interphase of the human (a) and the pygmy chimpanzee (b). Note that most *YRRM* (red) and *TSPY* (green) signals are mainly separated, while only a few yellow signal spots appear due to overlapping *YRRM* (red) and *TSPY* (green) signals.

main signals being in Yp11.2 and the secondary signals in Yq11.23. Surprisingly, the hybridization signals of *YRRM* and *TSPY* cosmids seem to coincide on the Y chromosome of great apes and the white hand gibbon (Schempp *et al.*, 1995).

In order to get a better resolution of *YRRM* and *TSPY* signals we performed two-colour FISH on Y chromosomes of the human and various higher primate species in mitotic stages as well as on released Y chromatin in interphase nuclei (Senger *et al.*, 1994) of man and pygmy chimpanzee. In all our experiments the *YRRM*-related cosmid A5F (Taylor *et al.*, 1991) was labeled with digoxigenin and detected with TRITC resulting in red fluorescence signals, while the *TSPY*-related cosmid 2.2133 (Taylor *et al.*, 1991) was biotinylated and detected with FITC resulting in green signals. Some examples of two-colour FISH on Y chromosomes of different species are given in Figure 3.2. Apparently, no further resolution of the reverse and complementary signal pattern of main and secondary signals of *YRRM* and *TSPY* cosmids can be achieved on human Y chromosomes in prometaphase (Fig. 3.2a). Overlapping *YRRM* main (red) and *TSPY* secondary signals (green) result in a yellow signal in Yq and vice versa in Yp. As in the human no further resolution of the *YRRM* and *TSPY* signal pattern could be found on the Y chromosome of the pygmy chimpanzee (Fig. 3.2b) and the pig-tail macaque (Fig. 3.2c). Our results of two-colour FISH of *YRRM* and *TSPY* cosmids on released Y chromatin in interphase are exemplified for the human and the pygmy chimpanzee in Figure 3.3. Even on moderately released Y chromatin structures as shown in Figure 3.3, *YRRM* and *TSPY* signals are mostly separated into red and green spots, although yellow signal spots due to superimposed red (*YRRM*) and green (*TSPY*) signals are still visible. In our maximum released Y chromatin stretches, however, the *YRRM* and *TSPY* signals are completely separated into red and green spots on human Yq and Yp (Conrad *et al.*, in preparation).

3.3 CONCLUSIONS

These results unequivocally lend support to the view that the *YRRM* and *TSPY* gene families have an interspersed organization on the Y chromosome of man and higher primates. Thus the distribution of *YRRM* and *TSPY* signals might well go back to a common organization of these genes next to each other on an ancient Y chromosome. Concerning possible functional role in spermatogenesis it was shown that expression of both *YRRM* and *TSPY* genes appears to be confined to germ cells of the spermatogonial and/or early spermatocyte stages in adult males (Chandley and Cooke, 1994). However, more detailed understanding of the possible role of *YRRM* and *TSPY* genes in male gametogenesis will require in the first stage their closer functional analysis.

ACKNOWLEDGEMENTS

I wish to thank Ann Chandley, Joachim Arnemann, Kay Taylor and Jonathan Wolfe for kindly providing the *YRRM* and *TSPY* probes. I am especially grateful to Ann Chandley for initiating these studies and for many helpful discussions. This work was supported by grants from the Deutsche Forschungsgemeinschaft.

REFERENCES

Arnemann, J., Epplen, J.T., Cooke, H.J. *et al* (1987) A human Y-chromosomal DNA sequence expressed in testicular tissue. *Nucleic Acids Research*, **15**, 8713—24.

Arnemann, J., Jakubiczka, S., Thüring, S., Schmidtke, J. (1991) Cloning and sequence analysis of a human Y chromosome-derived, testicular cDNA, TSPY. *Genomics*, **11**, 108—14.

Chandley, A.C., Gosden, J.R., Hargreave, T.B. *et al* (1989) Deleted Yq in the sterile son of a man with a satellited Y chromosome (Yqs). *Journal of Medical Genetics*, **26**, 145—53.

Chandley, A.C., Cooke, H.J. (1994) Human male fertility — Y-linked genes and spermatogenesis. *Human Molecular Genetics*, **3**, 1449—52.

Jakubiczka, S., Schnieders, F., Schmidtke, J. (1993) A bovine homologue of the human TSPY gene. *Genomics*, **17**, 732—5.

Ma, K., Sharkey, A., Kirsch, S., *et al* (1992) Towards the molecular localization of the AZF locus: mapping of microdeletions in azoospermic men within 14 subintervals of interval 6 of the human Y chromosome. *Human Molecular Genetics*, **1**, 29—33.

Ma, K., Inglis, J.D., Sharkey, A. *et al* (1993) A Y chromosome gene family with RNA-binding protein homology: candidates for the azoospermia factor AZF controlling human spermatogenesis. *Cell*, **75**, 1287—95.

Manz, E., Schnieders, F., Müller-Brechlin, A., Schmidtke, J. (1993) TSPY-related sequences represent a microheterogeneous gene family organized as constitutive elements in DYZ5 tandem repeat units on the human Y chromosome. *Genomics*, **17**, 726—31.

Reijo, R., Lee, T.-Y., Salo, P. *et al* (1995) Diverse spermatogenic defects in humans caused by Y chromosome deletions encompassing a novel RNA-binding protein gene. *Nature Genetics*, **10**, 383—92.

Schempp, W., Binkele, A., Arnemann, J. *et al* (1995) Comparative mapping of YRRM- and TSPY-related cosmids in man and hominoid apes. *Chromosome Research*, **3**, 227—34.

Senger, G., Jones, T.A., Fidlerová, H. *et al* (1994) Released Chromatin: linearized DNA for high resolution fluorescence *in situ* hybridization. *Human Molecular Genetics*, **3**, 1275—80.

Taylor, K., Ulinowski, Z., Wolfe, J. (1991) Contig assembly of cosmid clones from the Y chromosome. *Cytogenetics and Cell Genetics*, **58**, 2096.

Tiepolo, L., Zuffardi, O. (1976). Localization of the factors controlling spermatogenesis in the nonfluorescent portion of the Y chromosome long arm. *Human Genetics*, **34**, 119—24.

Tyler-Smith, C., Taylor, L., Müller, U. (1988). Structure of a hypervariable tandemly repeated DNA sequence on the short arm of the human Y chromosome. *Journal of Molecular Biology*, **203**, 837—48.

Zhang, J.S., Yang-Feng, T.L., Müller, U. *et al* (1992). Molecular isolation and characterization of an expressed gene from the human Y chromosome. *Human Molecular Genetics*, **1**, 717—26.

4

Subchromosomal painting libraries (SCPLs) from somatic cell hybrids

M. Rocchi, R. Antonacci, R. Marzella, P. Finelli, C. Cassano, A. Lonoce, C. Cino, A. Forabosco and N. Archidiacono

4.1 INTRODUCTION

Fluorescent in situ hybridization (FISH) has become one of the most powerful techniques in human cytogenetics. The characterization of chromosomal rearrangements has been greatly facilitated by the availability of several kind of probes suitable for FISH experiments:

1. large inserts accommodated in vectors like phage, cosmid, BAC, YAC etc.;
2. probes recognizing satellite DNAs (alphoid DNA, satellite III etc.);
3. whole chromosome painting libraries (WCPL), obtained from sorted chromosomes (Collins *et al.*, 1991), or from somatic cell hybrids retaining single whole human chromosomes (Lichter *et al.* 1990; Liu *et al.* 1993);
4. microdissection libraries, specific for single chromosome bands (Meltzer *et al.*, 1992; Guan *et al.*, 1995);
5. subchromosomal painting libraries (SCPLs), specific for a definite chromosomal region. SCPLs can be generated, via Alu-PCR, from somatic cell hybrids retaining fragments of human chromosomes (Archidiacono *et al.*, 1994).

We recently reported a large panel of SCPLs derived from 140 human-hamster somatic cell hybrids: about 300 different SCPLs, spanning the entire genome, were identified (Antonacci *et al.*, 1995).

This panel of SCPLs can be of great help in the study of clinical cytogenetics cases (see, for example, Zelante *et al.*, 1994, and also examples reported in this paper), and in the study of karyotype evolution of primates (Muller *et al.*, 1996).

Most of the chromosome fragments found in the hybrids were the result of *in vitro* rearrangements which occurred by chance (hamster chromosomes are prone to rearrangements). Some hybrids were generated in order to isolate specific derivative chromosomes present in patients carrying well-characterized chromosome rearrangements (usually X/autosome translocations). In these cases the presence in the hybrid of a specific chromosome region was known in advance. This large panel of hybrids retaining fragments of human chromosomes can also be used for fine subregional mapping of short DNA probes not suitable for FISH experiments (Hulsebos *et al.*, 1995).

The full exploitation of these hybrids as sources for SCPLs and for subregional mapping purposes faces two major problems:

1. the simultaneous presence, in most hybrids, of extra fragments and/or entire human chromosomes, in addition to the fragment of interest, makes the interpretation of FISH results difficult, so that co-hybridization experiment are often necessary. Co-hybridization experiments, however, are difficult to deal with if digital imaging is not available (as in most clinical cytogenetic laboratories);
2. cytogenetic characterization, alone, is not completely satisfactory if the hybrids are intended to fulfil high resolution mapping demands.

These two problems can be avoided by (i) generating hybrids retaining a single or very few fragments as the only human contribution, and (ii) characterizing them using YAC or STSs (Sequence Tagged Sites) probes whose genetic and/or physical mapping has been established.

In the present paper we present data on the generation of a panel of radiation hybrids, specific for chromosome 10, each hybrid retaining a single or a few fragments of this chromosome. In addition, we report preliminary data on the fine characterization, by YACs, of a panel of somatic cell hybrids retaining fragments of the X-chromosome. As a conclusion, some explanatory examples of the use of SCPLs in the fine characterization of chromosome rearrangements encountered in clinical practice are illustrated.

4.2 CHROMOSOME 10 SCPL PANEL

The monochromosomal hybrid GM10926D (Camden Cell Repository) retaining chromosome 10 as the only human contribution, was irradiated at 5 Krads (first experiment) or 10 Krads (second experiment), then fused to a TK⁻CHO cell line. HAT selection was applied and, after 15—20 days, clones were picked-up and transferred to a 24-well plate. When a hybrid clone reached confluency in the 1.5 cm well, it was trypsinized. Half the cells were seeded in a 3.5cm Petri dish with a round coverslip at the bottom, for *in situ* cytology; the remaining cells were kept in the same well. *In situ* preparations were performed as described (Pecile *et al.*, 1984). Slides were hybridized *in situ* using biotinylated total human DNA to discriminate hybrids retaining human material. This experiment also gave precious information on the homogeneity of the hybrid. Hybrids were considered homogeneous enough if the percentage of positive cells was higher than 50%. Many clones, however, showed positive signals in 90—100% of the nuclei examined. Homogeneous clones were then expanded. DNA was extracted from a confluent large flask (175cm^2). The extraction yielded up to 1mg of DNA per flask. The DNA from each hybrid (30—50ng per 50μl reaction) was amplified by Alu-PCR. Alu-PCR products were labeled with biotin by nick-translation and hybridized to normal human metaphases, in order to cytogenetically characterize the human chromosome content (for technical details see Archidiacono *et al.*, 1994).

Tables 4.1a and b summarize the fragments obtained with the two series of experiments. The chromosome fragments, representing potential SCPLs, are reported as solid bars spanning the length of the fragment. Fig. 4.1a shows a human metaphase hybridized with biotinylated Alu-PCR product of hybrid RH171, retaining only the region 10q11.2→10p14.

Radiation hybrids available in the literature are frequently obtained by high dosage irradiation, so that small and numerous fragments per hybrid are obtained. For the present purposes, a low dosage experiment (5Krads) was found more satisfactory. Large uninterrupted fragments, indeed, are more useful as SCPLs: very small fragments are comparable to large YACs already available. These hybrids can be considered 'renewable' material, as opposite to radiation hybrids with multiple minute fragments. They can be put back in culture and expanded for further DNA extraction, since it is quite easy to check

Table 4.1 Ideogram of the banding pattern of chromosome 10. On the left of the ideogram are reported the fragments scored in radiation hybrids experiments (1a) and in normal somatic cell hybrids (1b). Each fragment is represented by a solid bar describing its length and subchromosomal location. The hybrid code number is also reported.

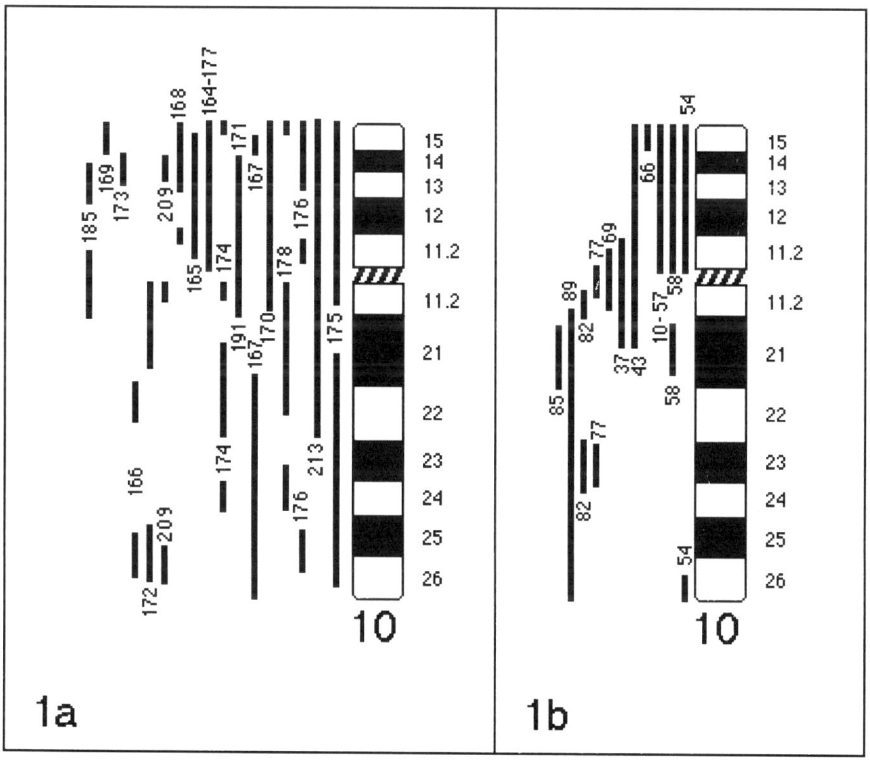

if the hybrid is still retaining the fragment identified in the primary clone. The small amount of DNA necessary for a single PCR reaction (30—50ng) enables about 30,000 reactions to be theoretically performed per mg of DNA. In addition, each 50μl amplification reaction can be used for about 15—20 FISH experiments.

4.3 CHROMOSOME X SCPL PANEL

Alu sequences are not regularly distributed along the human chromosomes (Korenberg and Rikowski, 1989). Therefore, Alu-PCR hybridization produces a banding pattern corresponding to R-banding (Baldini and Ward, 1991). As a consequence, the characterization of break-points located in pale interbands could be inaccurate, with potentially detrimental consequences especially if the hybrid is used for mapping studies. This applies particularly to some chromosomes: inter-bands 7q21 and 7q31, for instance, seem quite completely devoid of amplificable inter-Alu sequences. YACs and STSs, however, can be used to better characterize the extension of each fragment. YACs from Genethon are particularly suitable for the study since their genetic mapping (through STSs recognizing them), and occasionally their physical mapping, are established (Gyapay *et al.*, 1994).

In Table 4.2 a panel of fragments of the X-chromosome is diagrammatically reported. The extension of each segments, initially characterized by the Alu-PCR approach, has been refined by using a set of YAC probes distributed along the X-chromosome. The hybrids were tested as follows. Each hybrid cell line was harvested by trypsinization after two hours of colcemid. Hypotonic treatment was done using a 0.075M KCl solution, then the cells were fixed and washed three times in methanol:acetic acid 3:1. An appropriate amount of fixative containing human metaphases, obtained from PHA-stimulated peripheral blood lymphocytes, was added to the suspension. This mixture was used to prepare slides, according to standard procedures. The slides were then hybridized *in situ* with biotinylated PCR products from an individual YAC (Fig. 4.1b,c). The human metaphases were added for two reasons: the fine physical mapping of the YACs from Genethon was not always known; in addition they provide an internal positive control in case the hybrid is negative for a specific YAC. In these cases, indeed, one could not discard with certainty the possibility of a technical failure. The relative position along the X chromosome of YACs mapping in the same region was established by co-hybridizing the two differently-labelled YACs on elongated chromosomes (Fig. 4.1d). The results were always in agreement with the genetic mapping data, when available, derived from STS recognizing these YACs (Gyapay *et al.*, 1994).

Table 4.2 Ideogram of the chromosome X (left). Close to the ideogram, on the right, are reported the YAC probes with their location on the X chromosome (small bar). The chromosome fragments, with hybrid code, are represented by vertical solid bars. The table summarizes the results of the cytogenetic characterization of the hybrids and of FISH experiments of YAC probes on the hybrids.

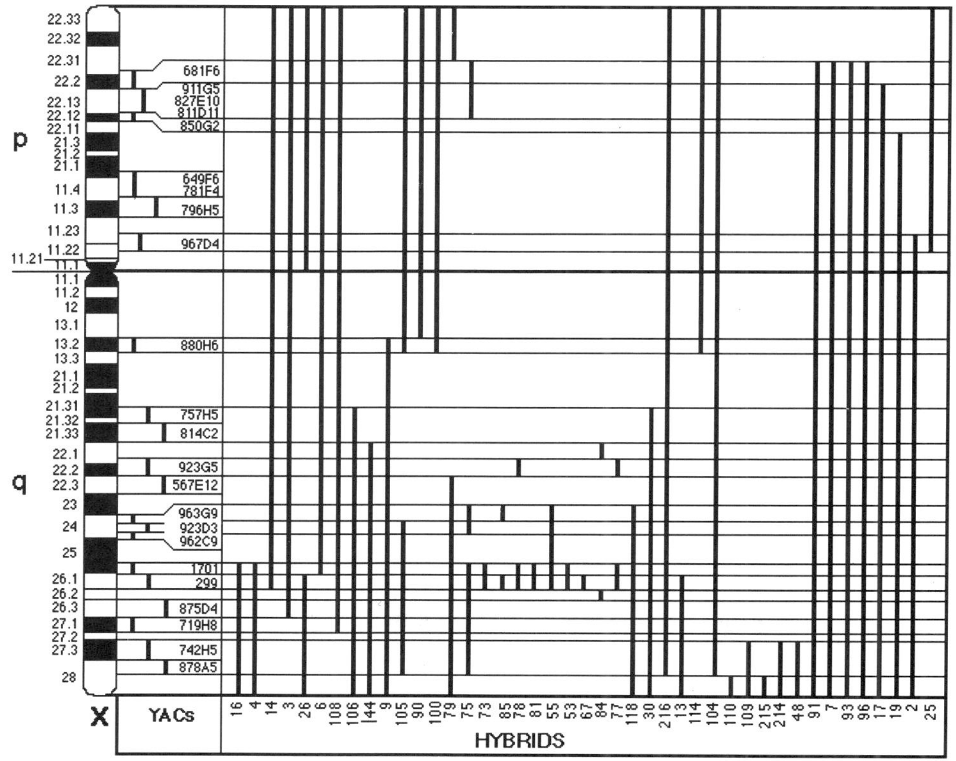

Fig. 4.1 ***Colour Plate 2*** **(a)** Partial DAPI-banded human metaphase hybridized with the radiation hybrid #171 retaining the region 10q11.2→10p14 (red signal). **(b)** A partial metaphase from hybrid #105, retaining the Xq28→Xq24::Xq13→pter region, hybridized with the YAC 923_d_3 (arrow). **(c)** Signal on the human X-chromosome mapping at Xq24. **(d)** Co-hybridization of YACs 923_d_3 (red signal) and 962_c_9 (green signal) on normal human X, indicating that 962_c_9 is telomeric.

Fig. 4.2 ***Colour Plate 2*** **(a)** Pericentric inversion of chromosome 16 (16pter→16p12::16q24→ 16p12::16q24→16qter). A SCPL specific for 16p has been used as a probe. Two pairs are shown with normal 16 on the left, the upper row lacking the hybridization signal. **(b)** A partial karyotype from a patient carrying a balanced translocation (2pter→2q24::6q22→6qter; 6pter→ 6q22::2q24→2qter), hybridized with SCPL #148 (red signals). From left: normal chromosome 2, rearranged chromosome 2, rearranged chromosome 6, normal chromosome 6 (upper row without hybridization signals). **(c)** An interstitial deletion of chromosome 4 (4pter→4q13.1::4q22.2→qter). The deleted chromosome 4 is on the right. Four different SCPLs have been used. Left-hand pair: co-localisation of SCPL #19 (red) and SCPL #81 (green), the overlapping region is yellow.

4.4 EXPLANATORY EXAMPLES

SCPLs are suitable for filling the gap between probes yielding a dotted FISH signal, like YACs, and WCPLs. Furthermore, in some cases, these latter probes are ineffective. WCPLs, for instance, would be useless in pericentric or paracentric inversions. In cases of patients bearing de novo unbalanced translocations or marker chromosomes, the use of YAC probes to identify the origin of the extra material would be fragmented and time consuming.

The three examples shown in Figure 4.2 illustrate the use of SCPLs in appropriate cases: a pericentric inversion (2a), a balanced translocation (2b) and an interstitial deletion (2c).

All pictures are pseudocoloured digital images obtained using a Zeiss Axioplan epifluorescence microscope equipped with a cooled CCD camera (Princeton Instruments, NJ). FITC, Cy3 and DAPI fluorescence, detected using a specific filter set combination (Chroma Technology, VT), were recorded separately as grey scale images. DAPI staining, giving a Q-like banding, has been used for chromosome identification. Pseudocolouring and merging of images were performed using the Adobe Photoshop software.

ACKNOWLEDGMENTS

The study was supported in part by AIRC and Telethon. The cytogenetic clinical cases reported in Fig. 4.2a and 4.2b have been studied in collaboration with Alessandria Cytogenetic Center, Rome.

REFERENCES

Antonacci, R., Marzella, R., Finelli, P. *et al.* (1995) A panel of subchromosomal painting libraries representing over 300 regions of the human genome. *Cytogenetics and Cell Genetics*, **68**, 25—32.

Archidiacono, N., Antonacci, R., Forabosco, A., and Rocchi, M. (1994) Preparation of human chromosome painting probes from somatic cell hybrids. In: Methods in Molecular Biology, Vol. XX: In situ Hybridization Protocols. K.H.A. Choo, ed. Humana Press Inc., Totowa, NJ, pp. 1—13.

Baldini, A. and, Ward, D.C. (1991) In situ hybridization of human chromosomes with Alu-PCR products: a simultaneous karyotype for gene mapping studies. *Genomics*, **9**, 770—4.

Collins, C., Lin Kuo, W., Segraves, R. *et al.* (1991) Construction and characterization of plasmid libraries enriched in sequences from single human chromosomes. *Genomics*, **11**, 997—1006.

Guan, X.-Y., Meltzer, P.S., Burgess, A.C. *et al.* (1995) Coverage of chromosome 6 by chromosome microdissection: Generation of 14 subregion-specific probes. *Human Genetics,* **95**, 637—40.

Gyapay, G., Morissette, J., Vignal, A. *et al.* (1994) The 1993—94 Genethon human genetic linkage map. *Nature Genetics*, **7** special issue, 246—339.

Hulsebos, T.J.M., Cerosaletti, K.M., Fournier, R.E.K. *et al.* (1995) Identification of the human bA2 cristallin gene; localization of the gene on human chromosome 2 and of the homologous gene on mouse chromosome 1. *Genomics*, **28**, 543—8.

Korenberg, J.R. and Rykowski, M.C. (1989) Human genome organization: Alu, lines, and the molecular structure of metaphase chromosome bands. *Cell*, **53**, 391—400.

Lichter, P., Ledbetter, S., Ledbetter, D. *et al.* (1990) Fluorescence in situ hybridization with Alu and L1 polymerase chain reaction probes for rapid characterization of human chromosomes in hybrid cell lines. *Proceedings of the National Academy of Science USA*, **87**, 6634—8.

Liu, P., Siciliano, J., Seong, D. *et al.* (1993) Dual Alu PCR primers and conditions for isolation of human chromosome painting probes from hybrid cell. *Cancer Genetics and Cytogenetics*, **65**, 93—9.

Meltzer, P.S., Guan, X.Y., Burgess, A. *et al.* (1992) Rapid Generation of Region Specific Probes by Chromosome Microdissection and Their Application. *Nature Genetics*, **1**, 24—8.

Muller, S., Koehler, U., Wienberg, J. *et al.* (1996) Comparative chromosome mapping of primate chromosomes with Alu-PCR generated probes from human/rodent somatic cell hybrids.. *Chromosome Research*, 4, 38—42.

Pecile, V., Archidiacono, N., and Rocchi, M. (1984) Treated glass cover-slips for in situ culture of amniotic cells: better cell adhesion and growth than on plastic. *Journal of Medical Genetics*, **21**, 136—7.

Zelante, L., Calvano, S., Dallapiccola B. *et al.* (1994) Patient with *de novo* 12p+ syndrome identified as dir dup (12p13.1-pter) using subchromosomal painting libraries from somatic cell hybrids. *Clinical Genetics*, **46**, 368—71.

PLATES 2, 3 and 4

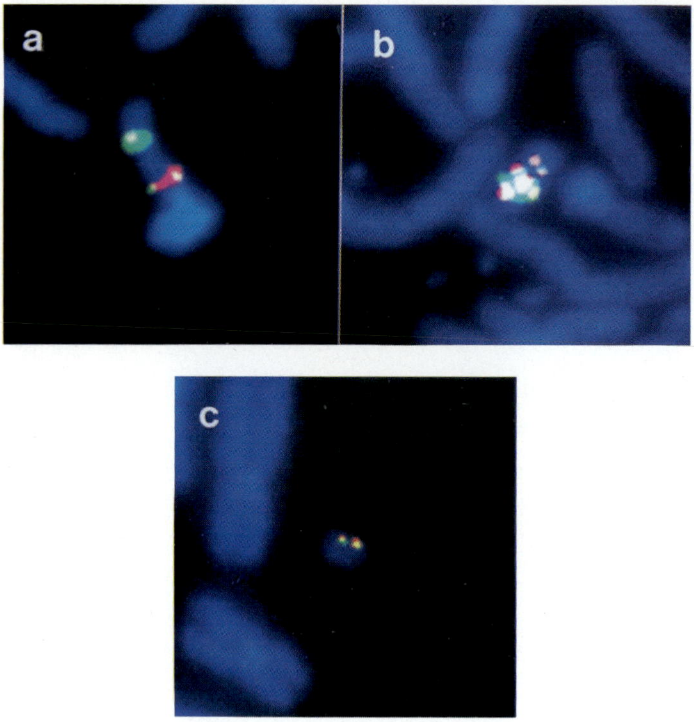

Plate 2 **Fig. 3.2** a–c (see p.53)

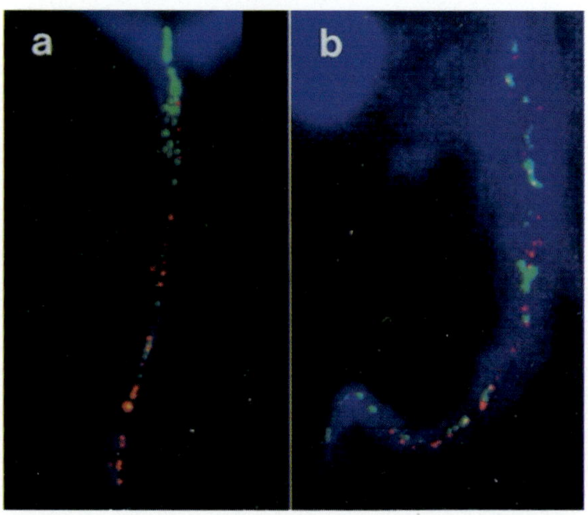

Plate 2 **Fig. 3.3** a–b (see p.53)

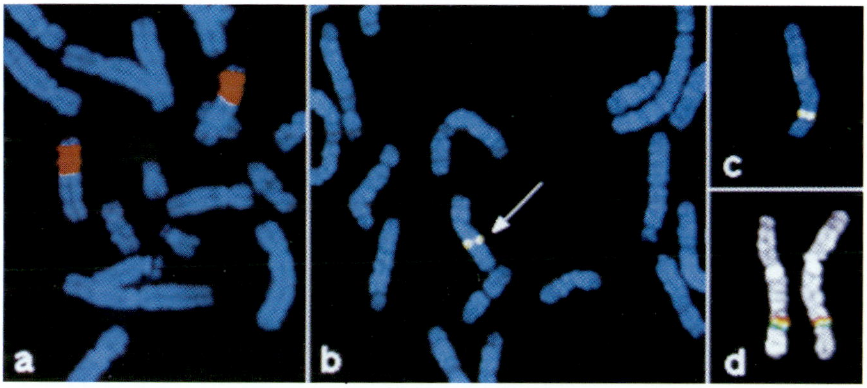

Plate 2 **Fig. 4.1** a–d (see p.62)

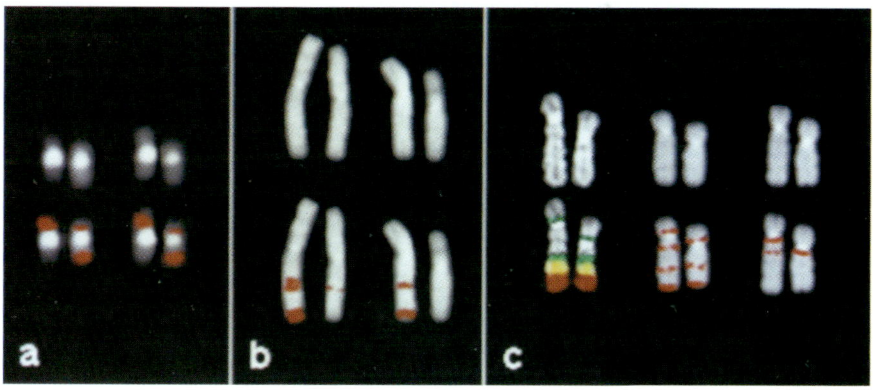

Plate 2 **Fig. 4.2** a–c (see p.62)

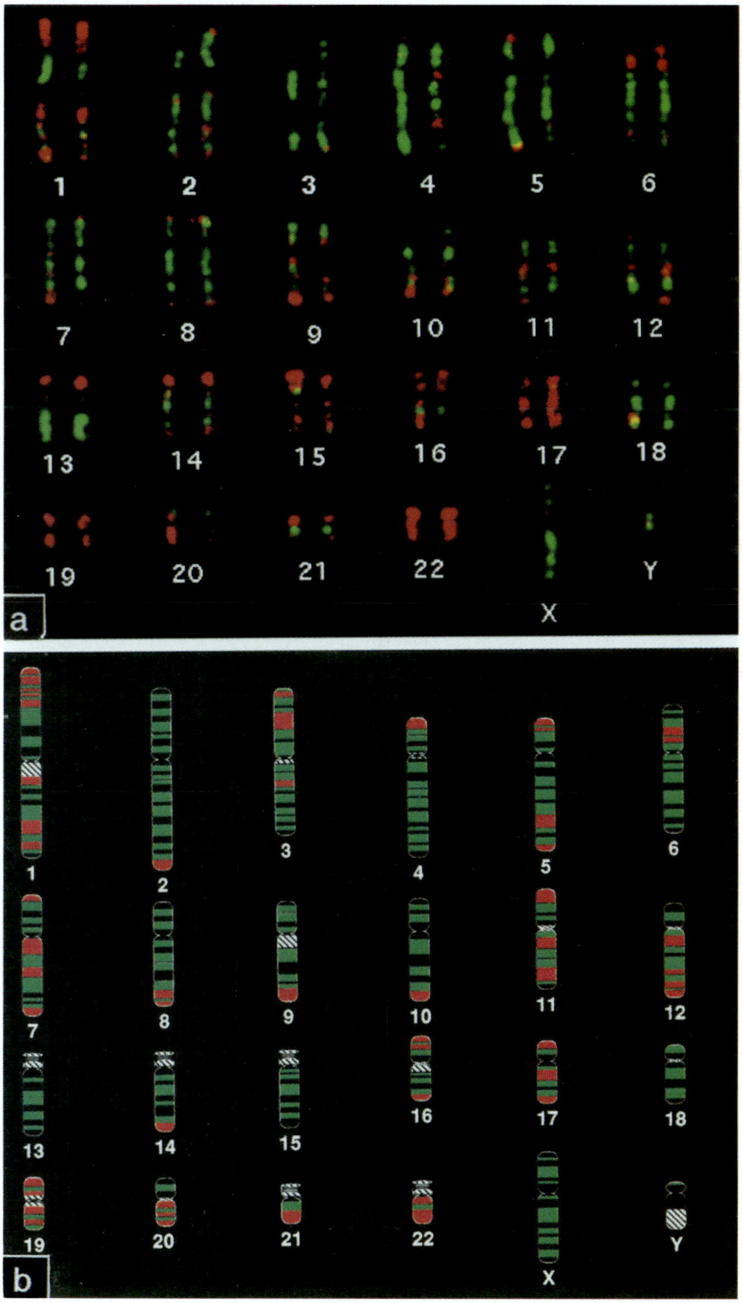

Plate 3 **Fig. 5.2** a–b (see p.73)

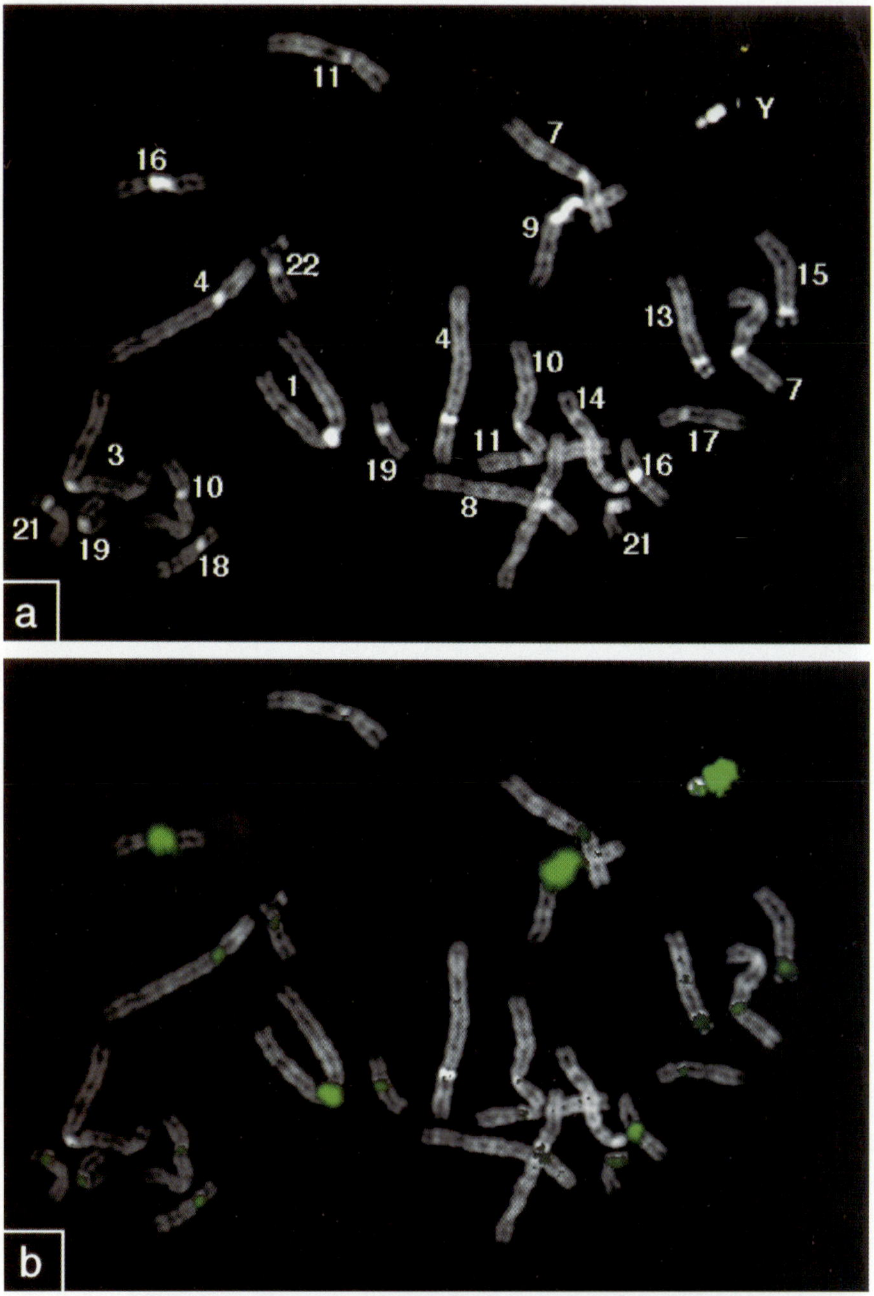

Plate 4 **Fig. 5.4** a–b (see p.76)

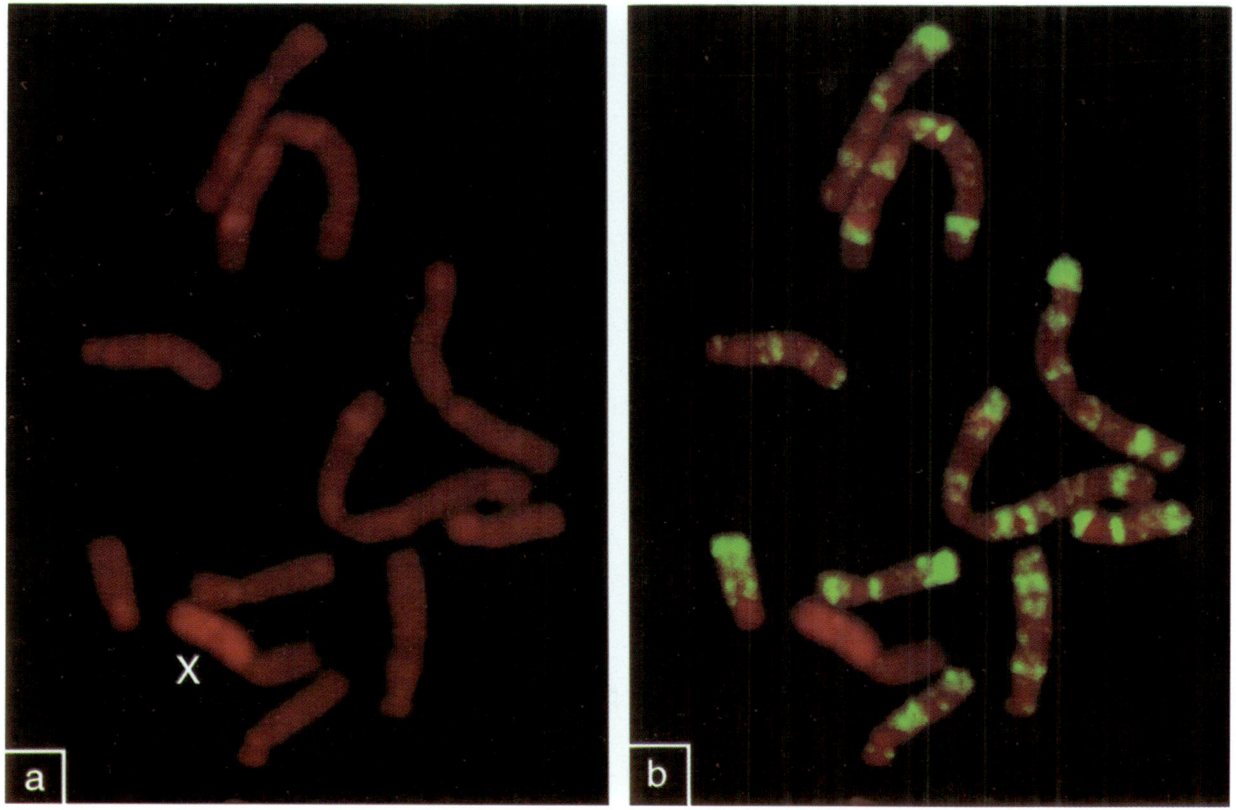

Plate 4 **Fig. 5.5** a–b (see p.76)

5

The relationship between gene density and chromosome banding patterns in mammalian nuclei

J.M. Craig and W.A. Bickmore

5.1 WHAT ARE CHROMOSOME BANDS?

Most species of eukaryotes examined so far exhibit some degree of chromosome banding (reviewed in Sumner, 1990). Of all eukaryotes, however, mammalian chromosomes respond to the widest range of banding techniques. Of all mammalian species, humans have been the most widely studied with respect to chromosome banding and physical gene mapping. However, the vast majority of banding patterns demonstrated in humans have also been shown in other mammalian species.

Mammalian chromosome bands are not only visible with staining protocols but exist as physical and functional entities in their own right. Chromosome banding protocols can therefore be dye-based, structure-based or function-based. The most common dye-based banding methods are C (centromere), Q (quinacrine), G (Giemsa) and R (reverse) banding. Bands that exhibit a stronger staining reaction are referred to as positive bands and weakly-staining bands, negative. Positive C bands contain constitutive heterochromatin. In man, this is found in and around the centromere of each chromosome and towards the end of the long arm of the Y chromosome. Euchromatin stains weakly with C banding, but can be banded with any of the other techniques (Q, G and R). Euchromatic Q-positive bands are identical to G-positive and R-negative bands.

Chromosomes at the pachytene stage of meiosis can be stained without any pretreatment to give a pattern of densely staining, condensed chromomeres separated by less condensed inter-chromomeric DNA. These chromomeres correspond to mitotic G bands. The most widely used function-based banding protocol is replication banding and is based on the fact that R bands replicate before G bands during S phase of the cell cycle (Ganner and Evans, 1971).

5.2 SUB-CLASSIFICATION OF CHROMOSOME BANDS

Chromosome banding techniques have been shown to produce more than just two sub-types of chromosome band. Different 'shades' of Q bands (Korenberg & Engels, 1978), G bands (Francke, 1994) and R bands (ISCN, 1985; Drouin & Richer, 1988) have all been recorded. There is also a banding technique, T banding, that identifies a distinct subset of intensely-staining R bands (Dutrillaux, 1973; Ambros & Sumner, 1987). It is thought to identify the GC-richest R bands, of which approximately half occur at telomeres in the human genome (Holmquist, 1992). In addition, Holmquist, (1992) split human R bands into four 'flavours' depending not only on GC content (all T bands were GC rich), but also on *Alu* interspersed repeat content. Dutrillaux *et al.*, (1976) also split S phase arbitrarily into eighteen time-segments and assigned each G and R band a relative replication time.

5.3 THE PROPERTIES OF CHROMOSOME BANDS: EVIDENCE FOR A VARIATIONS IN GENE DENSITY

The major properties of mammalian chromosome bands are listed in Table 5.1. Most properties are graded between G bands and T bands and are related to differences in genetic activity. Only a few percent of the mammalian genome has a coding function, and there is mounting evidence that genes are non-randomly distributed in the human genome. The location of coding sequences is fundamentally important for genome mapping and for understanding the relationship between chromosome structure and function.

Table 5.1 Properties of human euchromatic chromosome bands. For explanation and discussion see text.

	G bands	*R bands*	*T bands*
Replication time	late	early	very early
Base composition	AT-rich	GC-rich	very GC-rich
Gene density	low	high	highest
Condensation time	early	late	latest
Density of cancer exchanges, X-ray exchanges and meiotic chiasmata	low	high	highest
Nuclease sensivity	low	high	high
Levels of histone acetylation	low	high	high

5.3.1 Replication timing and base composition

In placental mammals, both constitutive (C band) and facultative (the inactive X chromosome) heterochromatin replicate late. Based on this link, Ganner and Evans, (1971) suggested that G bands were likely to be genetically more quiescent than R bands. There is a direct correlation between gene density, early replication and GC content in mammalian chromosomes (reviewed in Craig and Bickmore, 1993). In humans, the GC-richest subset of DNA has also been shown to be the most gene-rich (Bernardi *et al.*, 1985; Mouchiroud *et al.*, 1991).

5.3.2 Mapped genes

In a survey of genes mapped to single chromosome bands, Craig and Bickmore, (1993) found that 80% of 1023 mappped genes were found in R bands, even though R and G bands contain approximately the same amount of DNA at metaphase. Although a further 750 genes had been mapped in such a way at the time of going to press, the proportion of genes with R bands has remained constant. Furthermore, 57% of genes map to R bands, which acount for only 20% of genomic

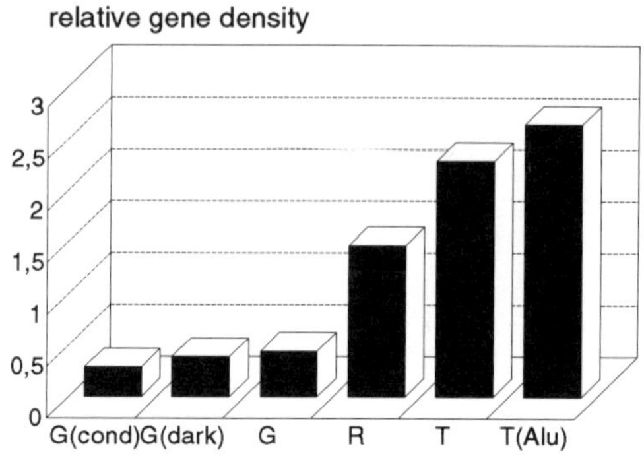

Fig. 5.1 Comparison of gene density for major and minor human chromosome band 'flavours'. Relative gene density was calculated as observed/expected gene density, assuming for expected gene density that genes are distributed evenly throughout the human genome. Band types are: G_{Cond}, G bands that do not condense from prophase to metaphase (from comparison of ISCN (1985) ideograms); G_{Dark}, the darkest staining G bands (Francke, 1994); G/R/T, G, R and T bands as defined in the text; T_{Alu}, *Alu*-rich T bands (Holmquist, 1992).

DNA. Relative (observed/expected) gene density can be calculated for each band subtype (Fig. 5.1). This reveals a gradient of gene density throughout the human genome, from darkly-stained condensed G bands, through G, R and T bands, to *Alu*-rich T bands. However, this sample represents only approximately 2% of all human genes.

5.3.3 Long range physical maps

In vertebrate genomes, the dinucleotide CpG occurs at a much lower frequency than GpC. Approximately 30% of non-methylated CpGs are clustered around the 5' end of genes (all housekeeping genes and approximately 40% of tissue-specific genes, Bird *et al.*, 1985; Bird, 1986; Larsen *et al.*, 1992). CpG islands are defined as having an observed/expected CpG:GpC ratio of >0.6 and a size of >200bp. Due to the lack of methylation within CpG islands, they are sites for methylation-sensitive restriction enzymes, especially those with one or more CpGs in their recognition sequence. Enzymes such as *Not*I (GCGGCCGC) have proved useful for long-range physical mapping of genes because almost all cleavage sites are expected to lie within CpG islands (Bickmore and Bird, 1992). Of 189 *Not*I fragments surveyed from 19 different chromosome bands, average spacings were 1483kb for G bands, 434kb for R bands and 381kb for T bands (Craig, 1995). Again, this sample of fragments accounts for only a small percentage of the human genome.

5.3.4 *In situ* hybridisation of cDNA and RNA

Yunis *et al.,* (1977) and Yunis and Tsai, (1978) hybridised cDNA, poly(A) mRNA and heterogeneous nuclear RNA (HnRNA) to metaphase chromosomes and found that 70—80% of signals were coincident with R bands. Obvious problems with this type of study are that in any particular cell type, all genes do not transcribe at the same level, with many not transcribing at all, biasing any sample of RNA. There is also a high probability of finding *Alu* sequences within HnRNA.

5.3.5 Chromosome condensation

As chromosomes condense from prophase to metaphase, R sub-bands have been shown to fuse more readily with each other than G bands, with many G bands not fusing at all (Drouin *et al.* 1991). It was also found that by prophase, G bands have almost attained maximal contraction, whereas R bands still have much contraction (fusion) forthcoming. It is possible that during interphase, gene-rich T bands need to be less condensed for a longer period of time than gene-poor G bands, to facilitate transcription.

5.3.6 'Open' chromatin

Nucleases such as DNase I preferentially digest R bands and T bands of intact chromosomes, with G bands and C bands being increasingly refractive to digestion (reviewed in Sumner, 1990). R bands have higher levels of acetylated histone H4 than G bands (Jeppesen and Turner, 1993) and it has been shown that in rat chromosomes, that the 'linker' histone, histone H1 is enriched in G bands (Breneman *et al.*, 1993). This 'open chromatin' state is found in CpG island chromatin, which contains hyperacetylated histones H3 and H4, has a reduced amount of the linker histone H1, is non-methylated, and contains nucleosome-free regions (Tazi and Bird, 1990). All these features have been associated with genetic activity (Allfrey, 1980; Hebbes *et al.*, 1988; Keshet *et al.*, 1986; Gross and Garrard, 1988; Turner, 1991).

5.3.7 Recombination and DNA breakpoints

R bands have a greater density of recombination events than G bands (reviewed in Holmquist, 1992; Craig and Bickmore, 1993). They also have a higher frequency of X-ray-induced and disease-associated breakpoints (Holmquist, 1992). High levels of recombination activity have been linked with the open chromatin structure found around active genes in the organisms as diverse as yeasts (Wu and Lichten, 1994; Oliver *et al.*, 1992; Egel, 1984) and humans (Hsieh and Lieber, 1992; Nickoloff, 1992). If R band DNA is in a more 'open' configuration, it will have a larger volume per unit length of DNA than G band DNA. It is therefore likely that this differential chromatin compaction effectively presents R band DNA as a larger target for biophysical events such as chromosome rearrangements or recombination.

5.4 EUCHROMATIC GENE DISTRIBUTION IN RELATION TO GENE FUNCTION

Larsen *et al.*, (1992) set up a database of sequenced genes, categorised according to function and whether they possess a CpG island. They categorised genes as having either widespread expression ('housekeeping' genes) or limited/tissue-specific expression ('tissue-specific' genes). Of 900 genes listed in the current database, 25% are 'housekeeping' and 75% are 'tissue-specific'. All 'housekeeping' and 38% of 'tissue-specific' genes have a CpG island. The proportion of all genes having a CpG island is thus 53%. The genomic distribution of this subset of genes is similar to all mapped genes. However, all but two of the 56 'housekeeping' genes map to R bands, as do 80% of the tissue-specific genes. G bands therefore contain mainly tissue-specific genes with and without CpG islands.

5.6 HOW MANY GENES CAN WE LOCATE?

It is difficult to know if the small number of human genes mapped to individual chromosome bands is representative of all human genes. At present, it is not possible to identify every human gene and the best we can do is to choose some feature that most genes share. The CpG island, present in greater than 50% of human genes, has proved useful in this respect (Craig and Bickmore, 1994). A number of methylation-sensitive restriction enzymes can be used to cleave DNA at CpG islands (Bickmore & Bird, 1992). DNA liberated by these enzymes can be fractionated according to size and appropriate size fractions hybridised back to chromosomes using *in situ* suppression hybridisation.

5.7 USING INTER-CPG ISLAND DNA AS A TOOL TO INVESTIGATE THE DISTANCE BETWEEN HUMAN GENES

If genomic DNA were to be cleaved exclusively at CpG islands, the size of the resulting fragments would be inversely proportional to gene density. The enzymes chosen to fulfill this task were *Bss*HII (GCGCGC), *Eag*I (CGGCCG) and *Sac*II (*Sst*II: CCGCGG), all of which are methylation sensitive. Each CpG island is expected to have on average, 1.68 sites for each of these enzymes (Bickmore & Bird, 1992). In practice, out of 40 gene sequences surveyed from GenBank, all but three contained sites for at least one of these enzymes (Bickmore & Bird, 1992). Most sites for these enzymes are within CpG islands and those that are not are often methylated in genomic DNA and therefore uncuttable. A combination of all three should effectively yield either tiny fragments of <500bp from within CpG islands, or larger inter-island fragments, which, on the evidence from published mapping data, would range from kb to Mb scale. The size of restriction fragments would then be directly proportional to the distance between CpG islands and hence genes.

5.7.1 *In situ* hybridisation of inter-CpG island size-fractions from a single human chromosome

To provide a starting point, DNA from a hamster—human somatic cell hybrid containing a single human chromosome 11 was cut with CpG island-specific restriction enzymes *Bss*HII, *Eag*I and *Sst*II (Craig and Bickmore, 1994). Restriction fragments were then fractionated according to size using pulsed field gel electrophoresis. This provided pools of DNA, each with a characteristic range of CpG island densities, ranging from 15—100kb to >1Mb. Fractions were then hybridised to metaphase chromosomes (Craig and Bickmore, 1994). Results showed that each type of chromosome band on human chromosome 11 has a characteristic gene density. T bands were shown to possess inter-island distances of less than 1Mb, with most being less than 500kb and many less than 100kb apart. CpG islands were spaced between 100kb and 1Mb apart in non-T R bands and >1Mb apart in G bands and C bands. The gene density of a band was also correlated with its time of replication, with gene-rich R bands replicating early and gene-poor G bands late. The fraction of DNA containing the shortest (15—100kb) inter-island fragments was also hybridised to more extended chromosomes, this time counterstained by late-replication (G) banding. All gene-rich bands were early replicating, although not all early-replicating bands were gene-rich (Craig and Bickmore, 1994).

5.7.2 *In situ* hybridisation of inter-CpG island size-fractions from the whole human genome

Would this relationship between banding patterns and gene density be reiterated throughout the whole human genome? To answer this question, total (male) genomic DNA was cut with the same three restriction enzymes (*Bss*HII, *Eag*I and *Sac*II) and size-fractionated as before. This time the extremes of gene density were concentrated on, and the inter-island fractions used for hybridisation were 15—100kb and >1Mb (split into two fractions of 1—3Mb and 3—6Mb). Each fraction was biotin-labelled by nick translation, hybridised to metaphase chromosomes and detected in a fraction-dependent manner.

(a) *In situ* hybridisation of the shortest genomic inter-CpG island fragments

Hybridisation of the the shortest inter-island fraction (15—100kb) from total human DNA was detected with avidin-Texas Red (Fig. 5.2). Chromosomes were late-replication-G banded. This well-contrasted banding pattern was achieved by releasing cells synchronised in mid-S phase with bromodeoxyuridine (BrdU) and using fluorescein

Fig. 5.2 ***Colour Plate 3.*** Results from the *in situ* hybridisation of the shortest human genomic inter-CpG island fragments, in relation to chromosome banding patterns. **5.2a** Genomic inter-island fragments of between 15 and 100kb were biotin-labelled, hybridised to peripheral blood lymphocyte chromosomes and detected with avidin-Texas Red. Late replicating G bands were detected with FITC-anti-BrdU (green) and electronically cut, pasted, and rotated where necessary to construct a karyotype. Bent chromosomes were straightened by cutting in a region devoid of signal and rotation of the cut segment. Each chromosome has its own characteristic pattern and in most cases, hybridisation patterns for each homologue are almost identical. Hybridisation sites correspond well with GC-rich T bands. Some chromosome regions and indeed some chromosomes e.g. 13, 18, X and Y appear to be relatively gene-poor. Hybridisation to the short arms of the acrocentric chromosomes is most likely from the ribosomal DNA clusters contained in these regions. **5.2b** Ideogram for comparison with hybridisation results. G bands are shown in green, non-T R bands in black, and T bands in Red. Adapted from Holmquist, 1992.

isothiocyanate (FITC)-conjugated anti-BrdU (green) for detection after FISH (Craig and Bickmore, 1994). Hybridisation signal was mostly over early-replicating (R band) DNA and was unevenly distributed both within a single chromosome and between chromosomes. Chromosomes in Figure 5.2a have been rearranged as a karyotype. Hybridisation patterns on homologous chromosomes are generally identical. Some chromosome regions and even some whole chromosomes appear gene-rich. Apart from the ribosomal DNA clusters on the short arms of the acrocentric chromosomes, known to contain unmethylated, GC-rich DNA (Bird *et al.*, 1985), there are regions of strong hybridisation on chromosomes 1, 9, 15, 16, 17, 19, 20 and 22. The bulk of the signal occurs over euchromatin corresponding to early replicating R bands, thus most euchromatin is painted either red or green with little overlap of the signal (yellow). As for chromosome 11, the very highest concentration of signal is to the GC-richest subset of R bands, the T bands (compare with the ideogram in Figure 5.2b). Thus chromosome 19, where all the R bands are T bands, is very heavily labelled despite its relatively small size, whilst the comparably sized chromosome 18, where none of the R bands are T bands, has little hybridisation signal. There were a few exceptions: there are several non-T R bands which hybridised strongly with the probe, for example those on the long arm of chromosomes 9 and 15. Hybridisation to the X and Y chromosomes is weak. This is expected for the Y chromosome with its dearth of mapped genes, but not expected for the X chromosome. This could be explained by the fact that unmethylated CpG islands from these chromosomes are present at half the copy number of those from autosomes, whether male or female DNA is used as the probe source (The female inactive X chromosome has methylated and therefore uncuttable CpG islands; Grant and Chapman, 1988; Riggs and Pfeifer, 1992). Even so, the signal on the X is still lower than expected from the number of mapped genes. This could be due either to a bias in the type of gene on the X chromosome or to the fact that the X chromosome has been studied more than any other so far.

'Tiny' fragments (<600bp) of genomic DNA produced by the methylation-sensitive restriction enzyme *Hpa*II, which are known to originate from CpG islands, produced an almost identical pattern of hybridisation pattern to that of the shortest inter-island fragments (Craig and Bickmore, 1994). This confirmed the link between gene density, GC richness and early replication.

(b) In situ hybridisation of the inter-CpG island fragments of between 1 and 3Mb

Hybridisation of genomic inter-island fragments of 1—3Mb were detected with avidin-FITC. Chromosomes were banded with DAPI (4,6-diamino-2-phenylindole) and actinomycin D and pseudocoloured for easier comparison of signal and banding patterns. Hybridisation patterns are shown next to enhanced, pseudocoloured DAPI-stained chromosomes in Figure 5.3. Only one homologue is shown for clarity. This arrangement best illustrates the genomic distribution of this fraction, which corresponds to brightly-staining G bands and constitutive heterochromatin at most centromeres and the pericentric heterochromatin of chromosomes 1, 16 and the long arm of Y. This mirrors the distribution of similar sized fragments on chromosome 11. Even though 10μg of Cot 1 DNA per slide was used to compete out repetitive DNA, signal is still present over centromeric heterochromatin. The signal over the heterochromatic regions is most likely due to Cot 1 DNA not being able to suppress the large proportion of the probe that is repetitive DNA.

(c) *In situ* hybridisation of the longest genomic inter-CpG island fragments

Hybridisation results from inter-island fragments of 3—6Mb are shown in Figure 5.4. Chromosomes, DAPI-banded as in Figure 5.3, are shown in Figure 5.4a and are shown together with hybridisation signals in Figure 5.4b. The majority of signal is over constitutive

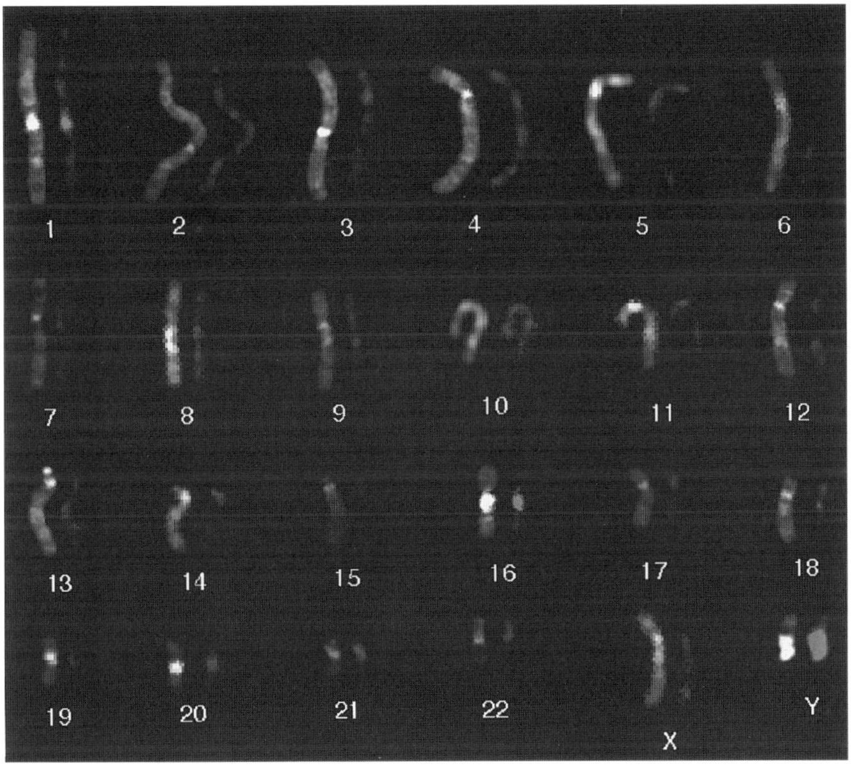

Fig. 5.3 Karyotype of results from the hybridisation of genomic inter-CpG island fragments of between 1 and 3Mb. Genomic inter-island fragments of between 1 and 3Mb were biotin-labelled, hybridised to peripheral blood lymphocyte chromosomes and detected with avidin-FITC (on the right in each pair). Chromosomes were counterstained and G-banded with DAPI and actinomycin D. For each chromosome, only one homologue of each pair is displayed. On the left, the banded chromosome (blue DAPI stain changed to white for a better contrast) is shown, and on the right, the hybridisation signal only. The hybridisation signal is discontinuous, with the brightest areas corresponding to bright G bands in most cases. Other bright spots of hybridisation occur over the heterochromatin of chromosomes 1, 16 and Y (but not 9) and over centromeres, which usually stain brightly with DAPI. This signal distribution reflects the gene-poor nature of these regions.

Fig. 5.4 ***Colour Plate 4*** Hybridisation of the longest genomic inter-CpG island fragments of 3—6Mb. Genomic inter-island fragments of between 3 and 6Mb were biotin-labelled, hybridised to peripheral blood lymphocyte chromosomes and detected with avidin-FITC (green). Chromosomes were G-banded with DAPI and actinomycin D (as for Figure 3.3). In this case, white chromosomes (identified in a) are shown overlaid with green signal (b) to show that hybridisation is mostly to heterochromatin on chromosomes 1, 9, 16 and Y and to most centromeres. There is very little signal over euchromatin. These result confirm that centromeres and other regions of constitutive heterochromatin are most likely gene-free.

Fig. 5.5 ***Colour Plate 4*** Clustering of gene-rich regions on hamster chromosomes. The shortest inter-CpG island fraction (fragments between 15 and 100kb) from the hamster-human somatic cell hybrid J1 was biotin-labelled, hybridised to hamster (CHO) chromosomes and detected with avidin-FITC (bright). Chromosomes were counterstained with DAPI/propidium iodide (red) which stains R bands brightly. As with the shortest human inter-island fragments, the brightest hybridisation signals correspond to R bands, with the exception of the brightly-staining X chromosome.

heterochromatin, the brightest signal over pericentric regions of chromosomes 1, 9, 16 and over the long arm heterochromatin of the Y, with less intense signals over centromeres. Very little signal is seen over euchromatin, but is notable over the G bands on chromosomes 4, 7, 8, and 13.

These results indicate that the regions of the genome most devoid of genes are bands containing constitutive heterochromatin and to a lesser extent, G bands. Thus in humans, there is a gradient of (potentially) genetically-active regions from gene-rich, GC-rich, early-replicating T bands; through other R bands; gene-poor, GC-poor, late-replicating G bands and finally constitutive heterochromatin.

5.7.3 CpG island distribution in other mammalian species

To determine whether this organisation is specific to the human genome, or generally true of mammalian genomes, other mammals must be examined in a similar manner.The somatic cell hybrids used earlier in this chapter have hamster genomes as their main component, and inter-island fractions were already available for hybridisation. When the shortest inter-island fraction (15—100kb) was hybridised to hamster metaphase chromosomes banded with propidium iodide and

DAPI, most signal was again found to be coincident with R bands (Fig. 5.5). Results from other rodents seem to confirm these findings (W. Bickmore, unpublished observations). The relationship between gene density and GC-rich chromosome bands would therefore appear to be widespread amongst mammals.

5.7 DISCUSSION

Can CpG island clustering be extrapolated to gene density itself? Of the 230 mapped and sequenced genes, 90% of genes with CpG islands and 77% of the genes without islands mapped to R bands. In using the CpG island as a global marker for gene position, we are therefore ignoring only a relatively small proportion of genes without islands. Therefore, although there is a slight bias for CpG island-associated genes to be located in R bands, a large proportion of the genes missed by FISH analysis will still have a similar genomic distribution.

Evidence presented here has provided an insight into gene distribution and function within mammalian chromosome bands. Each type or 'flavour' of chromosome band seems to have a characteristic set of properties, most of which are linked to the density of genes in that 'flavour'. 'Flavours' seem to be graded from the most GC-rich, 'open', early replicating, darkest staining R bands (the T bands) to the most condensed (at prophase and possibly throughout interphase), AT-rich, late replicating, darkly-staining G bands. These relationships are also likely to be a general feature of the genomes of all placental mammals.

It is possible that the mammalian prophase chromosome band, ranging in size from around 400kb to 6,500kb (corresponding to the thinnest and widest of 2,000 bands observed by Yunis, 1981), is relatively homogeneous with respect to gene density, GC content and replication time. These bands may also correspond to 'isochores', regions of DNA >300kb in length, each with a homogeneous base composition (Bernardi, 1989; Mouchiroud *et al.*, 1991; Saccone *et al.*, 1992; 1993). When chromosomes condense, bands coalesce, with smaller bands being engulfed by larger ones, to eventually produce metaphase chromosomes with six times fewer bands. Each band should then have GC content and gene density which are the average of those of the constituent bands. However, it looks like the more GC-rich the band, the greater the degree of band fusion (Drouin *et al.*,

1991). These results account for the finding that low resolution R bands are more heterogeneous in their base composition and gene density than G bands (Bernardi, 1989).

Not incompatible with these findings is the possibility that true high resolution G band chromatin, containing very few genes (<1 per Mb), remains relatively condensed throughout the cell cycle (Okada and Comings, 1974; Comings, 1978; Camargo and Cervenka, 1982). Any genes contained in G bands could be locally decondensed during G1 for expression. The highly condensed nature of G bands is particularly apparent on meiotic chromosomes where G bands appear as chromomeres. Chromomeres are thought to represent regions of centromeric and 'intercalary heterochromatin'. The latter are normally classed as euchromatic G bands, and this implies that these bands are more tightly condensed or packaged during interphase, mitosis and meiosis (Okada and Comings, 1974; Comings, 1978; Craig and Bickmore, 1993).

There have been indications about the nature of this cryptic 'intercalary heterochromatin'. The most striking was the discovery of a 1Mb tract of repetitive DNA in a mouse G band, which comprised around 50 copies of a 14kb repeat sequence (38.5% GC) together with AT-rich *L1* repeats (Nasir *et al.*, 1991). Other reports show a similar picture, with Mb stretches of such 'macrosatellites', both GC- or AT-rich, interrupted only by interspersed repetitive elements (Giacalone *et al.*, 1992; Nishioka *et al.*, 1992). It is therefore possible that a proportion of high resolution 'euchromatic' G bands could actually be completely gene-free. Because of the locus specific nature of these macrosatellite repeats, it is no surprise that they have not been detected by methods such as renaturation kinetics. More evidence for the existence of gene-free G bands comes from reports of duplication of whole G bands without any apparent phenotypic effects (Fernández *et al.*, 1994; Verma *et al.*, 1993). Another possibility is that G bands contain very large genes of >400kb which take up most of the band they are in. However, although dystrophin (2.4Mb) is located within Xp21, a G band, no other genes of similar size have been mapped to date.

All eukaryotes with chromosomes large enough to study show replication banding and meiotic chromomeres, even though their chromosomes may not respond to G and R banding techniques (reviewed by Sumner, 1990; Craig, 1995). The distribution of genes is not known for many eukaryotic species, although heterochromatin is always thought to be gene-poor. In yeasts, late replicating regions of DNA have been shown to be inactive (Ferguson and Fangman, 1992; Aparicio and Gottschling, 1994). This information suggests a fundamental link between genetic activity and chromosome banding patterns. However, the question of how can such a precise genome organisation have evolved and be maintained has been discussed at length elsewhere (Holmquist, 1992; Bernardi, 1993).

It is emerging that the 'bar-code' pattern of chromosome bands carries important information for nuclear activity. We are now beginning to be able to read and understand this code. The mapping and sequencing of mammalian and other eukaryotic genomes will therefore lead to a better understanding of chromosome organisation and evolution with respect to banding patterns.

ACKNOWLEDGEMENTS

The authors wish to thank all members of the Blood Lab at the MRC Human Genetics Unit, Edinburgh, Scotland for the help, support and encouragement they have provided during the last four years. J.C. was supported by a PhD studentship from the UK Medical Research Council. W.A.B. is a Lister Research Fellow.

REFERENCES

Allfrey, V.G. (1980) Molecular aspects of regulation of eukaryotic transcription—nucleosomal proteins and their postsynthetic modification in the control of DNA conformation and template function. In *Cell Biology: a Comprehensive Tratise*, vol. 3, (eds L. Goldstein and D.M. Prestcott), Academic Press, New York.

Ambros, P.F. and Sumner, A.T. (1987) Correlation of pachytene chromomeres and metaphase bands of human chromosomes, and distinctive properties of telomeric regions. *Cytogenetics and Cell Genetics*, **44**, 223—8.

Aparicio, O.M. and Gottschling, D.E. (1994) Overcoming telomeric silencing: a transactivator competes to establish gene expression in a cell cycle-dependent way. *Genes and Development*, **8**, 1133—46.

Bernardi, G. (1989). The isochore organisation of the human genome. *Annual Review of Genetics*, **23**, 637—61.

Bernardi, G. (1993) The vertebrate genome, isochores and evolution. *Molecular Biology and Evolution,* **10**, 186—204.

Bernardi, G., Olofsson, B., Filipski, J. *et al.* (1985) The mosaic genome of warm blooded vertebrates. *Science,* **228**, 953—8.

Bickmore, W.A. and Bird, A.P. (1992) The use of restriction endonucleases to detect and isolate genes from mammalian cells, in *Methods in Enzymology*, vol. 216 (ed R Wu), Academic Press, San Diego, pp 224—45.

Bird, A.P. (1986) CpG-rich islands and the function of DNA methylation. *Nature,* **321**, 209—13.

Bird A., Taggart M., Frommer M. *et al.* (1985) A fraction of the mouse genome that is derived from islands of nonmethylated CpG-rich DNA. *Cell,* **40**, 91—9.

Breneman J.W., Yau P., Teplitz R.L. and Bradbury E.M. (1993) A light microscope study of linker histone distribution in rat metaphase chromosomes and interphase nuclei. *Experimental Cell Research,* **206**, 16—26.

Camargo, M. and Cervenka, J. (1982). Patterns of DNA replication of human chromosomes. II. Replication map and replication model. *America Journal of Human Genetics,* **34**, 757—80.

Comings, D.E. (1978). Mechanisms of chromosome banding and implications for chromosome structure. *Annual Review of Genetics,* **12**, 25—46.

Craig, J.M. (1995) The functional significance of mammalian chromosome banding. PhD thesis, University of Edinburgh.

Craig, J.M. and Bickmore, W.A. (1993) Chromosome bands — flavours to savour. *BioEssays,* **15**, 349—54.

Craig, J.M. and Bickmore, W.A. (1994) The distribution of CpG islands in mammalian chromosomes. *Nature Genetics,* **7**, 376—82.

Drouin, R. and Richer, C.-L. (1988). High resolution R banding R banding at the 1250-band level II. Schematic representation and nomenclature of human RBG-banded chromosomes. *Genome,* **32**, 425—39.

Drouin, R., Lemieux, N. and Richer, C.-L. (1991). Chromosome condensation from prophase to late metaphase: Relationship to chromosome bands and their replication time. *Cytogenetics and Cell Genetics,* **57**, 91—9.

Dutrillaux, B. (1973) Nouveau systeme de marquage chromosomique des bandes T. *Chromosoma* **41**, 395—402.

Dutrillaux, B., Couturier, J., Richer, C-L. and Viegas-Péguinot, E. (1976). Sequence of DNA replication in 277 R- and Q-bands of human chromosomes using a BrdU treatment. *Chromosoma,* **58**, 51—61.

Egel, R. (1984) 2 tightly linked silent cassettes in the mating-type region of *Scizosaccharomyces pombe*. *Current Genetics,* **8**, 199—203.

Ferguson, B.M. and Fangman, W.L. (1992) A position effect on the time of replication origin activation in yeast. *Cell,* **68**, 333—9.

Fernández, J.L., Pereira, S., Campos, A. *et al.* (1994) An extra band within the human 9qh+ region that behaves like the surrounding constitutive heterochromatin. *Journal of Medical Genetics*, 31, 632—634.

Francke, U. (1994) Digitised and differentially shaded human chromosome ideograms for genomic applications. *Cytogenetics and Cell Genetics,* **65**, 206—19.

Ganner, E. and Evans, H.J. (1971) The relationship between patterns of DNA replication and of quinacrine fluorescence in the human chromosome complement. *Chromosoma,* **35**, 326—41.

Giacalone, J., Freides, J. and Franke, U. (1992). A novel GC-rich human macrosatellite VNTR in Xq24 is differentially methylated on active and inactive X chromosomes. *Nature Genetics,* **1**, 137—43.

Grant, S.G. and Chapman, V.M. (1988) Mechanisms of X chromosome regulation. *Annual Review of Genetics,* **22**, 199—233.

Gross, D.S. and Garrard, W.T. (1988) Nuclease hypersensitive sites in chromatin. *Annual Review of Biochemistry,* **57**, 159—97.

Hebbes, T.R., Thorne A.W. and Crane-Robinson C. (1988) A direct link between core histone acetylation and transcriptionally active chromatin. *EMBO Journal,* **7**, 1395—402.

Holmquist, G.P. (1992). Chromosome bands, their chromatin flavors, and their functional features. *American Journal of Human Genetics,* **51**, 17—37.

Hsieh, C-L. and Lieber, M.R. (1992). CpG methylated minichromosomes become inaccessible for V(D)J recombination after undergoing replication. *EMBO Journal*, **11**, 315—25.

ISCN (1985): *An International System for Human Cytogenetic Nomenclature*. (eds Harnden, D.G. and Klinger, H.P.), pp. 48—57. Published in collaboration with *Cytogenetics and Cell Genetics* (Karger, Basel 1985).

Jeppesen, P. and Turner, B.M. (1993) The inactive X chromosome in female mammals is distinguished by a lack of histone H4 acetylation, a cytogenetic marker for gene expression. *Cell,* **74**, 281—9.

Keshet, I., Lieman-Hurwitz, J. and Cedar, H. (1986) DNA methylation affects the formation of active chromatin. *Cell,* **44**, 535—43.

Korenberg, J.R. and Engels, W.R. (1978) Base ratio DNA content and quinacrine brightness of human chromosomes. *Proceedings of the National Academy of Science U.S.A.,* **75**, 3382—6.

Larsen, F., Gundersen, G., Lopez, R. and Prydz, H. (1992). CpG islands as gene markers in the human genome. *Genomics,* **13**, 1095—107.

Mouchiroud, D., D'Onofrio, G., Aïssani, B., *et al.* (1991). The distribution of genes in the human genome. *Gene,* **100**, 181—7.

Nasir J., Maconochie M.K. and Brown S.D.M. (1991). Co-amplification of L1 line elements with localised low copy repeats in Giemsa dark bands: implications for genome organisation. *Nucleic Acids Research,* **19**, 3255—60.

Nickoloff J.A. (1992). Transcription enhances intrachromosomal homologous recombination in mammalian cells. *Molecular and Cell Biology,* **12**, 5311—8.

Nishioka Y., Dolan B.M., Fiorellino A. and Prado V.F. (1992). Nucleotide sequence analysis of a mouse Y chromosomal DNA fragment containing BKM and LINE elements. *Genetica,* **87**, 7—15.

Okada T.A. and Comings D.E. (1974). Mechanisms of chromosome banding. III. Similarities between G-bands of mitotic chromosomes and chromomeres of meiotic chromosomes. *Chromosoma,* **48**, 65—71.

Oliver S.G., Van der Aart Q.J.M., Agostini-Carbone M.L. *et al.* (1992) The complete DNA sequence of yeast chromosome III. *Nature,* **357**, 38—46.

Riggs A.D. and Pfeifer G.P. (1992). X-chromosome inactivation and cell memory. *Trends in Genetics,* **8**, 169—74.

Saccone S., De Sario A., Della Valle G. and Bernardi G. (1992). The highest gene concentrations in the human genome are in T bands of metaphase chromosomes. *Proceedings of the National Academy of Science U.S.A.,* **89**, 4923—17.

Saccone S., De Sario A., Wiegant J., *et al.* (1993) Correlations between isochores and chromosomal bands in the human genome. *Proceedings of the National Academy of Science U.S.A.,* **90**, 11929—33.

Sumner A.T. (1990) *Chromosome Banding*, Unwin Hyman, London.

Tazi J. and Bird A. (1990) Alternative chromatin structure at CpG islands. *Cell,* **60**, 909—20.

Turner B.M. (1991) Histone acetylation and control of gene expression. *Journal of Cell Science*, **99**, 13—20.

Verma, R.S., Luke, S., Brennan, J.P. *et al.* (1993) Molecular topography of the secondary constriction region (qh) of human chromosome 9 with an unusual euchromatic band. *American Journal of Human Genetics*, **52**, 981—986.

Wu T-.C. and Lichten M. (1994) Meiosis-induced double-strand break sites determined by yeast chromatin structure. *Science,* **263**, 515—8.

Yunis J.J. and Tsai M.Y. (1978). Mapping of polysomal messenger RNA and heterogeneous nuclear RNA to the lightly staining G-bands of human chromosomes. *Cytogenetics and Cell Genetics*, **22**, 364—7.

Yunis J.J., Tien Kuo M. and Saunders G.F. (1977) Localisation of sequences specifying messenger RNA to light-staining G-bands of human chromosomes. *Chromosoma*, **61**, 355—44.

Yunis, J.J. (1981) Mid prophase human chromosomes. The attainment of 2000 bands. *Human Genetics*, **56**, 293—8.*Chromosoma,* **61**, 335—44.

PART TWO

MAINTENANCE AND BREAKDOWN OF STABILITY

Maurizio Gatti

Harald Biessmann

José Egozcue

Martine Muleris

6

Genes controlling chromosome structure in *Drosophila melanogaster*

F. Verni', M.P. Somma, R. Gandhi, M.L. Goldberg and M. Gatti

6.1 INTRODUCTION

The DNA of eukaryotic cells is condensed approximately 10,000-fold into metaphase chromosomes. This extremely tight DNA compaction is brought about with such an exquisite precision that metaphase chromosomes maintain, mitosis after mitosis, not only their relative size but also fine morphological features such as primary and secondary constrictions, differential condensation of euchromatin and heterochromatin, and chromosome banding. The early steps of DNA compaction, resulting in a 40-fold packing of this molecule, are rather well defined. They involve DNA—histone interactions producing the 11 nm nucleosomal fibre which is further coiled into a 30 nm solenoid (reviewed by McGhee and Felsenfeld, 1980). The higher order condensation of the solenoid is less understood. The most widely accepted model is that the solenoid fiber is folded into 60—100 kb loops anchored in a radial fashion to a 'scaffold' of nonhistone proteins (Marsden and Laemmli, 1979). This scaffold is thought to contain a loose network of discrete anchoring complexes which, upon histone depletion and fixation, may collapse into a continuous proteinaceous structure (for reviews see Paulson, 1988; Gasser *et al.*, 1989). Only one of the identified scaffold components, the enzyme topoisomerase II, has been shown to have a role in chromosome condensation; this enzyme is also required for decatenation of sister chromatids at the onset of anaphase (for reviews see Earnshaw, 1991; Holloway, 1995).

Our understanding of the structural elements distinguishing

morphological features of the chromosomes such as centromeres is also at an early stage. The use of autoimmune CREST sera has led to the identification of several proteins that bind mammalian centromeres, but only two of these polypeptides, CENP-A and CENP-B, are thought to play a role in centromeric chromatin packaging (reviewed by Rattner, 1991; see also Pluta *et al.*, 1992). CENP-A is a centromere-specific histone-like protein (Palmer *et al.*, 1991), while CENP-B is a helix-loop-helix DNA binding protein that specifically interacts with a 17 bp sequence (CENP-B box) of centromeric satellite DNAs of mouse, primates and man (Masumoto *et al.*, 1989; Sullivan and Glass, 1991; Muro *et al.*, 1992). Polypeptides that bind the CENP-B box have been recently isolated also in *Drosophila melanogaster*. These possible functional homologues of CENP-B accumulate in centromeric heterochromatin of this organism but it is currently unknown whether they are needed for heterochromatin packaging (Avides and Sunkel, 1994). To the best of our knowledge, no proteins have been identified to date that are specifically involved in finer structural features of metaphase chromosome, such as the differential condensation of heterochromatin or chromosome banding.

One approach to identify proteins implicated in the various levels of chromosome condensation and structure is genetic dissection. This approach has the obvious advantage that the organism 'tells' us which gene products are important for the process under study. Moreover, it allows the identification of proteins essential for chromosome condensation but not sufficiently abundant to be isolated by conventional biochemical methods.

Drosophila melanogaster is in many ways an ideal organism for mutational dissection of the processes of chromosome condensation. The sophisticated genetics of *D. melanogaster* allows the collection and the characterization of large numbers of mutants and facilitates the molecular cloning of the corresponding genes. Moreover, the finding that lesions in genes specifying essential mitotic functions often have a distinctive lethal syndrome, makes the isolation of mutants affecting chromosome structure very straightforward (Gatti *et al.*, 1983a; Gatti and Baker, 1989; see below). The cytological phenotypes of these mutants can be easily examined in dividing larval neuroblasts which exhibit morphologically well defined chromosomes that can be stained and longitudinally differentiated by a variety of procedures (for reviews see Gatti and Goldberg, 1991; Gatti *et al.* 1994).

Finally, *D. melanogaster* offers the possibility of testing whether the gene products identified by condensation-defective mutants also have roles in the control of gene expression. This can be done by examining whether these mutants have the ability to modify position effect variegation (PEV). PEV is the mosaic inactivation of a euchromatic gene relocated next to heterochromatin by a chromosomal rearrangement or a transposition event. This phenomenon is thought to be the consequence of the spreading from heterochromatin to euchromatin of multiprotein complexes that make the variegating gene inaccessible to transcriptional machinery (for reviews see Grigliatti, 1991; Reuter and Spierer, 1992; Weiler and Wakimoto, 1995). Strong support for this hypothesis has recently been provided by Wallrath and Elgin (1995), who showed that variegating transgenes inserted into *D. melanogaster* heterochromatin have an altered chromatin structure.

6.2 ISOLATION OF MUTANTS AFFECTING CHROMOSOME MORPHOLOGY

The first two mutants shown to affect mitotic chromosome morphology in *D. melanogaster*, *mus* 101^{tsl} and *mit(1)4*, were isolated by screening a collection of 168 X-linked temperature-sensitive *(ts)* lethals (Gatti *et al.*, 1983b; Smith *et al.*,1985). The approach of screening *ts* lethals to detect lesions in essential mitotic functions has been widely exploited in yeast (reviewed by Pringle and Hartwell,1981); it is based on the expectation that temperature-induced reduction of mitotic gene activity will result in an aberrant mitotic phenotype.

Although *ts* lethals are a reasonably good source of mutants affecting mitotic chromosome architecture, functions needed for proper chromosome condensation in *Drosophila* can also be identified by cytologically screening nonconditional larval/pupal lethals with defective imaginal discs. The rationale for screening such mutants is based on a hypothesis for the control of cell division during the life cycle of *D. melanogaster* (Baker *et al.*,1982; Gatti and Baker, 1989). It has been proposed that most, if not all, of the embryonic mitotic divisions needed to form a larva are accomplished using maternal products packaged into the egg. Divisions of cell populations that are progenitors of adult structures, such as the imaginal discs, the abdominal histoblasts and the neuroblasts, would instead be under the control of the zygotic genome. Thus, a zygote homozygous for a mutation in an essential mitotic gene, which is necessarily produced by a heterozygous mother, would not die until the larval/pupal transition, when the larval cells are replaced by the progenitor cells of the adult. However, due to the progressive depletion of the maternal product, the imaginal cells and the neuroblasts of these homozygous mutant larvae would progressively lose the ability to undergo a normal cell division. Hence, homozygous mutants for essential cell cycle genes are expected to die at the larval/pupal boundary and to exhibit small, morphologically defective imaginal discs.

Based on this hypothesis 59 late larval/early pupal lethals, including 41 with defective imaginal discs, were examined cytologically for mitotic defects in larval neuroblasts (Gatti and Baker, 1989). This analysis showed that 30 mutations, all but one with defective imaginal discs, cause disruptions in mitotic chromosome behavior. Remarkably, 21 of these 30 mutants exhibit more or less pronounced defects in chromosome condensation. The remaining 8 mutants affect other aspects of mitotic cell division such as progression through interphase, chromosome integrity, spindle formation and/or function, or cytokinesis. These results indicate that the phenotypes of larval/pupal lethality and poorly developed imaginal discs are together almost diagnostic of a mutation that disrupts mitotic division. Moreover, more than one half of the mutations that cause these phenotypes identify genes required for proper mitotic chromosome condensation.

Our model for the control of cell division during the *Drosophila* life cycle predicts that maternal products can partially sustain mitotic divisions of imaginal discs and neuroblasts. Indeed, there is evidence that some wild-type mitotic products persist at least until the end of the second larval instar and even during the third larval instar (Carmena *et al.* 1991; Ripoll *et al.*, 1992). Larval brain neuroblasts can therefore be regarded as a cell population that keeps dividing while a product essential for mitosis gets progressively depleted. This raises the concern that some of the observed terminal mitotic phenotypes are not the direct consequence of the absence of a mitotic product, but are instead indirect effects of a defect in cellular metabolism. More specifically, it is possible that sick, dying cells fail to properly condense their DNA in preparation for mitosis. However, two considerations suggest that most mutants with abnormally condensed mitotic chromosomes do in fact specify functions involved in chromatin packaging or maintenance. First, many mutants exhibit quite specific chromosomal phenotypes that permit them to be grouped into several categories. Second, most condensation defective mutants appear to suppress PEV in a dominant manner, suggesting that they identify chromatin components (see below).

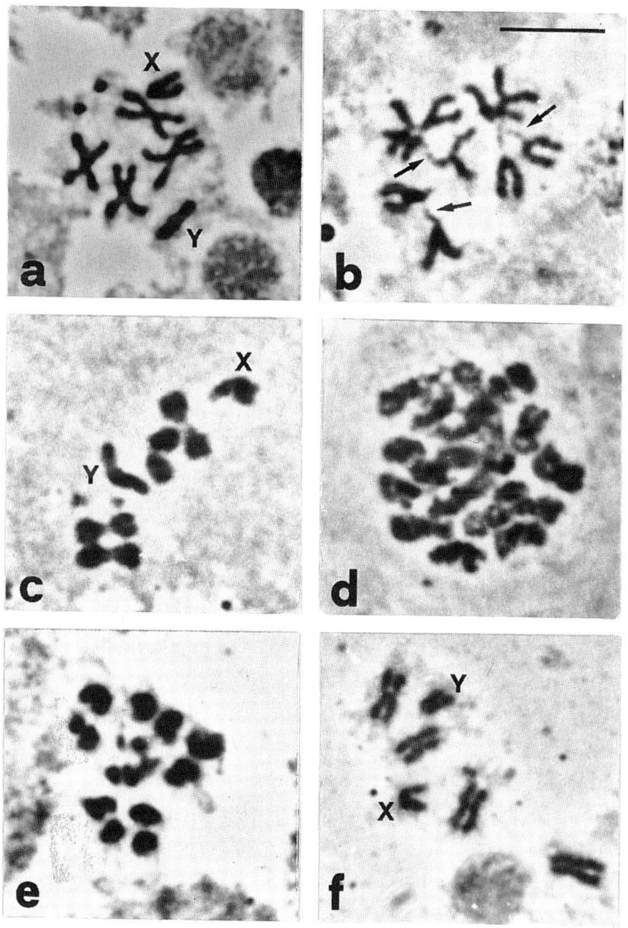

Fig. 6.1 (a) Normal male metaphase. (b) female metaphase of *mus-101*^{tsl}; note undercondensation of heterochromatin (arrows). (c) male metaphase of *mit(1)4* showing abnormal condensation of both euchromatin and heterochromatin. (d) irregularly condensed polyploid metaphase in *l(3)2612*. (e) overcondensed metaphase chromosomes in *l(1)d.deg10*. (f) *pasc* metaphase showing sister chromatid separation in the heterochromatic regions. Bar, 5μm.

6.3 THE CYTOLOGICAL PHENOTYPES OF MUTANTS AFFECTING CHROMOSOME MORPHOLOGY

To date, mutants at 28 loci have been identified that disrupt normal chromosome morphology in larval neuroblasts. Based on their cytological phenotypes, these mutants can be grouped into four broad classes.

6.3.1 Mutants defective in heterochromatin condensation

The only mutant in this category is a heat-sensitive lethal allele of *mus-101* (*mus-101*^{tsl}) (Gatti *et al.*, 1983 b; Smith *et al.*, 1985). *mus-101* is an X-linked locus originally identified by viable mutagen-sensitive mutations defective in postreplication repair (Boyd *et al.*, 1976; Boyd and Setlow, 1976). Upon exposure to restrictive temperature, homozygous and hemizygous *mus-101*^{tsl} individuals exhibit a drastic undercondensation of the heterochromatin of all chromosomes; remarkably, euchromatic arms remain normally condensed (Fig. 6.1b). This failure of heterochromatin condensation is not due to gross defects in DNA synthesis because undercondensed heterochromatic regions incorporate normal amounts of tritiated thymidine during S phase (Gatti *et al.*, 1983b). These results indicate that *mus-101*$^{+}$ encodes an essential product required for heterochromatin condensation. They also suggest that mutagen sensitivity and defective DNA repair may be secondary effects of a primary defect in heterochromatin packaging. However the finding that *mus-101*$^{+}$ function is also required for chorion gene amplification (Orr *et al.*, 1984), indicates that this locus is not solely involved in heterochromatin condensation.

It is worth noting that the cytological phenotype of *mus-101* is reminiscent of that observed in the ICF (Immunodeficiency, Centromeric instability and Facial anomalies) syndrome, a rare human disorder characterized by the undercondensation of the heterochromatic regions of chromosomes 1, 9 and 16 (Tiepolo *et al.*, 1978; Hulten, 1978; Maraschio *et al.*, 1988).

6.3.2 Mutants defective in condensation of both euchromatin and heterochromatin

Although the 21 mutants in this category have the common characteristic of affecting both euchromatin and heterochromatin, they display markedly different cytological phenotypes.

Neuroblast chromosomes in mutants at the *l(1)d.deg-9* and *mit(1)4* loci are swollen and fuzzy, and have sister chromatids that tend to be closely apposed (Smith *et al.*, 1985; Gatti and Baker, 1989). Moreover, heterochromatic regions are slightly elongated, and the Y chromosome often exhibits a decondensed region on the short arm (Fig. 6.1c). In *l(1)d.deg-9* about 10% of metaphases show chromosome breakage, whereas *mit(1)4* neuroblasts have a normal level of breakage (less than 0.5%). However, it should be noted that the map positions of these mutations are compatible with the possibility that they are allelic.

Mutants at *l(3)IX-14*, *l(3)g60A*, *l(3)7m-75* and *l(3)15m-25* are characterized by swollen and unevenly condensed metaphase chromosomes and by low levels of chromosome breakage (Gatti and Baker, 1989). *l(1)het-2* and *l(3)2004* behave similarly, but also exhibit 5—10% polyploid cells (compared with less than 0.1% in wild type) (Gatti and Baker, 1989). Finally, *l(1)d.deg-12* and *l(3)2612*, which display particularly swollen and irregularly condensed metaphase chromosomes (Fig. 6.1d), exhibit about 10% polyploid cells but no chromosome breakage (Gatti and Baker, 1989).

Abnormal chromosome condensation and polyploid cells are also caused by mutations in *abnormal anaphase resolution* (*aar*) (Gomes *et al.* 1993). In addition, mutants at this locus exhibit stretched and lagging chromatids during anaphase and delay in the metaphase-anaphase transition. *aar* encodes the 55 kd regulatory subunit of *Drosophila* phosphatase 2A, an enzyme involved in the regulation of $p34^{cdc2}$ as well as in dephosphorylation of p34-dependent phosphorylation sites in histone H1 (Mayer-Jaekel *et al.* 1993, and references therein).

In *l(3)XII-10* neuroblasts there is only a mild defect in chromosome condensation, resulting primarily in uneven condensation of chromatids. In addition, about 10% of the cells of this mutant have broken chromosomes and 5% are polyploid (Gatti and Baker, 1989). Relatively mild defects in chromosome condensation are also observed in *l(3)e20*, *l(3)k43* and *l(3)1902*, three mutants that display very few dividing cells in larval brains. However, the rare mitotic figures observed in these mutants show extensive chromosome breakage (Gatti and Baker, 1989). Interestingly, *l(3)1902* is a null allele of the *Enhancer of zeste* [*E(z)*] locus (Jones and Gelbart, 1990), also called *polycombeotic* (*pco*) (Phillips and Shearn, 1990), a member of the Polycomb-group genes involved in the maintenance of the repressed state of the *Antennapedia* (*Ant-C*) and *bithorax* (*BX-C*) homeotic gene complexes.

In mutants at the *l(3)8m-12*, *l(3)11m-254*, *l(3)12m137*, *l(3)IIb1* and *l(3)m45* loci, effects on chromosome condensation are very mild: chromosomes are only slightly swollen or unevenly condensed. Neither chromosome breakage nor polyploid cells have been observed in these mutants. How directly the wild-type functions identified by these mutants are concerned with chromatin condensation remains unclear.

Finally, a unique pattern of abnormal chromosome condensation has been observed in larval brain ganglia of mutants at *lethal(2)giant larvae [l(2)gl]* (Radhakrishnan and Sinha, 1987). *l(2)gl* is a recessive oncogene of *Drosophila melanogaster* which, when mutated, causes tumorous growth of imaginal discs and larval brains (Gateff, 1978). In about 50% of *l(2)gl* brain cells, mitotic chromosome are two-to four-fold longer than normal but do not appear to be otherwise irregularly condensed. *l(2)gl* has been cloned and shown to encode a 127 kd polypeptide that displays characteristics of a cytoskeletal protein (Strand *et al.*, 1994). Whether this protein has a direct role in chromosome condensation is unknown.

6.3.3 Mutants leading to chromosome overcondensation

Lethal mutations at the *l(1)d.deg-3*, *l(1)d.deg-10*, *dotted* and *podgy* loci arrest cell division at metaphase and produce polyploid cells (Ripoll *et al.*, 1987; Gatti and Baker, 1989). However, the degree of metaphase chromosome condensation observed in these mutants is much higher than that found in other *D. melanogaster* metaphase arrest mutants or in wild-type cells exposed to prolonged colchicine treatments (Fig. 6.1e). In addition to this extreme chromosome condensation, *l(1)d.deg3* and *l(1)d.deg10* exhibit both frequent sister chromatid separation and chromosome breakage that appears to occur at or near the centromere (Gatti and Baker, 1989). There are two possible explanations for the phenotypes of these mutants. They may be defective in functions that allow cells blocked in metaphase to revert to interphase, so that their chromosomes would keep to condense for a longer period than in wild-type. Alternatively, these mutants may specify lesions in functions that prevent excessive chromosome condensation, and their metaphase arrest phenotypes could be a downstream consequence of these primary defects.

Extremely condensed metaphase chromosomes and hyperploid/polyploid cells have been also observed in mutants that disrupt one of the *Drosophila* protein phosphatase 1 isoenzymes (PP1 87B) (Axton *et al.*, 1990). Interestingly, excessively condensed chromosomes are also seen in *Schizosaccaromyces pombe* carrying mutations in the *dis* genes that encode products highly homologous to PP1 (Ohkura *et al.*, 1989). It has been suggested that chromosome overcondensation observed in such PP1 mutants may either be a consequence of a failure of $p34^{cdc2}$ kinase inactivation leading to a mitotic block, or instead reflect a direct involvment of PP1 phosphatase in the control of chromatin packaging. The finding that mutants at the *PP1 87B* locus are dominant suppressors of PEV provides support for the latter hypothesis (Baksa *et al.*, 1993).

6.3.4 Mutants affecting heterochromatin organization

This class includes only *parallel sister chromatids* (*pasc*), a locus defined by mutations that disrupts sister chromatid apposition in heterochromatic regions without affecting heterochromatin condensation (Verni' *et al.*, in preparation). In wild type metaphase chromosomes, sister chromatids are well separated in euchromatin but closely apposed in the heterochromatic regions. This results in the characteristic morphology of the metacentric autosomes and of the telocentric X chromosomes which appear X-shaped and V-shaped, respectively. In *pasc* mutants, sister chromatids adhesion in the heterochromatic regions is disrupted. As a consequence, sister chromatids are aligned parallel to each other and remain connected only at centromeres by a tiny thread of chromatin (Fig. 6.1f). This cytological defect does not affect parameters measuring cell cycle progression, nor does it result in chromosome breakage or missegregation. A cytological phenotype very similar to that elicited by *pasc* mutations is observed in Roberts syndrome, a rare human recessive disorder characterized by growth retardation, craniofacial abnormalities and tetraphocomelia. However, in several cases of Roberts syndrome the defect in heterochromatin organization is accompained by a low frequency of chromosome loss during anaphase (reviewed by Van Den Berg and Francke, 1993).

The observation that *pasc* is required for sister chromatid adhesion in heterochromatic regions suggested that this locus may also control heterochromatin pairing during female meiosis. A role for heterochromatin in pairing of achiasmate homologues during the first meiotic division of *Drosophila* females has been recently suggested by Hawley and coworkers (Hawley *et al.*, 1993). Examination of the progeny of females heterozygous for *pasc* shows that mutations at this locus increase nondisjunction of achiasmate X chromosomes in a dominant manner, suggesting a general role of $pasc^+$ in heterochromatin cohesion (Verni' *et al.*, in preparation). Further support for a general role of $pasc^+$ in the control of heterochromatin organization is provided by the finding that mutant alleles of this gene act as dominant suppressors of both *white* and *Stubble* variegation (Verni' *et al.*, in preparation).

Molecular cloning of *pasc* (Gandhi *et al.*, in preparation) indicates that this gene gives rise to two alternatively spliced transcripts of 6.5 and 5 kb that appear to encode proteins that vary at their N-termini but are otherwise identical. These proteins show no substantial homology with known polypeptides. However, they contain stretches of acidic amino acids, a characteristic showed by several chromosome-bound proteins including the products of *Su(var)2-5* and *Su(var)3-7*, two genes that suppress PEV (reviewed by Reuter and Spierer, 1992).

6.4 CHROMOSOME CONDENSATION AND PEV SUPPRESSION

Hundreds of mutations, representing about 80 loci, have been identified to date that act as dominant suppressors *[Su(var)s]* or enhancers *[E(var)s]* of PEV (Grigliatti, 1991; Reuter and Spierer, 1992; Weiler and Wakimoto, 1995). *Su(var)* genes are thought to specify structural or regulatory chromatin components that have a dosage-dependent effect on the expression of variegating genes. Thus, many *Su(var)* genes could potentially be involved in the packaging of chromatin into metaphase chromosomes. However, to date only three *Su(var)* mutants have been examined for effects on chromosome morphology. As previously described, mutants at the *pasc* and *PP1 87B* [also called *Su(var)3-6*] loci, which have distinctive effects on chromosome morphology, also suppress PEV. Another gene involved in both PEV suppression and in the control of chromosome structure is *Su(var)2-5*. This gene encodes HP-1, a protein predominantly localized at the chromocentre of polytene chromosomes which shares a region of similarity termed the chromodomain with the Polycomb protein (Eissenberg *et al.*, 1990; Paro and Hogness, 1991). Examination of HP-1 distribution during embryonic divisions reveals a more complex cell cycle-dependent immunostaining pattern. The protein binds heterochromatin and, to a lesser extent, euchromatin of interphase nuclei. However, HP-1 dissociates from chromosomes during metaphase and anaphase and becomes associated with chromosomes again at telophase (Kellum *et al.*, 1995). Although the biological meaning of this dynamic binding pattern is not yet understood, cytological analysis of cell division in *Su(var)2-5* homozygous embryos indicates that HP-1 is required for chromosome condensation. In these embryos, chromosomes exhibit undercondensation of both euchromatin and heterochromatin and fail to segregate properly. It has been suggested that the defective chromosome segregation is a secondary consequence of incomplete chromosome condensation that causes chromatids to become tangled (Kellum and Alberts, 1995).

To investigate whether chromatin assembly and PEV modulation are controlled by the same genes we tested whether the mutants we have previously found to affect chromosome condensation also act as PEV modifiers. Work in progress strongly suggests that mutants at most loci controlling chromosome condensation suppress *white* variegation in *In(1)w*^{m4}. In particular, a clear suppression effect has been observed for *l(1)d.deg 12, l(3)8m12, l(3)12m137, l(3)15m25, l(3)2004, l(3)m45, l(3)XII-10, l(3)1902, l(3)2612, l(3)IX-14, l(3)11m254* and *l(3)K43*, whereas *l(3)e20* and *l(3)g60A* do not appear to affect variegation (Verni', Somma and Gatti, unpublished results). These findings provide strong support for the hypothesis that PEV suppressors identify functions implicated in chromatin packaging or maintenance. Moreover they indicate that the abnormal chromosome morphology observed in many late lethals is not a non-specific consequence of defects in cellular metabolism.

6.5 SUMMARY AND PERSPECTIVES

D. melanogaster offers great advantages for the isolation of mutations affecting chromosome structure. Large numbers of these mutants can be recovered by screening lethal mutations that die at the larval/pupal boundary with poorly developed imaginal discs. The roles of the loci corresponding to these mutations in both chromatin packaging and regulation of gene expression can be further defined by testing their ability to modulate PEV.

Cytological analysis and PEV allow initial identification of genes controlling chromosome structure. However, further insight into the roles of these genes in chromatin packaging requires molecular analysis. Most of the chromosome condensation mutants isolated to date were induced by ionizing radiations or chemical mutagens and are thus quite difficult to clone. However, mutants affecting chromosome morphology can also be isolated by single P elements mutagenesis (our unpublished results; M. Heck, personal communication), a recently introduced powerful technology that greatly facilitates gene cloning (reviewed by Gatti and Goldberg, 1991). Thus, molecular characterization of *Drosophila* chromosome condensation genes and the proteins they encode is likely to be relatively easy.

Finally, we believe that the molecular probes obtained in *Drosophila* will permit the isolation of orthologous genes in vertebrates including man. This is suggested by the finding that the three cloned genes controlling chromosome structure in *Drosophila* have mammalian counterparts. *PP1 87B* is homologous to mammalian phosphatases (Axton *et al.*, 1990, and references therein), and HP-1 is widely conserved in plants and animals including man (Singh *et al.*, 1991; Saunders *et al.*, 1993; Wreggett *et al.*, 1994). Moreover, we have recently obtained evidence that *pasc* has a human homologue (Somma, Verni' and Gatti, unpublished results).

ACKNOWLEDGEMENTS

This work was supported by grants from Progetto Finalizzato Ingegneria Genetica (CNR) and Telethon-Italy (grant no. E27) to M. Gatti, and by NIH grant GM 48430 to M. L. Goldberg.

REFERENCES

Avides, M.C. and Sunkel, C.E. (1994) Isolation of chromosome-associated proteins from *Drosophila melanogaster* that bind a human centromeric DNA sequence. *Journal of Cell Biology*, **127**, 1159—71.

Axton, J.M., Dombradi, V., Cohen, P.T.W. and Glover, D.M. (1990) One of the protein phosphatase I isoenzymes in *Drosophila* is essential for mitosis. *Cell*, **63**, 33—46.

Baker, B.S., Smith, D.A. and Gatti, M. (1982) Region-specific effects on chromosome integrity of mutations at essential loci in *Drosophila melanogaster. Proceedings of the National Academy of Science USA*, **79**, 1205—09.

Baksa, K., Morawietz, H., Dombradi, V. *et al.* (1993) Mutations in the protein phosphatase 1 gene at 87B can differentially affect suppression of position effect variegation and mitosis in *Drosophila melanogaster. Genetics*, **135,** 117—25.

Boyd, J.B., Golino, M.D., Ngugyen, T.G., Green, M.M. (1976) Isolation and characterization of X-linked mutants of *Drosophila melanogaster* which are sensitive to mutagens. *Genetics*, **84**, 485—506.

Boyd, J.B. and Setlow, R.B. (1976) Characterization of postreplication repair in mutagen-sensitive strains of *Drosophila melanogaster. Genetics*, **84**, 507—26.

Carmena, M., Gonzalez, C., Casal, J, and Ripoll, P. (1991) Dosage dependence of maternal contribution to somatic cell division in *Drosophila melanogaster. Development*, **113**, 1357—64.

Earnshaw, W.C. (1991) Large scale chromosome structure and organization. *Current Opinion in Structural Biology*, **1**, 237—44.

Eissenberg, J.C., James, T.C., Foster-Hartnett, D.M. *et al.* (1990) Mutation in a heterochromatin-specific chromosomal protein is associated with suppression of position effect variegation in *Drosophila melanogaster. Proceedings of the National Academy of Science USA*, **87**, 9923—27.

Gasser, S.M., Amati, B.B., Cardenas, M.E. and Hofmann, J.F.-X (1989) Studies on scaffold attachment sites and their relation to genome function. *International Review of Cytology*, **119**, 57—96.

Gateff, E. (1978) Malignant neoplasms of genetic origin in *Drosophila melanogaster. Science*, **200**, 1448—59.

Gatti, M., Pimpinelli, S., Bove, C. *et al* (1983a) Genetic control of mitotic cell division in *Drosophila melanogaster*, in *Genetics—New Frontiers. Proceedings of the XV International Congress of Genetics*, (eds V.L. Chopra, B.C. Joshi, R. P. Sharma and H.C. Bansal), Oxford and IBH Publishing Co., New Delhi, Vol.III, pp. 193—204.

Gatti, M., Smith, D.A. and Baker, B.S. (1983b) A gene controlling condensation of heterochromatin in *Drosophila melanogaster*. *Science*, **221**, 83—85.

Gatti, M. and Baker, B.S. (1989) Genes controlling essential cell-cycle functions in *Drosophila melanogaster. Genes and Development*, **3**, 438—53.

Gatti, M. and Goldberg, M.L. (1991) Mutations affecting cell division in *Drosophila*. *Methods in Cell Biology*, **35**, 543—86.

Gatti, M., Bonaccorsi, S. and Pimpinelli, S. (1994) Looking at *Drosophila* mitotic chromosomes. *Methods in Cell Biology*, **44**, 371—91.

Gomes, R., Karess, R.E., Ohkura, H. *et al.* (1993) Abnormal anaphase resolution (*aar*): a locus required for progression through mitosis in *Drosophila*. *Jounal of Cell Science*, **104**, 583—93.

Grigliatti, T. (1991) Position-effect variegation — An assay for nonhistone chromosomal proteins and chromatin assembly and modifying factors. *Methods in Cell Biology*, **35**, 587—627.

Hawley, R.S., Irik, H., Zitron, A.E. *et al.* (1993) There are two mechanims of achiasmate segregation in *Drosophila* females, one of which requires heterochromatic homology. *Developmental Genetics*, **13**, 440—67.

Holloway, S.L. (1995) Sister chromatid separation in vivo and in vitro. *Current Opinion in Genetics and Development*, **5**, 243—48.

Hulten, M. (1978) Selective somatic pairing and fragility at 1q12 in a boy with common variable immunodeficiency. *Clinical Genetics*, **14**, 194 (abstract).

Jones, R.S. and Gelbart, W.M. (1990) Genetic analysis of the *Enhancer of zeste* locus and its role in gene regulation in *Drosophila melanogaster*. *Genetics*, **126**, 185—99.

Kellum, R. and Alberts, B.M. (1995) Heterochromatin protein1 is required for correct chromosome segregation in *Drosophila* embryos. *Journal of Cell Science*, **108**, 1419—31.

Kellum, R., Raff, J.W. and Alberts, B.M. (1995) Heterochromatin protein 1 distribution during development and during the cell cycle in *Drosophila* embryos. *Journal of Cell Science*, **108**, 1407—18.

Maraschio, P., Zuffardi, O., Dalla Fior, T. and Tiepolo, L. (1988) Immunodeficiency, centromeric heterochromatin instability of chromosomes 1, 9 and 16 and facial anomalies: the ICF syndrome. *Journal of Medical Genetics*, **25**, 173—80.

Marsden, M.P.F. and Laemmli, U.K. (1979) Metaphase chromosome structure: evidence for a radial loop model. *Cell*, **17**, 849—58.

Masumoto, H., Masukata, H., Muro, Y. *et al.* (1989) A human centromere antigen (CENP-B) interacts with a short specific sequence in alphoid DNA, a human centromere satellite. *Journal of Cell Biology*, **109**, 1963—73.

Mayer-Jaekel, R.E., Ohkura, H., Gomes, R. *et al.* (1993) The 55 kd regulatory subunit of *Drosophila* protein phosphatase 2A is required for anaphase. *Cell*, **72**, 621—33.

McGhee, J.D. and Felsenfeld, G. (1980). Nucleosome structure. *Annual Review of Biochemistry*, **49**, 1119—56.

Muro,Y., Masumoto, H., Yod, K. *et al.* (1992) Centromere protein B assembles human centromeric α-satellite DNA at the 17-bp sequence, CENP-B box. *Journal of Cell Biology*, **116**, 585—96.

Ohkura, H., Kinoshita, N., Miyatani, S. *et al.* (1989) The fission yeast *dis2*$^{+}$ gene required for chromosome disjoining encodes one of two putative type 1 protein phosphatases. *Cell*, **57**, 997—1007.

Orr, W., Komitopoulou, K. and Kafatos, F.C. (1984) Mutants suppressing *in trans* chorion gene amplification in *Drosophila. Proceedings of the National Academy of Science USA*, **81**, 3773—77.

Palmer, D.K., O'Day, K., Le Trong, H. *et al.* (1991) Purification of the centromere-specific protein CENP-A and demonstration that it is a distinctive histone. *Proceedings of the National Academy of Science USA*, **88**, 3734—38.

Paro, R., and Hogness, D.S. (1991) The Polycomb protein shares a homologous domain with a heterochromatin-associated protein of *Drosophila*. *Proceedings of the National Academy of Science USA*, **88**, 263—67.

Paulson, J.R. (1988) Scaffolding and radial loops: the structural organization of metaphase chromosomes, in *Chromosomes and Chromatin* (eds K.W. Adolph), CRC Press, Boca Raton, Vol. 3, pp. 3—36.

Phillips, M.D. and Shearn, A. (1990) Mutations in *polycombeotic*, a *Drosophila* polycomb-group gene, cause a wide range of maternal and zygotic phenotypes. *Genetics*, **125**, 91—101.

Pluta, A.F., Saitoh, N., Goldberg, I. and Earnshaw, W.C. (1992) Identification of a subdomain of CENP-B that is necessary and sufficient for localization to the human centromere. *Journal of Cell Biology*, **116**, 1081—93.

Pringle, J.R. and Hartwell, L.H. (1981) The *Saccharomyces cerevisiae* cell cycle, in *Molecular biology of the Yeast Saccharomyces: Life, Cycle and Inheritance* (eds J.N. Strathern, E.W. Jones and J.R. Broach), Cold Spring Harbor Laboratory, Cold Spring Harbor, New York, pp. 97—142.

Radhakrishnan, V. and Sinha, P. (1987) Mitotic chromosome organization of neuroblastomas induced by the recessive oncogene *lethal(2)giant larvae*^{4} of *Drosophila melanogaster*. *Chromosoma*, **96**, 77—82.

Rattner, J.B. (1991) The structure of mammalian centromere. *BioEssays*, **13**, 51—56.

Reuter, G. and Spierer, P. (1992) Position effect variegation and chromatin proteins. *BioEssays*, **14**, 605—12.

Ripoll, P., Casal, J. and Gonzalez, C. (1987) Towards the genetic dissection of mitosis in *Drosophila*. *BioEssays*, **7**, 204—10.

Ripoll, P., Carmena, M. and Molina, I. (1992) Genetic analysis of cell division in *Drosophila*. *Current Topics in Developmental Biology*, **27**, 275—307.

Saunders, W.S., Chue, C., Goebl, M. *et al.* (1993) Molecular cloning of a human homologue of *Drosophila* heterochromatin protein HP1 using anti-centromere autoantibodies with anti-chromo specificity. *Journal of Cell Science*, **104**, 573—82.

Singh, P.B., Miller, J.R., Pearce, J. *et al.* (1991) A sequence motif found in a *Drosophila* heterochromatin protein is conserved in animals and plants. *Nucleic Acids Research*, **19**, 789—93.

Smith, D.A., Baker, B.S. and Gatti, M. (1985) Mutations in genes encoding essential mitotic functions in *Drosophila melanogaster*. *Genetics*, **110**, 647—70.

Strand, D., Raska, I. and Mechler B.M. (1994) The *Drosophila lethal (2)giant larvae* tumor suppressor protein is a component of the cytoskeleton. *Journal of Cell Biology*, **127**, 1345—60.

Sullivan, K.F. and Glass, C.A. (1991) CENP-B is a highly conserved mammalian centromere protein with homology to the helix-loop-helix family of proteins. *Chromosoma*, **100**, 360—70.

Tiepolo, L., Maraschio, P., Gimelli, G. *et al.* (1978) Concurrent instability at specific sites of chromosomes 1, 9 and 16 resulting multibranched structures. *Clinical Genetics*, **14**, 313—14.

Van Den Berg, D.J. and Francke, U. (1993) Roberts syndrome: a review of 100 cases and a new rating system for severity. *American Journal of Medical Genetics*, **47**, 1104—23.

Wallrath, L.L. and Elgin, S.C.R. (1995) Position effect variegation in *Drosophila* is associated with an altered chromatin structure. *Genes and Development*, **9**, 1263—77.

Weiler, K.S. and Wakimoto, B.W. (1995) Heterochromatin and gene expression in *Drosophila*. *Annual Review of Genetics*, in press.

Wreggett, K.A., Hill, F., James, P.S. *et al.* (1994) A mammalian homologue of *Drosophila* heterochromatin protein 1 (HP1) is a component of constitutive heterochromatin. *Cytogenetics and Cell Genetics*, **66**, 99—103.

7

Retrotransposons at *Drosophila* telomeres and terminal chromosome deficiencies

H. Biessmann, M.F. Walter, E. Kurenova and J.M. Mason

7.1 PREFACE

Telomeres are complex structures at the ends of linear chromosomes that perform several functions, at least two of which are vital: protection of the chromosome end from degradation and fusion ('capping'), and complete replication of DNA sequences at chromosome ends ('elongation'). Furthermore, telomeres are likely to be involved in associations with other telomeres and with structures in the nucleus, such as the nuclear scaffold and lamina. These associations may play a role in nuclear organization or meiotic chromosome pairing. In many eukaryotes, telomeres contain a stretch of simple repeats at the extreme end of the chromosome (telomeric repeat) and more complex repetitive sequences, often minisatellites, proximal to it. Several proteins have been identified that bind preferentially to these sequences (reviewed by Biessmann and Mason, 1992).

As recognized by Muller (1938), the predominant feature of the telomere is its capping function. However, there are some cases in which broken chromosomes are 'healed', and stable terminal chromosome deficiencies can be recovered. The healing of broken chromosomes was originally reported in maize by McClintock (1941), but more recently healing of broken chromosomes leading to terminal chromosome deficiencies has been observed in several other organisms including *Drosophila* (Mason et al., 1984; Biessmann and Mason, 1988;

Traverse and Pardue, 1988; Biessmann et al., 1990) and man (Wilkie et al., 1990). Telomere replication poses special problems for the linear chromosomes of eukaryotes, because all known DNA polymerases require a primer with a free 3'-end hydroxyl group, and synthesize in the 5' to 3' direction. Removal of the RNA primer of the lagging strand will result in a 5' terminal gap in one of the daughter strands and thus the DNA molecule will lose sequences from the end with each round of replication (Watson, 1972; Olovnikov, 1973). Therefore, every eukaryotic organism must possess a mechanism to counterbalance this inevitable terminal erosion in order to prevent loss of important genetic information.

Here we summarize the current knowledge of the DNA structure of *Drosophila melanogaster* telomeres and describe the unusual elongation mechanism by terminal attachment of specific retroposons. We also discuss the observation that terminal chromosome deficiencies occur under certain circumstances and may be quite frequent in natural populations.

7.2 *DROSOPHILA* TELOMERE ELONGATION

7.2.1 Mechanisms of telomere elongation

Two main mechanisms have been described to ensure the complete replication of telomeres. First, in many eukaryotes short tandem telomeric repeats are present at chromosome ends, e. g. $(TTGGGG)_n$ in the ciliate *Tetrahymena*, and $(TTAGGG)_n$ in humans. In these organisms telomere elongation is achieved by the addition of more of such repeats to the ends (Fig. 7.1a), catalyzed by telomerase, a reverse transcriptase that contains an RNA molecule encoding the species-specific template sequence for the GT-rich strand of the telomeric repeat (reviewed by Blackburn, 1992). A very different telomere elongation mechanism has been proposed for *Drosophila melanogaster* (reviewed by Mason and Biessmann, 1995), which depends on the attachment of specific retroposons, the HeT-A and TART elements, to chromosome ends (Fig. 7.1b). These retroposons belong to the evolutionarily widespread family of polyadenylated non-LTR transposons that includes the long interspersed elements, LINEs, in mammals. They are capable of attaching themselves with an oligo(A) tail to broken chromosome ends (see Fig. 7.6).

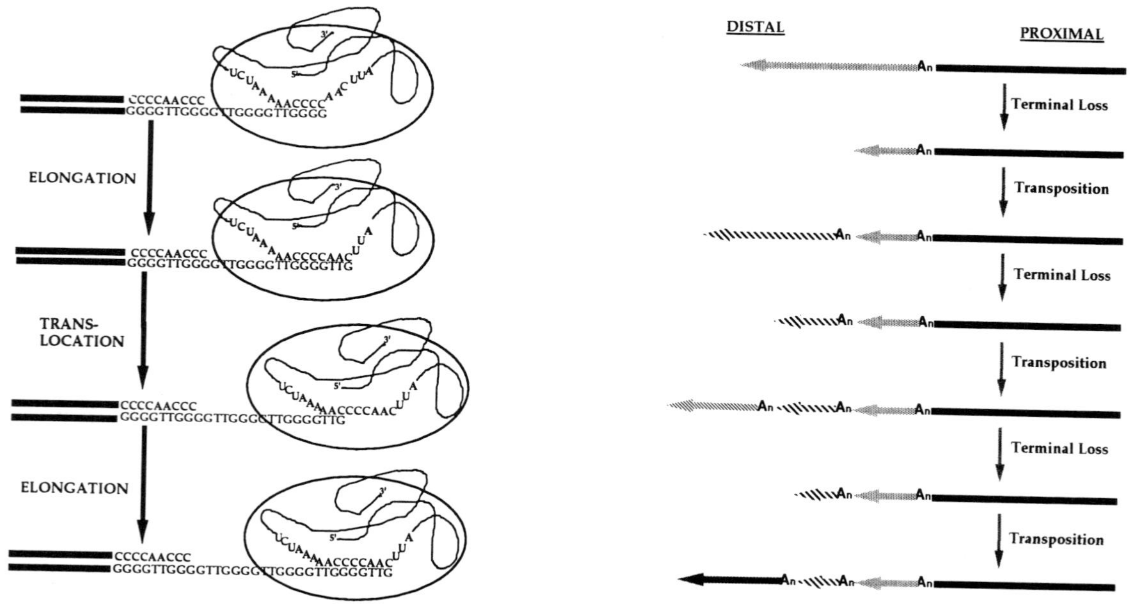

Fig. 7.1 Two telomere elongation mechanisms. **7.1a** Telomere elongation in e. g. *Tetrahymena* by telomerase synthesizing short telomeric repeats from an internal RNA template. **7.1b** Telomere elongation in *Drosophila* by frequent transpositions of HeT-A and TART retroposons to the chromosome ends. The arrows represent HeT-A or TART elements, their oligo(A) tails pointing proximally, attaching to the previous end (Reprinted in part with permission from: Mason and Biessmann, The unusual telomeres of *Drosophila*; published by Elsevier, *Trends in Genetics* 11, 58—62, 1995).

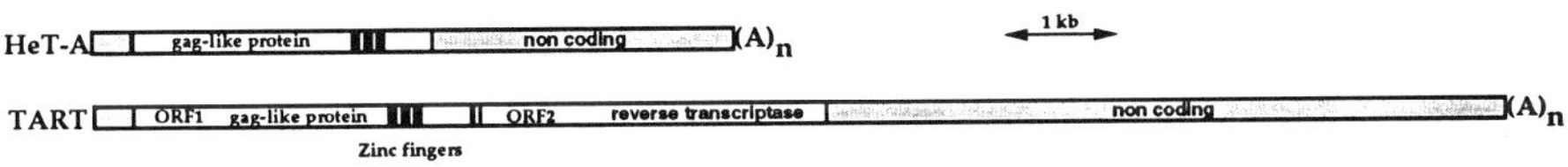

Fig. 7.2 Structural comparison of the two known telomere-specific LINE-like retroposons HeT-A and TART from *D. melanogaster*.

7.2.2 Structure of HeT-A and TART elements

HeT-A is approximately 6 kb in length (Fig. 7.2) and contains an unusually long, 2.4 kb, 3' non-coding region (Biessmann *et al.*, 1994; Danilevskaya *et al.*, 1994). Its RNA transposition intermediate has recently been identified (Danilevskaya *et al.*, 1994). The single 2.8 kb open reading frame (ORF) encodes a gag-like protein with three zinc finger nucleic acid binding motifs. While one recently transposed isolate contains an uninterrupted ORF (Biessmann *et al.*, 1992), the ORF in other isolates contain a —1 frameshift, interestingly always in the same codon (Danilevskaya *et al.*, 1992; Biessmann *et al.*, 1994). However, as in some retroviruses, this frameshift may be suppressed to produce a single, functional polypeptide (Danilevskaya *et al.*, 1994). A second ORF encoding a reverse transcriptase has not been found, even on very recently transposed HeT-A elements, suggesting that the required reverse transcriptase is encoded elsewhere in the genome and provided in *trans*. This would make HeT-A elements dependent on an external reverse transcriptase and distinguish them from the other retroposons. The TART element (Levis *et al.*, 1993) is about 12 kb long and differs from HeT-A, because, in addition to the ORF encoding a gag-like protein with three zinc fingers, it carries a second ORF encoding a reverse transcriptase. TART also has a long 3' non-coding region, measuring 5.1 kb, which shares no significant homology with HeT-A.

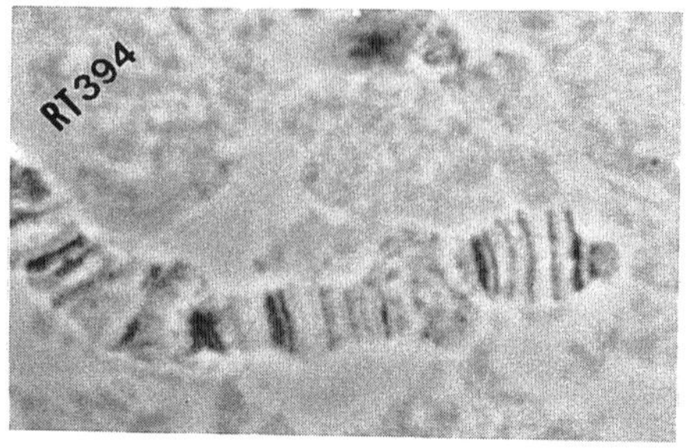

Fig. 7.3 X chromosome tip of a female heterozygous for the terminal deficiency RT394 and a wild type X chromosome.

7.2.3 HeT-A and TART retroposons, mobility and possible transposition mechanism

Terminal deficiencies of the X chromosome of *Drosophila* were recovered after irradiation of females carrying a *mutator-2* mutation (Mason *et al.*, 1984). Some deficiency breakpoints were positioned in the *yellow* gene which provides convenient probes for monitoring the ends of the broken chromosomes (Biessmann and Mason, 1988) (Fig. 7.3).

In order to study molecular events at the ends of the terminal deficiencies, individual deficiency/duplication males were taken at random from a stock to establish independent lines with attached-X females (Biessmann *et al.*, 1992). Every fourth generation, for 17 generations, lines were sampled by producing terminal deficiency/duplication females, preparing their DNA for genomic Southern analysis, and probing with a unique *yellow* DNA fragment (Fig. 7.4).

In general, DNA is lost from the distal end at 75 bp per generation (Fig. 7.5), providing evidence that two of the vital functions of the telomere, compensation for the loss of DNA at the end caused by incomplete replication, and protection from fusion with other DNA, can be separated. Since the terminal DNA fragments of these deficiencies have no repetitive DNA sequences attached to their distal ends (Biessmann *et al.*, 1990), we suggest that there may be non-DNA components, possibly proteins, at the distal ends of RT chromosomes that confer 'capping' function. In addition, 24 cases were discovered in which the terminal DNA fragment of an X chromosome had suddenly

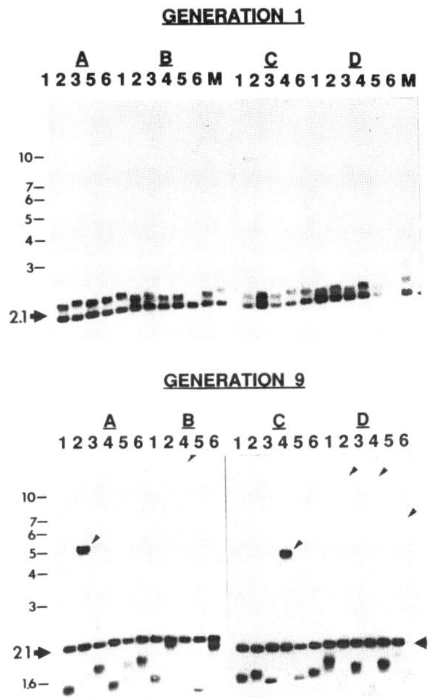

Fig. 7.4 Molecular events at the tip of the terminal deficiency X chromosome RT394 as monitored by genomic Southern blots. Four different sublines (A—D) were started from individual males and kept by crossing to attached—X females. Each lane represents DNA from daugthers of a single male. DNA was cut with NruI and probed with a *yellow* gene fragment that recognizes an internal 2.1 kb fragment (arrow) and the variable terminal fragment of the deficiency. Progressive shortening of the terminal DNA fragments and new DNA additions (arrowheads) that are evidenced by the sudden size increase of the terminal fragment are shown in this comparison of generation 1 and 9 after the start of the sublines. (Reproduced in part with permission from Biessmann et al., Frequent transpositions of *Drosophila melanogaster* HeT-A transposable elements to receding chromosome ends; published by Oxford University Press, *EMBO Journal,* 11, 4459—69, 1992).

increased in size, from 3 to 12 kb. We cloned seven of these new rearrangements, together with the adjacent *yellow* sequences, into lambda phages and showed that in each case the observed size increase of the terminal fragment was due to the *de novo* attachment of a HeT-A element to the chromosome end. No TART element transposed to the terminus of the X chromosome in this experiment, although TART

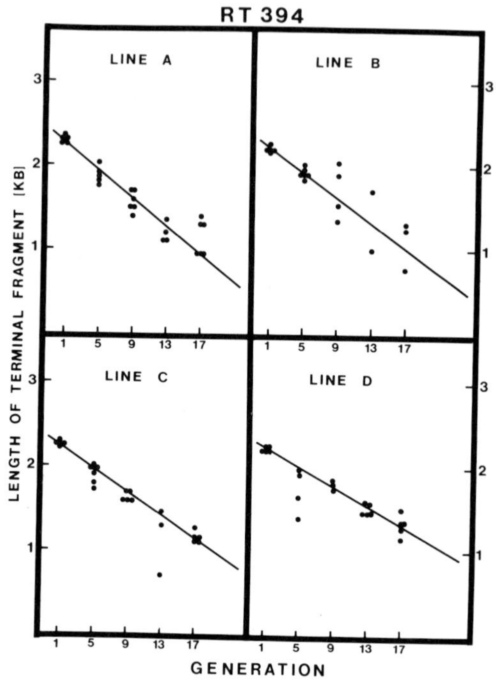

Fig. 7.5 Determination of the rate by which the X chromosomes recede in four independent sublines of the terminal deficiency stock RT394. Terminal DNA fragment sizes (except the ones that have acquired a new DNA fragment at the chromosome end) were plotted against the generation after starting the lines in which these single males were taken. In each of the four sublines the terminal DNA fragments of the X chromosome become progressively shorter at an average rate of 73 bp per generation. Terminal loss was determined separately for each of the four sublines. A, 84 bp/generation; B, 75 bp/generation; C, 75 bp/generation; D, 56 bp/generation.

elements have been shown to transpose to the 3R chromosome end (Sheen and Levis, 1994). This may be due to a much lower transposition rate of TART in the stock used.

The frequency of HeT-A transposition in the population has been calculated as about 1% per X chromosome per generation. Since the average estimated size of HeT-A elements is 6 kb, 60 bp would be added per generation. This observed frequency of HeT-A transpositions would be sufficient to counter-balance the loss that occurs due to incomplete replication.

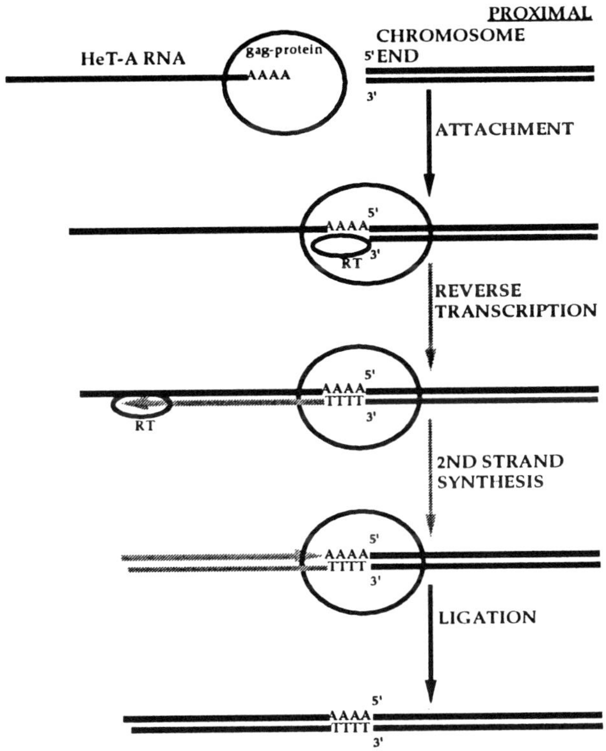

Fig. 7.6 Model for transposition of a telomere-specific HeT-A or TART retroposon onto a chromosome end. The gag-like protein is required in *cis* and may be responsible for targeting the RNA intermediate to chromosome ends. No specific sequences are required at the targeted end. Second strand synthesis is assumed to be dependent on cellular DNA repair enzymes. (Reproduced with permission from: Mason and Biessmann, The unusual telomeres of *Drosophila*; published by Elsevier, *Trends in Genetics* 11, 58—62, 1995).

Figure 7.6 shows a model for transposition of a telomere-specific retroposon. The gag-like protein encoded by the element is probably required in *cis* and may be responsible for targeting the RNA intermediate to chromosome ends. No specific sequences are required at the targeted end. Reverse transcriptase may use the free 3'OH for elongation. Second strand synthesis is assumed to be dependent on cellular DNA repair enzymes.

7.2.4 Speculation on the evolution of the *Drosophila* telomere elongation mechanism

How might such a mechanism have evolved? In an ancestor, the primary pathway of telomere elongation may have been lost, and transposons may have rescued some individuals. It is possible that a telomere-specific retroposon 'invaded' a dipteran ancestor and was used for this cellular process. It has been proposed that the P-element (Daniels *et al.*, 1990; Engels, 1992), as well as the I element (Bucheton *et al.*, 1992) recently 'invaded' *Drosophila melanogaster*. These elements may be spread among species by horizontal gene transfer (Kidwell, 1992), as has also been proposed for the mariner element present in a variety of insects (Robertson, 1993). Since the closest relative of the HeT-A and TART retroposons is the *Drosophila* transposable element jockey (Biessmann *et al.*, 1992; Danilevskaya *et al.*, 1992; Levis *et al.*, 1993), it is conceivable that both may have evolved from a jockey-like ancestor. Alternatively, both elongation mechanisms may have evolved independently.

7.3 *DROSOPHILA* SUBTERMINAL REGIONS

7.3.1 Organization of subterminal regions

Subterminal regions are located proximally adjacent to the DNA sequences that are performing chromosome elongation. The functions of these subterminal regions are currently not understood. In association with yet unidentified proteins, they may mediate interactions among telomeres, and between telomeres and the nuclear lamina or the nuclear scaffold, thus playing a role in maintaining the three-dimensional organization in the nucleus.

The region proximally adjacent to HeT-A and TART elements in *Drosophila* appears to be similar in composition and structure to the so-called 'subtelomeric' region of other eukaryotes. In all three *Drosophila* chromosomal subterminal regions studied: X (Karpen and Spradling, 1992), 3R (Levis *et al.*, 1993), and 2L (Walter *et al.*, 1995), minisatellites have been found that are different from each other but share some limited sequence similarities. Such minisatellites appear to be a common feature of all *Drosophila* subtelomeric regions.

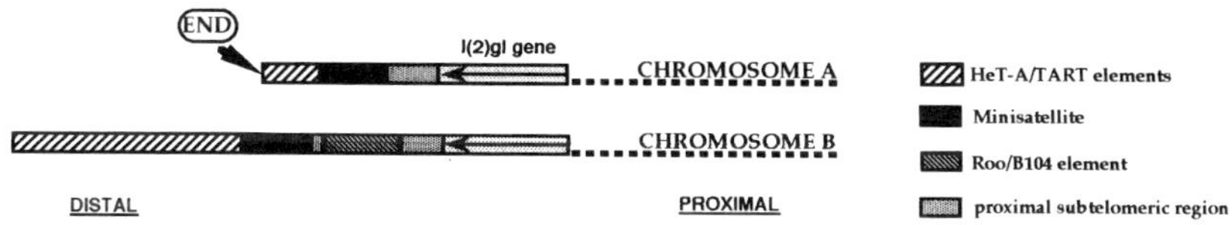

Fig. 7.7 Structural organization and polymorphism of subterminal elements at the tip of 2L of two different wild type chromosomes. The distinct subtelomeric regions which have been identified by mapping, cloning and sequencing are indicated. Superposition in chromosome A of apparent cutting sites of five of the seven restriction enzymes used in the analysis determine the position of the physical end (END) of this chromosome. The proximal subtelomeric region of chromosome B contains the insertion of a 8.5 kb roo/B104 element.

7.3.2 Characterization of DNA components and polymorphisms at the 2L telomere

In order to identify DNA components that may be important functional parts of the telomere, we have characterized the DNA structure of the entire tip of the left arm of chromosome 2 of *Drosophila* (Walter *et al.*, 1995). The recessive tumor-suppressor gene *lethal(2)giant larvae (l(2)gl)* is the most distal single copy gene at 2L (Mechler *et al.*, 1985), and provides convenient probes for isolating the subtelomeric region. Analysis of the subtelomeric region of chromosome arm 2L has revealed three regions of distinct DNA components: terminal HeT-A and TART retroposons, a subterminal minisatellite, and a 3.5 kb single copy region between the minisatellite and the *l(2)gl* gene (Fig. 7.7).

Two major configurations of naturally occurring polymorphisms were found (designated chromosomes A and B) that account for at least 90% of all 2L chromosome tips in our stock (Fig. 7.7). At their distal ends, both chromosomes A and B carry tandem HeT-A arrays with their oligo(A) tails facing proximally towards the centromere, consistent with the proposed function of HeT-A elements in telomere elongation. Proximal to the HeT-A element array lies a 7.5 kb region

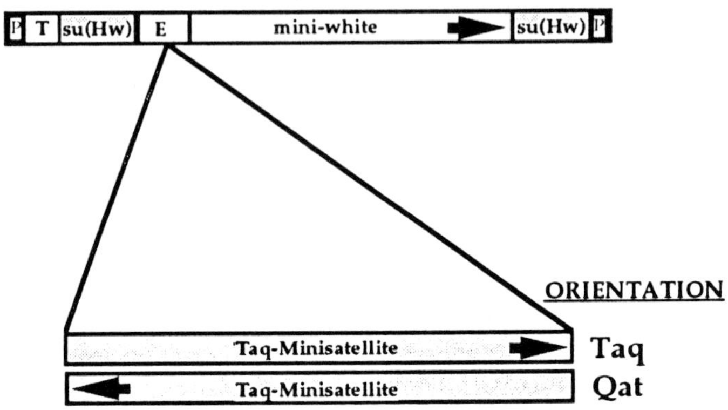

Fig. 7.8 P-element vector used in germ line transformations to test for silencing of the mini-*white* reporter gene by the subterminal minisatellite in its natural orientation at 2L (Taq), or in the opposite orientation (Qat). External heterochromatin effects at the integration site are buffered by flanking su(Hw) binding sites.

of tandemly arranged 460 bp repeats containing internal 150 bp subrepeats. *In situ* hybridization localizes this minisatellite exclusively to the tip of 2L (see Fig. 7.9b). Chromosome A carries the insertion of a 8.5 kb long transposable element (B104/roo). The physical end of the chromosome A lies about 16 kb distally to the 3' end of the *l(2)gl* gene, chromosome B has a much longer 2L telomere which extends at least 37 kb distal to *l(2)gl*.

7.3.3 Telomeric silencing

Tandem arrays of repeated sequences in *Drosophila* can produce effects on gene activity similar to position effect variegation that is induced by placing euchromatic genes next to heterochromatin (Dorer and Henikoff, 1994). Similarly, the *white*$^{+}$ genetic marker in P-element vectors that have inserted into telomeric regions have been found to variegate (Hazelrigg *et al.*, 1984; Gehring *et al.*, 1984, Wallrath and Elgin, 1995), cmnsistent with the possibility that the tandem array of 460 bp repeats found at the tip of 2L induces heterochromatin formation.

Table 7.1 Eye color phenotypes of transgenic lines transformed with the P-element vector shown in Fig. 7.8

Array size of satellite	*Orientation*	*Independent Lines*	*Eye color*
6. 0 kb	Taq	2	yellow
6. 0 kb	Qat	3	red
4. 6 kb	Taq	1	pink
1. 2 kb	Taq	1	orange
0 kb	—	3	red

In order to ask whether these tandem repeats at 2L are sufficient to interfere with gene activity, or whether some other component of the tip region is important, we have constructed a transformation vector carrying a mini-*white* reporter gene, and flanking su(Hw) binding sites to buffer the reporter gene from external heterochromatic influences at the integration sites (Fig. 7.8). The minisatellite or a derivative has been placed upstream of the *white* reporter gene and the construct was reintroduced into the genome of w^- flies by P-element mediated transformation.

When the full 6 kb repeat array is used in one orientation with respect to the *white*, the reporter gene is active and produces red eyes (Table 7.1). When the full length repeat is facing in the opposite orientation, the gene is partially inactivated and pale yellow eyes are produced. Partial deletions of the array in the latter orientation allow increased gene activity. The amount of eye pigment is inversely related to the size of the array. Variegated eye color has not been observed in any of these transgenic animals, as might be expected from heterochromatic reversibility and partial inactivation.

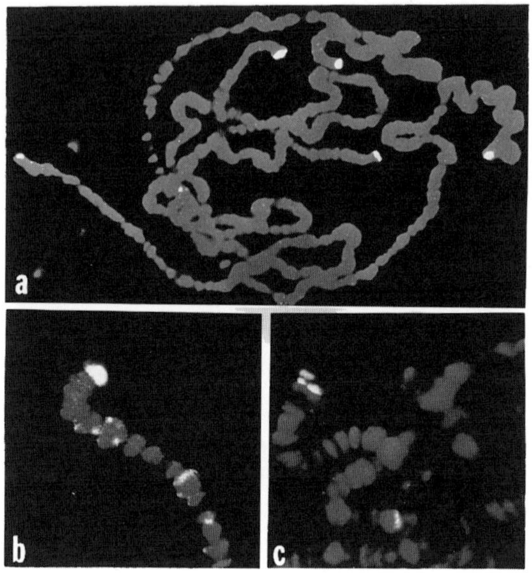

Fig. 7.9 *In situ* hybridizations with digoxigenin-labeled DNA probes. A probe from the HeT-A element labels all tips of the polytene chromosomes of wild type Oregon RC (a). Repeat units from the 2L subterminal minisatellite label only the telomere of 2L (b). Hybridization of HeT-A to the $l(2)gl^{U231}$ allele which is caused by a terminal chromosome deletion (c). The $l(2)gl^{U231}$ allele shows a shorter 2L chromosome as compared to the synapsed *Curly* balancer chromosome which carries HeT-A. Probably due to new transpositions, HeT-A elements are also present at the recessed tip of the $l(2)gl^{U231}$ chromosome, even though it has lost the minisatellite (not shown here). A second chromosome tip is also labeled with HeT-A and shows ectopic pairing with 2L. The left arm of the second chromosome (in panels b and c) is also labeled with a probe from the *decapentaplegic* locus which marks 22F1—2.

7.4 TERMINAL DEFICIENCIES

The telomere at 2L is of particular interest, because among second chromosomes from wild populations in the former USSR (Golubovsky, 1978) and the USA (Green and Shepherd, 1979) very high frequencies (1—2%) of *l(2)gl* mutations were detected, most of which are likely to be terminal deficiencies (Mechler *et al.*, 1985). These data suggest that terminal genes may be prone to loss, but more extensive quantitative studies will be required to estimate the frequency of naturally occurring and induced terminal chromosome deficiencies at all chromosome ends.

In order to investigate which DNA components had been lost from the tip of 2L in four putative terminally deficient chromosomes, $l(2)gl^{9}$, $l(2)gl^{GB26}$, $l(2)gl^{U231}$, $l(2)gl^{U258}$, *in situ* hybridizations to polytene chromosomes were done with probes from the telomeric HeT-A and TART retroposons, and from the subterminal regions previously identified at the wild type 2L. *In situ* hybridization showed that all four deficiency chromosomes have lost the subterminal minisatellite as well as the proximal subterminal region. HeT-A and TART elements are present at the recessed tip of $l(2)gl^{U231}$ and $l(2)gl^{U258}$, but not of $l(2)gl^{9}$ and $\mathrm{l(2)gl}^{GB26}$. An example is shown in Fig. 7.9. These results support the notion that terminal chromosome deficiencies occur in wild populations and remain targets for the transposition of HeT-A and TART.

How could these terminal deficiencies arise? Terminal chromosome deficiencies that break within euchromatin are not recovered in *Drosophila* after irradiation of sperm. However, in the presence of a homozygous mutation of the *mutator2 (mu2)* locus, terminal deficiencies of all chromosomes are recovered at a high frequency (0.2—0.3%) after irradiating females with low doses of X-rays (Mason *et al.*, 1984). Mobilizing a P[w^{+}; ry^{+}] element that had inserted into the subtelomeric region of chromosome 3R also resulted in terminal chromosome deficiencies which break distal to the P[w^{+}; ry^{+}] insert (Levis, 1989). However, chromosome breaks that occur in the subtelomeric region itself do not require the presence of *mu2* (J. M. Mason, unpublished observation). Thus, terminal deletions at the *l(2)gl* locus may arise in two different ways: first, a break may occur in the absence of *mu2* in the subterminal region, reducing telomere length, and eventually receding into the *l(2)gl* gene. Second, a break may occur in the presence of *mu2* directly in the euchromatic *l(2)gl* gene. Independent of their origin, both types of terminally deleted chromosomes are targets for the telomeric HeT-A retroposon.

Telomere shortening during mitotic cell divisions caused by incomplete DNA replication is well established. It is assumed that the stretch of telomeric repeats, associated with proteins, performs a vital cellular function, and that a minimal 'crisis' length of telomeric repeats is required for this function (Counter *et al.*, 1992; Allsopp *et al.*, 1992). The hypothesis of marginotomy (Olovnikov, 1973) proposes that aging of individuals may be tied to cellular senescence caused by shortened telomeres, based on the idea that quantitative differences in telomere repeat length are important. This hypothesis may not apply to *Drosophila* telomeres which are elongated by retroposons and do not carry short tandemly repetitive telomeric sequences. We have only shown that HeT-A elements are responsible for elongation, and there is no evidence that they are involved in any other nuclear function, although we have speculated in the past that their long 3' non-coding region may have some function in forming heterochromatic structures (Biessmann *et al.*, 1992). One might argue that at least one HeT-A element needs to remain at the tip for function. However, chromosomes that lack HeT-A and TART elements are stably maintained and have no deleterious effect on viability, nor do they cause cellular senescence.

The question arises whether any other component of the *Drosophila* telomere (including the subtelomeric region) could serve as a 'critical' component. Terminal deletions at 2L have lost the subterminal minisatellite, and two of the four tested also lack all HeT-A and TART elements. We conclude that neither HeT-A nor the subterminal minisatellite are essential for chromosome function and stability. Perhaps *Drosophila* has evolved a proliferation checkpoint system that tolerates terminally deleted chromosomes as long as they are 'capped' by a yet unidentified, sequence-independent protein.

ACKNOWLEDGEMENTS

Experimental work summarized here was partly supported by grants from the National Science Foundation (DMB89—03731) and the National Institutes of Health (GM46211) to H. B. We thank Larry Champion, Babak Kasravi, Cathy Jang and Jeff Donath for their contributions to the project.

REFERENCES

Allsopp, R., Vaziri, H., Patterson, C. *et al.* (1992) Telomere length predicts replicative capacity of human fibroblasts. *Proceedings of the National Academy of Science USA*, **89**, 10114—8.

Biessmann, H., Carter, S. B. and Mason, J. M. (1990) Chromosome ends in *Drosophila* without telomeric DNA sequences. *Proceedings of the National Academy of Science USA,* **87**, 1758—61.

Biessmann, H., Champion, L. E., O'Hair, M. *et al.* (1992) Frequent transpositions of *Drosophila melanogaster* HeT-A transposable elements to receding chromosome ends. *EMBO Journal,* **11**, 4459—69.

Biessmann, H., Kasravi, B., Bui, T., Fujiwara, G. *et al.* (1994) Comparison of two active HeT-A retroposons of *Drosophila melanogaster. Chromosoma,* **103**, 90—8.

Biessmann, H. and Mason, J. M. (1988) Progressive loss of DNA sequences from terminal chromosome deficiencies in *Drosophila melanogaster. EMBO Journal,* **7**, 1081—6.

Biessmann, H. and Mason, J. M. (1992). Genetics and molecular biology of telomeres. *Advances in Genetics,* **30**, 185—249.

Biessmann, H., Mason, J. M., Ferry, K. *et al.* (1990) Addition of telomere-associated HeT DNA sequences 'heals' broken chromosome ends in *Drosophila. Cell,* **61**, 663—73.

Biessmann, H., Valgeirsdottir, K., Lofsky, A. *et al.* (1992) HeT-A, a transposable element specifically involved in healing broken chromosome ends in *Drosophila melanogaster. Molecular Cell Biology,* **12**, 3910—8.

Blackburn, E. H. (1992) Telomerases. *Annual Review of Biochemistry,* **61**, 113—29.

Bucheton, A., Vaury, C., Chaboissier, M. C. *et al.* (1992) I elements and the *Drosophila* genome. *Genetica,* **86**, 175—90.

Counter, C. M., Avilion, A. A., LeFeuvre, C. E. *et al.* (1992) Telomere shortening associated with chromosome instability is arrested in immortal cells which express telomerase activity. *EMBO Journal,* **11,** 1921—9.

Daniels, S. B., Peterson, K. R., Strausbaugh, L. D., *et al.* (1990) Evidence for horizontal transmission of the P transposable element between *Drosophila* species. *Genetics,* **124**, 339—55.

Danilevskaya, O. N., Petrov, D. A., Pavlova, M. N. *et al.* (1992) A repetitive DNA element, associated with telomeric sequences in *Drosophila melanogaster,* contains open reading frames. *Chromosoma,* **102**, 32—40.

Danilevskaya, O. N., Slot, F., Pavlova, M. and Pardue, M. L. (1994) Structure of the *Drosophila* HeT-A transposon: A retrotransposon-like element forming telomeres. *Chromosoma,* **103**, 215—24.

Danilevskaya, O. N., Slot, F., Traverse, K. L. *et al.* (1994) *Drosophila* telomere transposon HeT-A produces a transcript with tightly bound protein. *Proceedings of the National Academy of Science USA,* **91**, 6679—82.

Dorer, D. R. and Henikoff, S. (1994) Expansions of transgene repeats cause heterochromatin formation and gene silencing in *Drosophila*. *Cell,* **77**: 993—1002.

Engels, W. R. (1992) The origins of P elements in *Drosophila melanogaster*. *Bioessays,* **14**, 681—6.

Gehring, W. J., Klemenz, R., Weber, U. and Kloter, U. (1984) Functional analysis of the *white*$^+$ gene of *Drosophila* by P-factor-mediated transformation. *EMBO Journal,* **3**, 2077—85.

Golubovsky, M. D. (1978) The 'lethal giant larvae' — the most frequent second chromosome lethal in natural populations of *D. melanogaster*. *Drosophila Information Service,* **53**, 179.

Green MM and Shepherd SH (1979) Genetic instability in *Drosophila melanogaster*: the induction of specific chromosome 2 deletions by MR elements. *Genetics,* **92,** 823—32.

Hazelrigg, T., Levis, R. W. and Rubin, G. M. (1984) Transformation of *white* locus DNA in *Drosophila:* Dosage compensation, *zeste* interaction, and position effects. *Cell,* **36**, 469—81.

Karpen, G. H. and Spradling, A. C. (1992) Analysis of subtelomeric heterochromatin in the Drosophila minichromosome *Dp1187* by Single-P element insertional mutagenesis. *Genetics,* **132**, 737—53.

Kidwell, M. G. (1992) Horizontal transfer. *Current Opinion in Genetics and Development,* **2**, 868—73.

Levis, R. W. (1989) Viable deletions of a telomere from a *Drosophila* chromosome. *Cell,* **58**, 791—801.

Levis, R. W., Ganesan, R., Houtchens, K. *et al.* (1993). Transposons in place of telomeric repeats at a Drosophila telomere. *Cell,* **75**, 1083—93.

Mason, J. M. and Biessmann, H. (1993) Transposition as a mechanism for maintaining telomere length in *Drosophila*, in *Chromosome Segregation and Aneuploidy* (ed B. K. Vig), Springer-Verlag, Berlin, pp.143—9.

Mason, J. M. and Biessmann, H. (1995) The unusual telomeres of *Drosophila*. *Trends in Genetics,* **11**, 58—62.

Mason, J. M., Strobel, E. and Green, M. M. (1984) mu-2: mutator gene in *Drosophila* that potentiates the induction of terminal deficiencies. *Proceedings of the National Academy of Science USA,* **81**, 6090—4.

Mechler, B. M., McGinnis, W. and Gehring, W. J. (1985) Molecular cloning of *lethal(2)giant larvae*, a recessive oncogene of *Drosophila melanogaster*. *EMBO Journal,* **4**, 1551—7.

McClintock, B. (1941) The stability of broken ends of chromosomes in *Zea mays*. *Genetics,* **26,** 234—82.

Muller, H. J. (1938) The remaking of chromosomes. *The Collecting Net,* **8**, 182—95.

Olovnikov, A. M. (1973) A theory of marginotomy. *Journal of Theoretical Biology,* **41**, 181—90.

Robertson, H. M. (1993) The mariner transposable element is widespread in insects. *Nature,* **262**, 241—5.

Sheen, F.-M. and Levis, R.W. (1994) Transposition of the LINE-like retrotransposon, TART, to *Drosophila* chromosome termini. *Proceedings of the National Academy of Science USA,* **91**, 12510—14.

Traverse K.L., Pardue M.L. (1988) A spontaneously opened ring chromosome of *Drosophila* has acquired He-T sequences at both new telomeres. *Proceedings of the National Academy of Science USA,* **85,** 8116—20.

Wallrath L.L. and Elgin S.C.R. (1995) Position effect variegation in *Drosophila* is associated with an altered chromatin structure. *Genes and Development,* **9,** 1263—77.

Walter, M. F., Jang, C., Kasravi, B. *et al.* (1995) DNA organization and polymorphism of a wild type *Drosophila* telomere region. *Chromosoma*, **104,** 229—41.

Watson, J. D. (1972) Origin of concatameric T7 DNA. *Nature New Biology,* **239**, 197—201.

Wilkie, A. O., Lamb, J., Harris, P. C. *et al.* (1990) A truncated human chromosome 16 associated with alpha thalassaemia is stabilized by addition of telomeric repeat (TTAGGG)n. *Nature,* **346**, 868—71.

8

Interspecific micronucleus model for the study of induced chromosome aberrations in human male germ cells

J. Egozcue, L. Tusell, R. Álvarez, A. Genescà, M.R. Caballín, L. Barrios and R. Miró

8.1 INTRODUCTION

The analysis of structural chromosome aberrations has long been used in biological dosimetry to determine the effect of different types of clastogenic agents (Lloyd *et al.*, 1980, de Boer and Tates, 1983, Beninson *et al.,* 1986, Bauchinger and Braselmann, 1990). These analyses may be carried out through the direct study of induced chromosome aberrations or sister chromatid exchanges, or indirectly evaluated, as in the micronucleus test.

However, while the determination of the effect of mutagens on somatic cells is a well established practice, extrapolation of the results to germ cells has always been considered as very difficult due to the different sensitivity of the successive stages of meiosis or of the gametes themselves, and to the difficulty inherent in the study of the germ cell line.

In recent years, the study of the effect of mutagenic factors on human spermatozoa has become possible through the interspecific human–hamster fertilization system, that allows the direct visualization of human sperm chromosomes (Martin *et al.*, 1986, Benet *et al.*, 1986). Using this system, Kamiguchi *et al.*, (1990a,b) first demonstrated that after treatment of human spermatozoa with different types of ionizing radiation the number of sperm with structural chromosome aberrations increased linearly with dose.

However, the study of human sperm chromosomes using this system is slow, time-consuming and expensive, because the number of high quality metaphases that can be obtained in each experiment is rather low.

Micronuclei have also been used as indicators of mutagenicity in somatic cells (Schmid, 1976, Heddle *et al.*, 1983, McGregor *et al.*, 1987). Scoring of micronuclei is much easier and faster by far than analysing whole chromosome complements. Since it is well known that there is a close relationship between induced chromosome aberrations and induced micronuclei in human somatic cells exposed to radiation (Hayashi *et al.*, 1984, Erexson *et al.*, 1989, 1991, Bauchinger and Braselmann, 1990), the micronucleus test has become a well established method in biological dosimetry (Ramalho *et al.*, 1988).

Recently, Kamiguchi *et al.*, (1991) incorporated the micronucleus test into the interspecific human–hamster fertilization system. According to these authors, the frequency of micronuclei in two-cell human–hamster embryos increased when human spermatozoa had been previously treated with γ-rays. Their results suggested that the increase in micronuclei was directly related to the effect of γ-rays, but this was just an assumption that could not be demonstrated.

However, since the method was promising, we decided to evaluate the possibility of incorporating the human–hamster micronucleus test into our radiation protection laboratory, to determine the effect of occupational or accidental exposure to different types of radiation on human spermatozoa.

To that end, it was necessary to establish a reproducible technique to obtain two-cell human–hamster embryos, to establish the criteria necessary to score micronuclei in this system, to determine whether the micronuclei obtained were of human or hamster origin, to establish a dose-effect curve for radiation-induced micronuclei, to determine if scoring of micronuclei was representative of the chromosome abnormalities induced by radiation in sperm derived chromosomes and to analyse the contents of micronuclei to determine whether it was possible to differentiate between clastogenic and aneugenic effects.

8.2 MATERIALS AND METHODS

The technique of obtaining two-cell human–hamster embryos was that described by Kamiguchi *et al.*, (1991) adapted to our human–hamster interspecific fertilization system (Genescà *et al.*, 1990). As a mutagen, we used γ-rays at different doses, as described in each case.

Briefly (Tusell *et al.*, 1995a), semen samples from healthy donors were collected in sterile, non-toxic containers, allowed to liquefy, divided into aliquots, irradiated at 35.5°C under a source of ^{60}Co and frozen in liquid nitrogen until use, because it is known that cryopreservation does not increase the frequency of chromosome abnormalities in human spermatozoa (Martin *et al.*, 1991). After thawing, motile sperm were selected by a swim-up procedure (Andolz *et al.*, 1987), treated with a calcium ionophore and washed.

Adult female hamsters were superovulated with PMSG followed by hCG, oocytes collected from the oviducts, freed from cumulus cells with 0.1% hyaluronidase, treated with 0.1% trypsin to remove the zona pellucida and co-incubated with selected sperm.

Controls of penetration were carried out periodically under phase contrast, and insemination stopped by washing when 50—70% of oocytes showed one single swollen sperm head in their cytoplasm.

Fertilized oocytes were cultured for about 20 h until they reached the first cleavage division, and were either scored under phase contrast or swollen in 1% sodium citrate, fixed with 3:1 ethanol—acetic acid and stored at —20 °C until used.

8.3 SCORING CRITERIA

For scoring micronuclei, several criteria have to be taken into account (Tusell *et al.*, 1996). These affect the type of fertilized oocytes to be used and the characteristics of the micronuclei.

Only monospermic oocytes can be used, to avoid overestimating the frequency of micronuclei. This implies selecting under Nomarski's interferential contrast only those hybrid zygotes with a second polar body and two visible pronuclei.

Micronuclei are only scored in two-cell embryos in which the main nuclei of the blastomeres can be visualized, and in which the second polar body is also visible, to avoid confusing it with a micronucleus. Furthermore, micronuclei must be free in the cytoplasm, and not be connected to the main nucleus (Fig. 8.1).

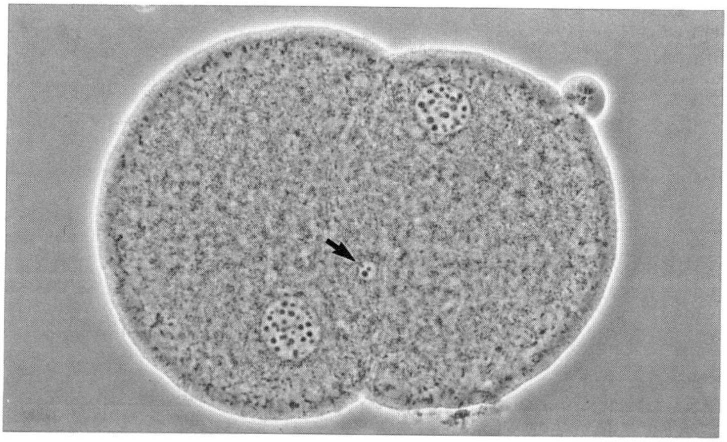

Fig. 8.1 Two-cell hybrid human–hamster embryo showing one micronucleus (arrow). Phase contrast.

8.4 ORIGIN OF MICRONUCLEI

Since the origin of micronuclei remained to be determined, we applied the fluorescent *in situ* hybridization techniques using whole genomic probes to our two-cell hybrid embryos. Human male and hamster female DNAs were isolated from peripheral blood leukocytes, sonicated and biotin labelled. *in situ* hybridization with human and hamster probes was carried out as described by Tusell *et al.*, (1995a). To obtain a good yield of micronuclei, sperm were irradiated with a 4 Gy dose of γ-rays.

To discard the possibility of cross-hybridization, slides containing identifiable, separated metaphases of human and hamster origin were treated with the different probes. Results demonstrated that cross-hybridization did not occur after routine pre-hybridization with herring sperm DNA.

Once the specificity of the probes had been established, slides with two-cell embryos containing micronuclei were hybridized to either human or hamster probes. Analysis of preparations hybridized to human probes demonstrated that 99.1% of micronuclei were of human origin (Fig. 8.2). Analysis of preparations hybridized to hamster probes demonstrated that 97.7% of micronuclei were also of human origin; this lower proportion of human micronuclei obtained when using hamster probes can be explained because when hamster probes are used, micronuclei of human origin lack a fluorescent signal, and may go unnoticed in spite of DAPI counterstaining, especially if they are small in size. This results in an overestimation of hamster micronuclei, and in an underestimation of human micronuclei, indicating that the margin of error is very low.

Thus, the micronucleus test in two-cell human-hamster hybrids can be used to ascertain the effect of γ-rays and of other mutagenic agents on human spermatozoa, with a slight correction needed to exclude the small proportion of micronuclei of hamster origin.

8.5 DOSE–EFFECT CURVE FOR RADIATION-INDUCED MICRONUCLEI

To establish a dose–effect curve for radiation-induced micronuclei, we performed the micronucleus test in two-cell human–hamster hybrid embryos after exposure of human spermatozoa to doses of 0, 0.10, 0.25, 0.50, 1.00, 2.00 and 4.00 Gy of γ-rays. The results were then compared to those obtained when analysing chromosome breaks and fragments in human sperm chromosomes using classical metaphase spreads.

A linear relationship between the different doses of radiation and the induction of micronuclei was demonstrated, although at the highest dose (4 Gy) this relationship showed a slight saturation effect (Tusell *et al.*, 1996).

To determine if scoring micronuclei could be used for the quantification of chromosome damage induced in human spermatozoa, we compared the frequency of micronuclei per two-cell embryo to the frequency of breaks and fragments in human sperm chromosome preparations. In the case of chromosome lesions, the frequency of breaks and fragments was also dose-dependent.

A good coincidence was found, as shown by the following results: when comparing the incidence of micronuclei per two-cell embryo (corrected for the estimated incidence of micronuclei of hamster origin) with the incidence of breaks and fragments per sperm complement, the results were 0.07 v. 0.07 for controls, 0.09 v. 0.07 at 0.25 Gy, 0.16 v. 0.12 at 0.50 Gy, 0.24 v. 0.25 at 1 Gy, 0.40 v. 0.45 at 2 Gy and 0.71 v. 0.81 at 4 Gy.

Fig 8.2 ***Colour Plate 5*** **8.2a** Fluorescent *in situ* hybridization of two-cell embryos with human genomic DNA. Yellow micronucleus (left) is of human origin. Segregation of human (yellow) and hamster (red) chromatins can be observed in the nucleus. **8.2b** Fluorescent *in situ* hybridization of two-cell embryos with hamster genomic DNA. Red micronuclei (arrows) are of human origin. Segregation of human (red) and hamster (yellow) chromatins can be observed in the nucleus.

Fig 8.3 ***Colour Plate 5*** Fluorescent *in situ* hybridization of two-cell embryos with human centromere (red) and unspecific telomere (green) probes. **8.3a** Nucleus from a blastomere in a human–hamster embryo; red signals correspond to human centromeres while green signals correspond to human and hamster telomeres. **8.3b** Nucleus (N), micronucleus (mn) and polar body (PB) in a human–hamster embryo. The micronucleus shows one centromere signal — big arrow — and two telomere signals — small arrows.

8.6 CHARACTERIZATION OF CHROMOSOME ABNORMALITIES IN INDUCED MICRONUCLEI

Since micronuclei may contain acentric fragments (Heddle and Carrano, 1977) as well as centric fragments or whole chromosomes with damaged centromeres (Schmid, 1976, Fenech and Morley, 1989, Farooqi *et al.*, 1993), and even groups of chromosomes that produce large micronuclei (Yamamoto and Kikuchi, 1980), we decided to carry out a FISH study using a combination of human specific centromere probes and of unspecific telomere probes to determine the content of the micronuclei induced by the fertilization of hamster oocytes with human spermatozoa previously treated with γ-rays (Tusell *et al.*, 1995b). In this work, human spermatozoa were irradiated under a ^{60}Co source at 2 and 4 Gy as indicated before.

Fluorescent *in situ* hybridization was carried out on fixed two-cell embryos using biotin labelled DNA probes specific for human centromeres (Miller *et al.*, 1991) as well as digoxygenin labelled telomere probes (Miller and Nüsse, 1993) that detect both human and hamster telomeres. However, since as shown earlier most micronuclei are of human origin, the lack of specificity of telomere probes was not of much importance. Most micronuclei contain just one fragment, and the aim of the study was to determine whether the fragment was centric or acentric, and included telomeres at one or both ends. Single and double FISH experiments were carried out.

In this series of experiments, the frequency of micronuclei was also dose—dependent. In control embryos, about two thirds of micronuclei were telomere positive and centromere negative, and about one third of spontaneously occurring micronuclei were centromere positive. This is within the normal range.

When using irradiated sperm, the incidence of micronuclei per embryo increased dramatically (0.1 for controls, 0.5 at 2 Gy and 0.7 at 4 Gy). Since at any dose most micronuclei are centromere negative, this indicates a predominantly clastogenic effect of γ-rays, because most micronuclei contain acentric fragments.

When using double FISH, most centromere positive micronuclei were also telomere positive (Fig 8.3), indicating that they probably contained whole chromosomes; the presence of whole chromosomes in micronuclei through a lagging effect has already been demonstrated by Viaggi *et al.*, (1987) and may be related to a weak aneugenic effect of ionizing radiation. The proportion of centromere positive-telomere negative micronuclei was always under 10 %.

8.7 CONCLUSIONS

1. The interspecific human—hamster micronucleus test is faster, cheaper and as reliable as the analysis of sperm chromosome complements to determine the incidence of radiation-induced chromosome anomalies in human spermatozoa.
2. The test can be used to quantify induced chromosome aberrations in cases of occupational or accidental exposures to radiation.
3. The use of centromere and telomere probes allows us to differentiate between clastogenic and aneugenic agents, although in some cases both types of effects may occur at the same time.
4. The micronucleus test can detect the effect of clastogens as well as that of aneugens that produce chromosome lagging, but not the effect of aneugens producing non-disjunction.
5. The test cannot be used to estimate the dose received at the testicular level or the possible damage to spermatogenic cells, although it is useful to evaluate the residual lesions left after exposure.

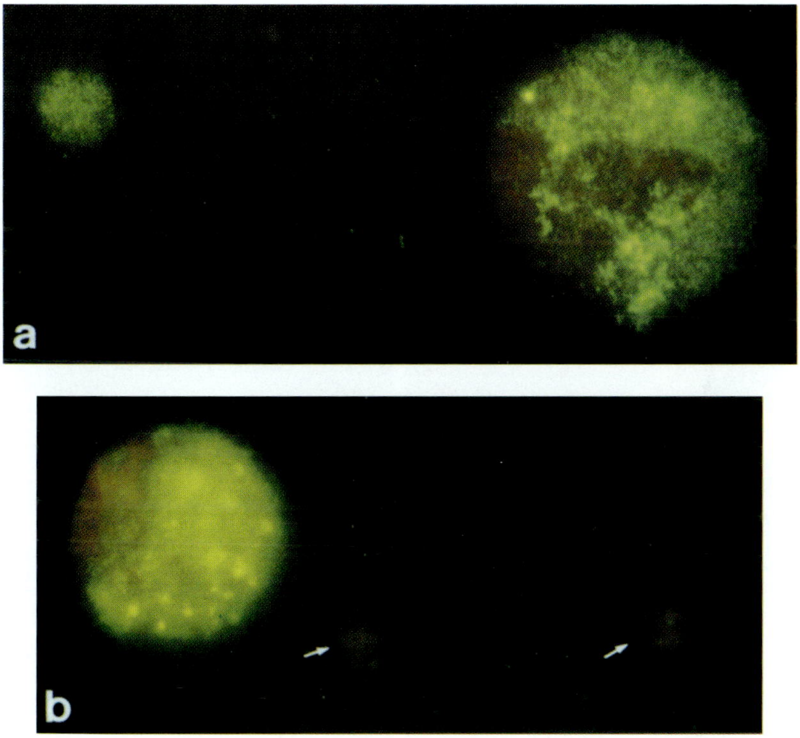

Plate 5 **Fig. 8.2** a–b (see p.127)

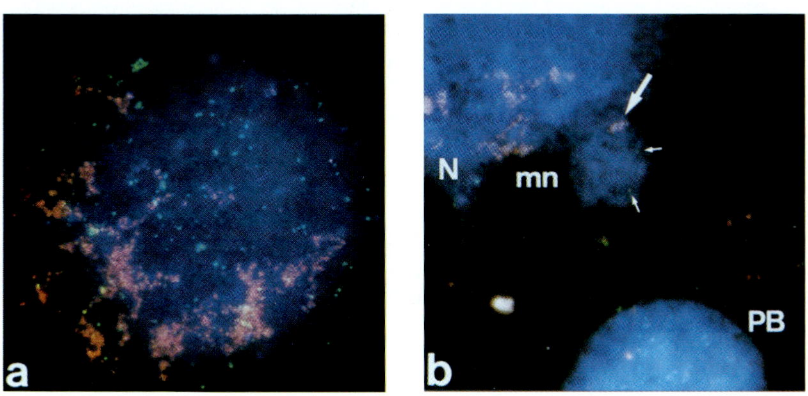

Plate 5 **Fig. 8.3** a–b (see p.127)

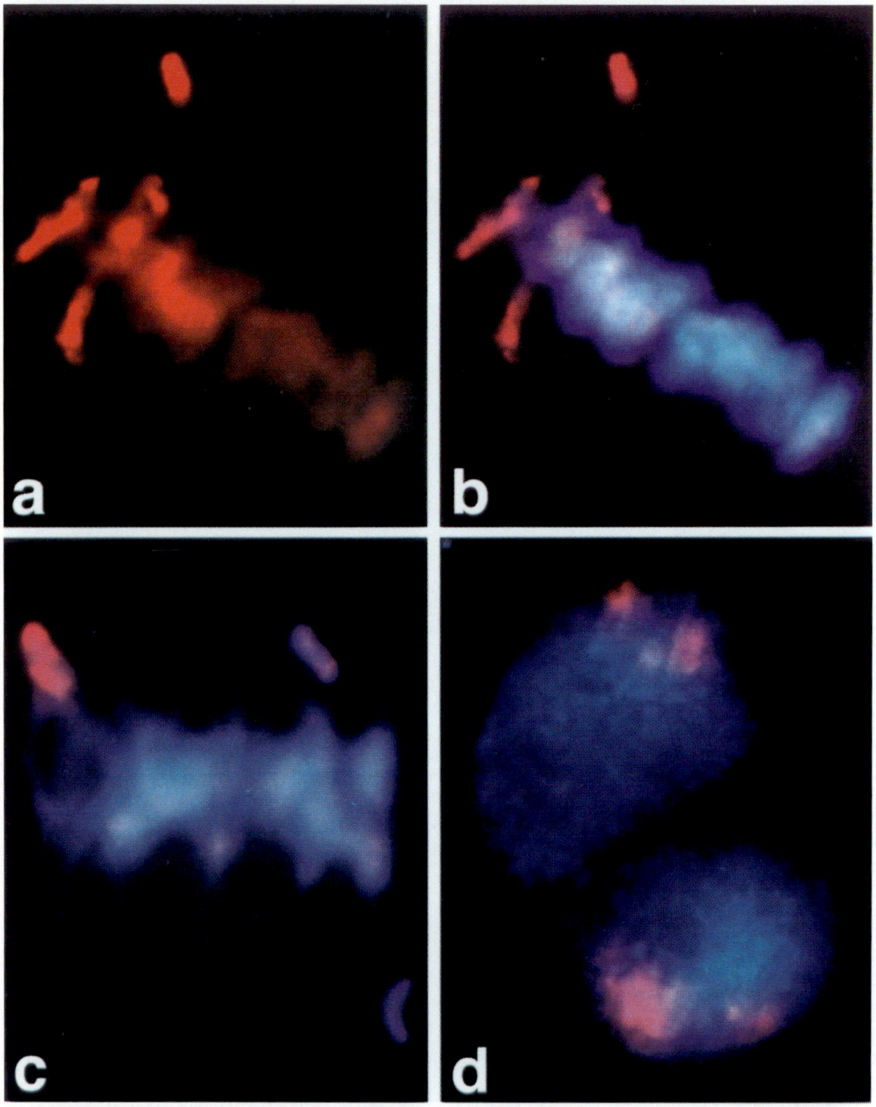

Plate 5 **Fig. 19.2** a–d (see p.326)

ACKNOWLEDGEMENTS

We thank Mark Grossmann for help with the manuscript. This work received financial support from the Radiation Protection Program of the EC (PL 920112), from CIRIT (GRQ 92–2025) and from the Consejo de Seguridad Nuclear.

REFERENCES

Andolz, P., Bielsa, M.A., Genescà, A. *et al.* (1987) Improvement of sperm quality in abnormal semen samples using a modified swim-up procedure. *Human Reproduction*, **2**, 99–101.

Bauchinger, M. and Braselmann, H. (1990) Use of micronuclei in biological dosimetry of absorbed radiation dose, in *Chromosomal Aberrations. Basic and Applied Aspects*, (eds G. Obe and A.T. Natarajan), Springer, Berlin, pp. 202–11.

Benet, J., Genescà, A., Navarro, J. *et al.* (1986) G-banding of human sperm chromosomes. *Human Genetics*, **73**, 181–2.

Beninson, D., Lloyd, D.C., Natarajan, A.T. *et al.* (1986) *Biological Dosimetry: Chromosomal Aberration Analysis for Dose Assessment.* IAEA Technical Reports Series No. 260, Vienna.

Boer, P.D. and Tates, A.D. (1983) Radiation–induced nondisjunction, in *Radiation–Induced Chromosome Damage in Man*, (eds. T. Ishihara and M.S. Sasaki), A.R. Liss, New York, pp. 299–325.

Erexson, G.L., Klingerman, A.D., Halperin, E.C. *et al.* (1989) Micronuclei in binucleated lymphocytes of mice following exposure to gamma radiation. *Environmental Molecular Mutagenesis,* **13**, 128–32.

Erexson, G.L., Klingerman, A.D., Bryant, M.F. *et al.* (1991) Induction of micronuclei by X-radiation in human, mouse and rat peripheral blood lymphocytes. *Mutation Research*, **253**, 193–8.

Farooqi, Z., Darroudi, F. and Natarajan, A.T. (1993) The use of fluorescence *in situ* hybridization for the detection of aneugens in cytokinesis-blocked mouse splenocytes. *Mutagenesis*, **8**, 329–34.

Fenech, M. and Morley A.A. (1989) Kinetochore detection in micronuclei: an alternative for measuring chromosome loss. *Mutagenesis*, **4**, 98–104.

Genescà, A., Miró, R., Caballín, M.R. *et al.* (1990) Sperm chromosome studies in individuals treated for testicular cancer. *Human Reproduction*, **3**, 286–90.

Hayashi, M., Sofuni, T. and Ishidate, M.Jr. (1984) Kinetics of micronucleus formation in relation to chromosomal aberrations in mouse bone marrow. *Mutation Research*, **127**, 129–37.

Heddle, J.A. and Carrano, A.V. (1977) The DNA content of micronuclei induced in mouse bone marrow by gamma irradiation: evidence that micronuclei arise from acentric chromosomal fragments. *Mutation Research*, **44**, 63–9.

Heddle, J.A., Hite, M., Kirkhart, B. *et al.* (1983) The induction of micronuclei as a measure of genotoxicity. A report of the US Environmental Protection Agency Gene-tox Program. *Mutation Research*, **123**, 61–118.

Kamiguchi, Y., Tateno, H. and Mikamo, K. (1990a) Dose-response relationship for the induction of structural chromosome aberrations in human spermatozoa after in vitro exposure to tritium β-rays. *Mutation Research*, **228**, 125–31.

Kamiguchi, Y., Tateno, H. and Mikamo K. (1990b) Types of structural chromosome aberrations and their incidences in human spermatozoa X-irradiated in vitro. *Mutation Research*, **228**, 133–40.

Kamiguchi, Y., Tateno, H. and Mikamo, K. (1991) Micronucleus test in 2-cell embryos as a simpler assay system for human sperm chromosome aberrations. *Mutation Research*, **252**, 297–303.

Lloyd, D.C., Purrott, R.J. and Reeder, E.J. (1980) The incidence of unstable chromosome aberrations in peripheral blood lymphocytes from unirradiated and occupationally exposed people. *Mutation Research*, **72**, 523–32.

Martin, R.H., Chernos, J.E. and Rademaker, A.V. (1991) Effect of cryopreservation on the frequency of chromosome abnormalities and sex ratio in human sperm. *Molecular Reproduction and Development*, **30**, 159–63.

Martin, R.H., Hildebrand, K., Yamamoto, J. *et al.* (1986) An increased frequency of human sperm chromosomal abnormalities after radiotherapy. *Mutation Research*, **174**, 219–25.

McGregor, J.T., Heddle, J.A., Hite, M. *et al.* (1987) Guidelines for the conduct of micronucleus assays in mammalian bone marrow erythrocytes. *Mutation Research*, **189**, 103–12.

Miller, B.M. and Nüsse, M. (1993) Analysis of micronuclei induced by 2-chlorobenzylidene malonitrile (CS) using fluorescence *in situ* hybridization with telomeric and centromeric DNA probes and flow cytometry. *Mutagenesis*, **8**, 35–41.

Miller, B.M., Zitzelsberger H.F., Weier, H.-Ul.G. *et al.* (1991) Classification of micronuclei in murine erythrocytes: immunofluorescent staining using CREST antibodies compared to *in situ* hybridization with biotinylated gamma satellite DNA. *Mutagenesis*, **6**, 297–302.

Ramalho, A., Sunjevaric, I. and Natarajan, A.T. (1988) Use of the frequencies of micronuclei as quantitative indicators of X-ray-induced chromosomal aberrations in human peripheral lymphocytes: comparison of two methods. *Mutation Research*, **207**, 141–6.

Schmid, W. (1976) The micronucleus test for cytogenetic analysis, in *Chemical Mutagens*, vol 4 (ed A. Hollaender), Plenum, New York, pp. 31–53.

Tusell, L., Álvarez, R., Genescà, A. *et al.* (1995a) Human origin of micronuclei in human × hamster two-cell embryos. *Cytogenetics Cell Genetics*, **70**, 41–4.

Tusell, L., Álvarez, R., Genescà, A. *et al.* (1995b) Centromere detection in radiation-induced micronuclei of two-cell human–hamster embryos. *Proceedings of the 25th Annual Meeting of the Environmental Mutagen Society*, Noordwijkerhout, p. 177.

Tusell, L., Álvarez, R., Caballín, M.R. *et al.* (1996) Induction of micronuclei in human sperm–hamster egg hybrids at the two-cell stage after in vitro gamma-irradiation of human spermatozoa. *Environmental Molecular Mutagenesis*, in press.

Viaggi, S., Bonatti, S. and Abbondandolo, A. (1987) New evidence for the presence of chromosomes in micronuclei of human–chinese hamster cells. *Mutagenesis*, **2**, 367–70.

Yamamoto, K.I. and Kikuchi, Y. (1980) A comparison of diameters of micronuclei induced by clastogens and by spindle poisons. *Mutation Research*, **71**, 127–31.

9

Characterization of amplified DNA sequences in human cancers

M. Muleris, A. Almeida, M. Gerbault-Seureau, A.M. Dutrillaux, B. Malfoy and B. Dutrillaux

9.1 INTRODUCTION

Gene amplification can be defined as an increase in the copy number of a restricted region of the genome. Originally discovered as a mechanism of drug resistance in cultured cell lines, it also occurs during tumourigenesis, more frequently in solid tumours than in hemopoietic malignancies (for review, Alitalo and Schwab, 1986; Schwab and Amler, 1990). Two types of chromosomal structures, the homogeneously staining regions (hsr) and double minute chromosomes (dmin), are currently regarded as hallmarks of gene amplification. hsr consist of chromosomal regions that fail to exhibit bands. They can vary in size and do not usually reside at the native locus of the amplified gene (Wahl, 1989). dmin are small paired acentric elements that replicate autonomously once per cell cycle and segregate randomly between daughter cells at mitosis (Wahl, 1989). Thus, unequal segregation of these structure provides means to increase the copy number of certain DNA sequences during a single cell cycle. Submicroscopic circular molecules, called episomes, were found in cells with amplified genes and might represent precursors of dmin (Carroll *et al.*, 1988; Van Hoff *et al.*, 1988; Wahl, 1989).

Gene amplification is part of the multistep progression of cancer that involves both activation of oncogenes and inactivation of tumour suppressor genes. Numerous amplified oncogenes have now been identified in cancer (Alitalo and Schwab, 1986; Schwab and Amler, 1990; Bishop, 1991; Brison, 1993). It has recently become evident that the currently known oncogenes account for only a fraction of all amplification events in primary tumours (Bruderlein *et al.*, 1990; Saint-Ruf *et al.*, 1990; Shiloh *et al.*, 1992) suggesting that a large number of unknown genes may play an important role in cancer progression.

Detection of gene amplification by molecular genetic techniques has usually been performed by hybridization of specific probes to tumour DNA (slot blot or Southern analysis). However, these techniques are limited to the screening of known genes, one gene or chromosome region at a time, leaving the rest of the genome unexplored. Other techniques based on reannealing of amplified sequences (Roninson, 1983), or chromosome microdissection (Meltzer *et al.*, 1992; Zhang *et al.*, 1993) have led to the identification and cloning of several amplified genes. A recent technique, comparative genomic hybridization (CGH), enables a rapid detection of regions of gains or losses of DNA sequences in the tumour genome and the mapping of these sequences on normal chromosomes (Kallioniemi *et al.*, 1992; Joos *et al.*, 1993; du Manoir *et al.*, 1993). We applied a modified CGH technique (mCGH) to identify and map amplified DNA sequences in a series of gliomas and primary breast cancers. Among tumours that have been characterized by cytogenetic analysis in our laboratory, 9 gliomas and 21 primary breast carcinomas that showed hsr or dmin were selected for mCGH analysis. mCGH results were interpreted in relation to the karyotypic data of each tumour. Identification of some amplified genes or DNA regions were specified using chromosome painting, fluorescence *in situ* hybridization (FISH) of specific probes on tumour metaphases and Southern analysis.

9.2 METHOD

The original CGH technique is based on the hybridization of differentially-labelled tumour DNA and normal reference DNA to normal human metaphase chromosomes, in the presence of unlabelled repeated sequences (Cot-1). The ratio of the hybridization intensity of tumour DNA (detected with a green fluorochrome) to that of normal DNA (detected in red) is measured along each chromosome using a digital image analysis system. Changes in the green to red fluorescence intensity ratio map chromosomal regions that are over — or under — represented in tumour DNA. Because our investigation was directed toward the identification of highly amplified genes involved in hsr or dmin, the technique was slightly modified, as follows. Simultaneous hybridization of tumour DNA, normal DNA and Cot-1 was performed according to the original protocol (Kallioniemi *et al.*, 1992) but only tumour DNA was revealed with a green fluorochrome. Chromosomes were counterstained with propidium iodide according to our usual procedure (Lemieux *et al.*, 1992). This treatment resulted in R-banded chromosomes with a uniform green-red staining on which amplified sequences appeared with a more intense green labelling. Analysis was performed with a conventional fluorescence microscope. No digital image analysis was used.

The combined mCGH and cytogenetic approach enabled us to discriminate between two types of mCGH signals. One corresponded to an over-representation of chromosomes or whole chromosome arms and resulted from either duplications or unbalanced translocations as demonstrated by cytogenetic analysis. The second one was limited to a chromosome segment and was likely to arise through a DNA amplification process.

9.3 ANALYSIS OF DNA AMPLIFICATION IN GLIOMAS

9.3.1 Material

Conventional cytogenetics performed on a series of gliomas enabled us to select 9 tumours with hsr (one case) or dmin (all other tumours). They were glioblastoma multiforme (7 tumours of which 3 were recurrent), anaplastic astrocytome and oligodendroglioma grade III (1 tumour each). Analyses were performed on fresh (2 cases) or xenografted (5 cases) tumours or both (2 cases).

9.3.2 Results and discussion

CGH analyses exhibited a single amplified locus in 8 tumours and 4 amplified loci in one tumour. Recurrent sites of amplification were detected in 7p11-p12 (5 cases) and 1q32.1 (2 cases). Signals were also observed in 4q11-q12, 5p15.1, 7q31, 8q24.1 and 9p2 in one tumour each (Muleris *et al.*, 1994a) (Fig. 9.1). Southern blotting demonstrated that the 7p11-p12 site is that of the EGFR gene, previously reported to be recurrently amplified in glioblastomas (for review, Fuller and

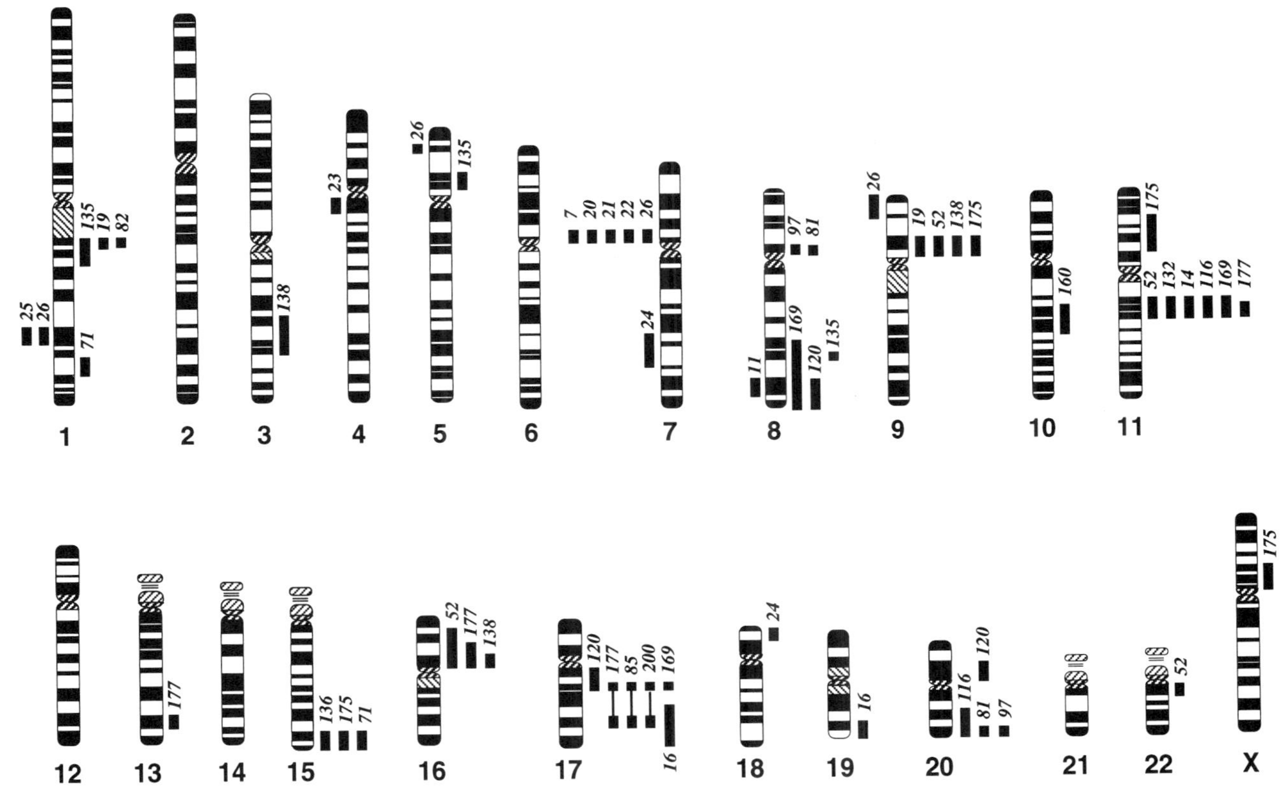

Fig. 9.1 Chromosomal localization of the amplified sites detected in 9 gliomas (left side of chromosomes) and 21 primary breast carcinomas (right side). Tumour numbers are indicated at the top.

Bigner, 1992). FISH with EGFR probe could be performed on tumour metaphases from all 5 cases. It exhibited a labeling on the dmin in 4 instances, and on episomes that have not been detected by conventional cytogenetics in the fifth tumour. Among other genes previously reported to be amplified in a subset of gliomas, only PDGFRA, corresponding to the mCGH site on 4q11-q12, was found to be involved in our series. FISH located it on dmin.

mCGH signals on 8q24 and 7q31 were demonstrated to correspond to MYC and MET amplification, respectively. Amplification of these oncogenes has been reported in a single cell line each (Trent *et al.*, 1986; Wullich *et al.*, 1993). Screening of the other tumours by Southern blotting revealed a weak MET amplification which was not evidenced by mCGH in one other tumour. FISH using the MYC probe exhibited signals on dmin in the fresh tumour and in hsr in the derived xenografted tumour. Other sites of gene amplification were observed in 1q32, 5p15.1 and 9p2 but no target genes can be proposed at present. Amplification of 5p15.1 and 9p2 sequences was found in the same tumour and cohybridization with chromosome 5 and 9 libraries using different dyes showed a bicolor staining of the dmin. Amplification of a sequence located in 1q32.1 was observed in 2 tumours in our series. It was also reported by Speicher *et al.* (1993) in one case.

9.4 ANALYSIS OF DNA AMPLIFICATION IN PRIMARY BREAST CARCINOMAS

9.4.1 Material

mCGH analyses were performed on 21 primary breast carcinomas in which one or several hsr were unambiguously detected by cytogenetic analysis of short-term cultured cells. The histological diagnoses, according to the WHO classification (1981), were ductal invasive (20 tumours) and lobular invasive (1 tumour). Histological grading distribution was T1N0 (2 tumours), T2N0 (11 tumours), T2N1 (7 tumours) and T4N0 (1 tumour) according to the TNM classification (1982) and SBR grade I (tumour 97), II and III (9 tumours each) (Bloom and Richardson, 1957). Two patients (160 and 177) had received treatment prior to surgery.

9.4.2 Results and discussion

The correlation between mCGH and cytogenetic findings was generally good, but not absolute. Principally in cases with very complex karyotypes, some duplications of chromosome arms demonstrated by mCGH had not been recognized by conventional cytogenetics. Twenty one different chromosomal regions were found to be involved in DNA amplification (Fig 9.1). Recurrent sites of amplification were observed in 11q13 (6 times), 17q21.1 (5 times), 9p13 and 17q23 (4 times), 1q21.1, 15q26, 16p11.2 and 20q13.3 (3 times), 8p1, 8q22 and 8q24.1 (twice) (Muleris *et al.*, 1994b).

Several of the regions detected harboured an oncogene or a growth factor receptor previously known to be amplified in breast carcinomas : 8p12 (FGFR1), 8q24 (MYC), 11q13 (FGF3, FGF4, EMS1, CCND1), 15q26 (FES, IGF1R) and 17q21.1 (ERBB2). These loci only account for 38 % (18/47) of all the amplification sites detected, showing that studies limited to the currently known oncogenes provide a restricted view of the target genes for amplification. Although ERBB2 has been reported to be located on 17q11.2–q12 (Human Gene Mapping 11, 1991), we found it more distal, on 17q21. We could determine that the target for amplification of band 15q26 was IGF1R since it was amplified and overexpressed in all 3 tumours showing a CGH signal, whereas FES oncogene was only occasionally coamplified but never overexpressed (Almeida *et al.*, 1994).

mCGH analyses resulted in the identification of 5 new regions of recurrent amplification which may harbour previously unknown genes, 1q21.1, 9p13, 16p11.2, 17q23 and 20q13.3. Amplification at the 2 last sites was also reported by Kallioniemi *et al.* (1994). In 15 tumours out of 21, two or more amplified loci were detected suggesting the complexity of the amplification process in breast carcinomas.

9.4.3 Organization of the amplified DNA sequences within hsr

To characterize the organization of the amplified sequences within hsr, 3 tumours showing two hsr each and for which metaphase spreads were available were selected for further FISH experiments (Muleris *et al.*, 1995). Two of them (tumours 16 and 97) had near-diploid karyotypes with few chromosomal alterations, the third (tumour 200)

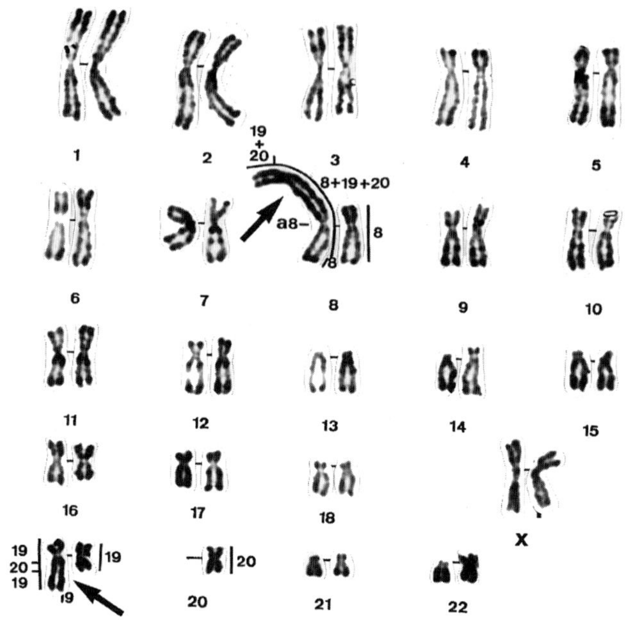

Fig 9.2 Karyotype of breast carcinoma 97. Data from painting chromosomes 8, 19 and 20, and FISH with the alphoid probe specific for chromosome 8 (a 8) are indicated beside chromosomes. hsr are indicated by arrows. The karyotype is 45, XX, hsr (8)(p1), hsr (19)(p1? or q1?), –20.

was hyperdiploid, with a more complex karyotype. In all hsr but one, two or more chromosomal regions were found to be amplified. In the tumour (case 97) with only a slightly modified karyotype, the only anomaly in addition to the two hsr was the loss of one chromosome 20. mCGH demonstrated two sites of amplification on 8p1 and 20q13.3. Painting with chromosome 8, 19 and 20 libraries showed that the two hsr had a common origin and comprised DNA sequences from these chromosomes with different patterns of amplification. In the smaller hsr, a block of amplified chromosome 20 sequences lies between chromosome 19 sequences, whereas in the larger hsr, sequences from chromosomes 8, 19 and 20 were coamplified (Fig. 9.2). This suggests that the rearrangement between chromosome 19 and 20q13.3 sequences probably preceeded the recombination with sequences from 8p, generating the hsr(8). In this very simple karyotype, the three missing normal chromosomes were involved in hsr formation.

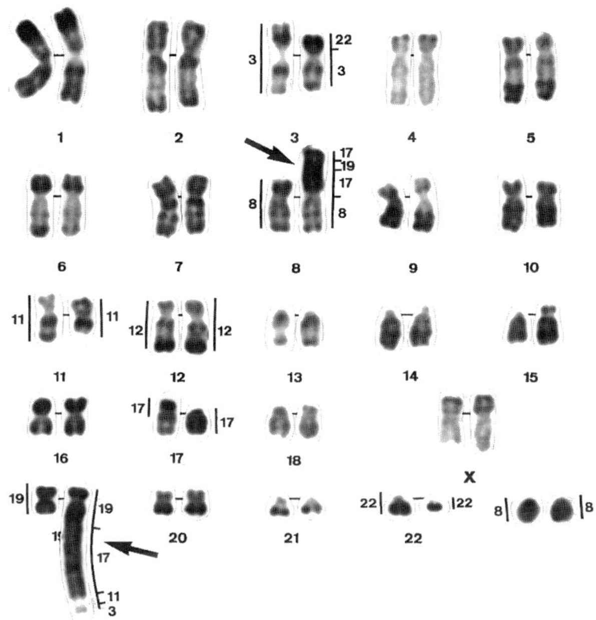

Fig. 9.3 Karyotype of breast carcinoma 16. Data from painting chromosomes 3, 8, 11, 12, 17, 19 and 22 are indicated beside chromosomes. hsr are indicated by arrows. The karyotype is 48, XX, der (3)t(3;22)(p14;q12.1), hsr (8p1?), + r(8) x 2, del (11)(q14), dup (12)(p1?), add (17)(q21), del (17)(p11), hsr (19)(p1? or q1?), del (22)(q12.1).

In tumour 16, two mCGH sites, on 17q22–q25 and 19q13.3, were detected. Painting with the corresponding chromosome libraries demonstrated that amplified chromosome 17 sequences constitute the main part of the largest hsr, whereas on the smallest one, amplified sequences from chromosome 19 hybridized as a central block in the middle of the hsr between two blocks originating from chromosome 17 sequences (Fig. 9.3). Southern blotting demonstrated that ERBB2 was not involved in the amplification. It is likely that the hsr(19), which is composed of 17q22–q25 amplified sequences only, occurred first. Secondary complex rearrangements subsequently transposed chromosome 17 and 19 sequences onto the 8p arm, and translocations then occurred between hsr(19) and chromosomes 3 and 11. The reconstruction of hsr formation is much more difficult for case 200 since 4 chromosome sites are involved (data not shown, Muleris *et al.*, 1995).

9.5 CONCLUSIONS

CGH led to the identification of new regions of DNA amplification that are likely to harbour genes that have an important role in tumour progression. However, this study also highlights some limitations of CGH and technical difficulties that we would like to point out.

The sensitivity of CGH depends on both the copy number change and the size of the affected region. CGH is highly sensitive in detecting small copy increases of large chromosomal regions, such as entire chromosomes or chromosome arms, whereas small amplicons need to be highly amplified to be detectable. This could explain why in breast tumour 97, the amplification of a chromosome 19 sequence, demonstrated by chromosome painting, could not be demonstrated using mCGH. The sensitivity of the technique could also be limited by normal cell contamination in tumour samples. For two gliomas, it was possible to analyse the fresh tumour and the corresponding xenograft. In one instance, Southern analysis demonstrated EGFR amplification in both tumours whereas it could be demonstrated by mCGH only in the xenograft (Muleris *et al.*, 1994a). It has been established that tumours are heterogeneous and transplantation in athymic mice selects the tumorigenic clones.

One difficulty in the interpretation of mCGH data is the distinction between gene amplifications and small increases in copy number due to chromosome rearrangements. In fact, this distinction is arbitrary since it is based on the size of the affected region. Reports establishing the direct relationship between CGH results and gene amplification remain limited, and it is necessary to combine different techniques to validate CGH data. Our experience shows that this distinction is possible but not absolute. In a breast cancer with a very complex karyotype including two hsr (tumour 138), three mCGH sites limited to small chromosomal regions (3q23, 9p1, 16p11.2) were detected. Further chromosome paintings showed that the chromosome 3 library hybridized to the largest hsr whereas chromosome 9 and 16 libraries stained parts of rearranged chromosomes that were not hsr (Muleris *et al.*, 1994b).

Another difficulty is that chromosome libraries are not always efficient in detecting amplification of short sequences. In breast tumour 200, the 50—100 fold amplification of ERBB2, demonstrated by a strong staining of the two hsr when FISH with the specific probe was used, was only barely detected by chromosome 17 painting

(Muleris *et al.*, 1994b). It suggests that ERBB2 sequences were not present in the chromosome 17 library we used. A similar observation was made in a glioma, in which the chromosome 7 library did not stain the numerous episomes that were demonstrated to harbour a 45–fold amplification of EGFR gene (located on 7p12) when FISH with the specific probe was used (Muleris *et al.*, 1994a).

CGH is a useful tool for a rapid identification and mapping of amplified sequences. In providing a whole genome overview in a single hybridization, it provides access to every clinical specimen regardless of mitotic activity. However, the ability to detect balanced rearrangements (reciprocal translocations and inversions) and to distinguish different subclones in order to trace clonal evolution remains the advantage of conventional karyotyping. Thus, CGH is complementary to other molecular cytogenetic and molecular genetic techniques for a complete characterization of amplification.

ACKNOWLEDGEMENTS

This work was supported by l'Association pour la Recherche sur le Cancer and la Ligue Nationale Française contre Cancer. We thank the pathologists of the Institut Curie and Prs M. Poisson and J.Y. Delattre for the obtention of tumour material.

REFERENCES

Alitalo, K. and Schwab, M. (1986) Oncogene amplification in tumour cells. *Advances in Cancer Research*, **47**, 235–81.

Almeida, A., Muleris, M., Dutrillaux, B. *et al.* (1994) The insulin-like growth factor I receptor gene is the target for the 15q26 amplicon in breast cancer. *Genes, Chromosomes and Cancer*, **11**, 63–5.

Bishop, J.M. (1991) Molecular themes in oncogenesis. *Cell*, **64**, 235–48.

Bloom, H.J.G. and Richardson, W.W. (1957) Histological grading and prognosis in breast cancer. *British Journal of Cancer*, **11**, 359–66.

Brison, O. (1993) Gene Amplification and tumour progression. *Biochimica Biophysica Acta*, **1155**, 25–41.

Bruderlein, S., Van der Bosch, K., Schlag, P. *et al.* (1990) Cytogenetics and DNA amplification in colorectal cancer. *Genes, Chromosomes and Cancer,* **2**, 63–70.

Carroll, S.M., DeRose, M.L., Gaudray, P. *et al.* (1988) Double minute chromosomes can be produced from precursors derived from a chromosomal deletion. *Molecular Cell Biology*, **8**, 1525–33.

Fuller, G.N. and Bigner, S.H. (1992) Amplified cellular oncogenes in neoplasms of the human central nervous system. *Mutation Research*, **276**, 299–306.

Hoff, D.D. von, Needham-Vandevanter, D.R., Yucel, J. *et al.* (1988) Amplified human MYC oncogenes localized to replicating submicroscopic circular DNA molecules. *Proceedings of the National Academy of Science USA*, **85**, 4804–8.

Human Gene Mapping 11 (1991) *Cytogenetics and Cell Genetics*, **58**, 1–2200.

Joos, S., Scherthan, H., Speicher, M.R. *et al.* (1993) Detection of amplified DNA sequences by reverse chromosome painting using genomic tumor DNA as probe. *Human Genetics*, **90**, 584–9.

Kallioniemi, A., Kallioniemi, O.P., Sudar, D. *et al* (1992) Comparative genomic hybridization for molecular cytogenetic analysis of solid tumors. *Science*, **258**, 818–21.

Kallioniemi, A,, Kallioniemi, O.P., Piper, J. *et al* (1994) Detection and mapping of amplified DNA sequences in breast cancer by comparative genomic hybridization. *Proceedings of the National Academy of Science USA*, **91**, 2156–60.

Lemieux, N., Dutrillaux, B., Viegas-Pequignot, E. (1992) A simple method for simultaneous R- or G-banding and fluorescence in situ hybridization of small single-copy genes. *Cytogenetics and Cell Genetics*, **59**, 311–2.

Manoir, S. du, Speicher, M.R., Joos, S. *et al.* (1993) Detection of complete and partial chromosome gains and losses by comparative genomic in situ hybridization. *Human Genetics*, **90**, 590–610.

Meltzer, P.S., Guan, X.Y., Burgess, A. *et al.* (1992) Rapid generation of region specific probes by chromosome microdissection and their application. *Nature Genetics*, **1**, 24–8.

Muleris, M., Almeida, A., Dutrillaux, A.M. *et al.* (1994a) Oncogene amplification in human gliomas: a molecular cytogenetic analysis. *Oncogene*, **9**, 2717–22.

Muleris, M., Almeida, A., Gerbault-Seureau, M. *et al.* (1994b) Detection of DNA amplification in 17 primary breast carcinomas with homogeneously staining regions by a modified comparative genomic hybridization. *Genes, Chromosomes and Cancer*, **10**, 160–70.

Muleris, M., Almeida, A., Gerbault-Seureau, M. *et al.* (1995) Identification of amplified DNA sequences in breast cancer and their organization within homogeneously staining regions. *Genes, Chromosomes and Cancer*, 14, 155–63.

Roninson, I.B. (1983) Detection and mapping of homologous, repeated and amplified DNA sequences by DNA renaturation in agarose gels. *Nucleic Acids Research*, **11**, 5413–31.

Saint-Ruf, C., Gerbault-Seureau, M., Viegas-Pequignot, E. *et al.* (1990) Proto-oncogene amplification and homogeneously staining regions in human breast carcinomas. *Genes, Chromosomes and Cancer*, **2**, 18–26.

Schwab, M. and Amler, L.C. (1990) Amplification of cellular oncogenes: a predictor of clinical outcome in human cancer. *Genes, Chromosomes and Cancer*, **1**, 181–93.

Shiloh, Y., Mor, O., Manor, A. *et al.* (1992) DNA sequences amplified in cancer cells: an interface between tumor biology and human genome analysis. *Mutation Research*, **276**, 329–37.

Speicher, M.R., du Manoir, S., Schršck, E. *et al.* (1993) Molecular cytogenetic analysis of formalin-fixed, paraffin-embedded solid tumors by comparative genomic hybridization after universal DNA-amplification. *Human Molecular Genetics*, **2**, 1907–14.

Trent, J., Meltzer, P., Rosenblum, M. *et al.* (1986) Evidence for rearrangement, amplification, and expression of c-myc in a human glioblastoma. *Proceedings of the National Academy of Science USA*, **83**, 470–3.

Tumor Node Metastasis–Union Internationale Contre le Cancer. TNM–Atlas (1982) Illustrated Guide to the TNM/pTNM Classification of malignant tumours, 2nd Ed., B. Spiesse, P. Hermanek, O. Scheibe, G. Wagner, eds. New York, NY Springer Verlag, Berlin.

Wahl, G.M. (1989) The importance of circular DNA in mammalian gene amplification. *Cancer Research*, **49**, 1333–40.

WHO (1981) Classification histologique internationale des tumeurs types histologiques des tumeurs du sein N2, 2nd Ed. Organisation Mondiale de la Santé, Genéve.

Wullich, B,, Mÿller, H.W., Fischer, U. *et al.* (1993) Amplified met gene linked to double minutes in human glioblastoma. *European Journal of Cancer*, **14**, 1991–95.

Zhang, J., Trent, J.M. and Meltzer, P.S. (1993) Rapid isolation and characterization of amplified DNA by chromosome microdissection: identification of IGF1R amplification in malignant melanoma. *Oncogene*, **8**, 2827–31.

PART THREE

CONTROL MECHANISMS

Peter Cook

Pablo Hernández

Nuno Neves

Denise Barlow

10

Transcription factories and chromosome structure

A. Pombo, J. McManus, T.A. Hughes, F.J. Iborra, D.A. Jackson and P.R. Cook

10.1 INTRODUCTION

Condensation is the hallmark of mitosis. Why, then, does the chromatin fibre not condense into the most compact form, a sphere? Why are chromatids cylindrical, and not spherical? What are the basic principles that determine the cylindrical shape?

Many models assume that chromosomes are constructed from a hierarchy of helices: the double helix first being coiled into a nucleosome (~11 nm dia.), nucleosomes into solenoids (~30 nm dia.), solenoids into chromatid fibres (225—250 nm dia.), and so on to give a cylinder. However, despite what our textbooks say, and despite the irrefutable evidence that solenoids are found *in vitro* in buffers containing <100 mM monovalent ion, there remains no evidence that solenoids exist *in vivo* (van Holde and Zlatanova, 1995). For example, solenoids cannot be seen in vitrified sections of metaphase CHO or HeLa cells that have been neither fixed nor stained, and it is unlikely that solenoids did exist but were missed because microtubules with the same mass/unit-length can be seen in the cytoplasm (McDowall *et al.*, 1986). There is also no agreement about higher-order coiling (Manton, 1950): gyres may appear constant in number but variable in sense in hypotonically-swollen human chromosomes (Ohnuki, 1968), of opposite sense in a minority of the sister chromatids of isolated HeLa 'scaffolds' (Boy de la Tour and Laemmli, 1988), or right-handed in polytene *Drosophila* chromosomes (Hochstrasser and Sedat, 1987). Coupled with the gap in the hierarchy at the level of the solenoid, such variability smacks more of biological diversity and/or differences in the conditions at the time of analysis instead of some general structural principle.

We will summarize an alternative model that explains why chromosomes are cylindrical (Cook, 1995). It takes as its starting point the observations of the early microscopists that chromosomes are formed from strings of chromomeres that condense at the end of interphase, to split and segregate during mitosis, and then decondense in the daughter cells (Wilson, 1928). It is to be contrasted with most current models that involve some form of helical coiling or ill-defined interactions between DNA and/or proteins (see, for example, DuPraw 1966; Bak *et al.*, 1977; Paulson and Laemmli, 1977; Rattner and Lin, 1985; Zatsepina *et al.*, 1989; Manuelidis, 1990; Rattner, 1992).

10.2 STUDIES USING 'PHYSIOLOGICAL' CONDITIONS

The study of nuclear and chromosome structure is bedevilled by the problem of artefacts. Chromatin is densely packed and poised in a metastable state so that even small changes in tonicity cause it to aggregate into an unworkable mess. As a result, unphysiological conditions are almost invariably used to isolate chromosomes, nuclei and nuclear substructures like matrices and scaffolds; then argument often rages as to what extent such isolates reflect structures found *in vivo* (reviewed by Cook, 1988). To sidestep these problems, we have developed conditions that allow us to analyse structure under conditions close to the physiological. The practical problems caused by chromatin aggregation at an isotonic salt concentration are overcome by encapsulating cells in agarose microbeads (50—150 μm dia.) before permeabilizing them in a 'physiological' buffer (Jackson *et al.*, 1988). The chromatin — now protected by agarose — can be pipetted freely; it is accessible to probes like antibodies and enzymes, its DNA is intact and polymerases retain their activity. The basic structural features of HeLa nuclei have been analysed using 'physiological' conditions and the approaches illustrated in figure 10.1.

The results can be summarized as follows. Lamins — which are members of the intermediate-filament family of proteins and which form an exo-skeleton that underpins the nuclear membrane (Gerace and Burke, 1988) — are also part of an internal nucleoskeleton (Jackson and Cook, 1988; Hozák *et al.*, 1995). Chromatin loops with an average contour length of 86 kbp are attached to this skeleton;

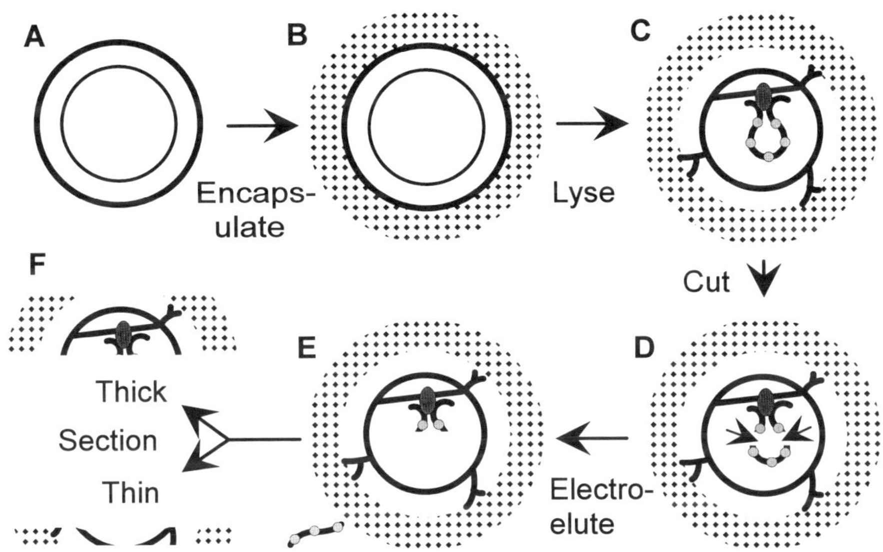

Fig. 10.1 A procedure for analysing chromatin structure using 'physiological' conditions. (A) HeLa cells are (B) encapsulated in an agarose bead (dotted surroundings). After permeabilization, the cytoskeleton, lamina, internal nucleoskeleton, associated transcription factory (oval) and DNA loop covered with nucleosomes (circles) all become accessible to molecular probes. (D) Added endonucleases can now diffuse through the agarose and cut chromatin loops (arrows) so that (E) most chromatin can be removed electrophoretically. (F) Skeletons, whether in the nucleus or cytoplasm, are best visualized by electron microscopy of thick sections. [From Cook (1995) with the permission of The Company of Biologists Ltd.] This procedure has been used to characterize: (i) An internal lamin-containing nucleoskeleton, once obscuring chromatin is removed (Jackson and Cook, 1988; Hozák *et al.*, 1995). (ii) The contour length of loops, from the average length and percentage of remaining DNA fragments (if fragment length is 8.6 kbp and 10% remains, contour length is $8.6 \times 1/(10/100) = 86$ kbp). It does not change during mitosis, so the molecular ties holding loops persist (Jackson *et al.*, 1990). (iii) Sequences remaining after elution; they are mainly promoters, enhancers and transcribed sequences, implying that engaged polymerases — which can still 'run-on' along residual fragments — mediate attachment to the skeleton (Jackson and Cook, 1985; 1993). (iv) Sites of transcription. Permeabilized cells (either before or after cutting and elution) are allowed to make RNA in the presence of Br-UTP, and then sites (ie factories) containing the incorporated analogue are immunolabelled using antibodies against Br-RNA (Jackson *et al.*, 1993; see also Wansink *et al.*, 1993).

importantly, this length does not change during mitosis, implying that the molecular ties must persist (Jackson *et al..*, 1990). Surprisingly, loops — whether part of the natural chromosome or a transfected 'minichromosome' — are attached through promoters/enhancers and transcribed sequences to polymerizing complexes on the skeleton (Jackson and Cook, 1985; 1993). The active polymerases that mediate attachment are concentrated into discrete 'foci' (Jackson *et al.*, 1993; Wansink *et al.*, 1993). (The term 'polymerase' is used to describe the large cluster of polypeptides in the active complex.) The foci have variable sizes and shapes which makes counting difficult, but we nevertheless estimate that there are ~2000 in a HeLa cell, or roughly the number of mitotic bands or chromomeres (unpublished results). The bright ones contain >50 active polymerases, plus associated templates and components of the splicing apparatus. We have christened them transcription 'factories' by analogy with the replication factories that are also fixed to the skeleton and contain all the machinery necessary to duplicate >20 replicons simultaneously (Nakamura *et al.,* 1986; Hozák *et al.,* 1993; 1994b).

10.3 NUCLEOLAR ORGANIZING REGIONS AND TRANSCRIPTION FACTORIES

We will assume that the remnants of transcription factories containing RNA polymerases II and III are equivalent to most of the mitotic chromomeres seen during mitosis. (We are currently collecting evidence for this assumption.) However, excellent evidence already exists that the transcription factories found in nucleoli (which contain RNA polymerase I) are equivalent to the mitotic 'chromomeres' of the nucleolar organizing region.

Nucleoli contain several 'fibrillar centres' surrounded by a 'dense fibrillar component'; in turn, these are embedded in a 'granular component' (Spector, 1993). Fibrillar centres — which are at nodes on the skeleton — store the required enzymes and transcription seems to occur as rDNA slides through polymerases on their surface. Nascent rRNA is then extruded into the dense fibrillar component and — after termination — it moves to the granular component to complete its maturation. Therefore these factories, which are several hundred nanometers in diameter, contain the active machinery on the surface of storage cores — the fibrillar centres (Hozák *et al.,* 1994a).

Each fibrillar centre is often associated with one, or a few, active ribosomal cistrons and the total number of centres is directly related to the transcription rate. For example, the ~234 in a fibroblast fall to ~156 on serum-starvation (Jordan and McGovern, 1981) and the ~9 in a peripheral blood lymphocyte rise to ~80 as it is stimulated to divide (Haaf *et al.*, 1991). In other words, increasing transcription increases surface area and so the number of polymerases accessible to promoters. When nucleoli disassemble during mitosis, most nucleolar components disperse (Hernandez-Verdun and Gautier, 1994) but — remarkably — all polymerase I and most of the transcription factor, UBF, remain bound as the remnants of the fibrillar centres aggregate into the nucleolar organizing regions on the chromosomes (Scheer and Rose, 1984; Roussel *et al.,* 1993).

It is attractive to suppose that factories containing polymerases II and III are built similarly. Increased transcription would disaggregate cores, increasing surface area and the number of accessible polymerases. When transcription ceases during mitosis, most core material would disperse and the rest would aggregate with other remnants to give the chromomeres of mitosis. Groups of small factories rich in genes would condense into Giemsa-light bands. And as the RNA polymerases and transcription factors that tie the chromatin into loops remain bound, those loops would retain their contour length.

10.4 A DIGRESSION ON CELLULAR ADHESION

But why do chromomeres condense into a cylinder and not a sphere? The principles involved can be exemplified by reference to the development of tissue layers during embryogenesis.

When disaggregated neural cells from frog embryos are randomly intermixed, they aggregate into a sphere and then sort into a semblance of the original tissue, with ectoderm surrounding the endoderm (Fig. 10.2A—D; Townes and Holtfreter, 1955). This process is simply explained if cells differ only in their adhesiveness for one another (Steinberg, 1964; Armstrong, 1989). First, the cells will aggregate; then, as random movement creates contacts, contacts between the more-adhesive cells will tend to persist, until eventually a sphere of less-cohesive cells forms around a core of the more-adhesive ones.

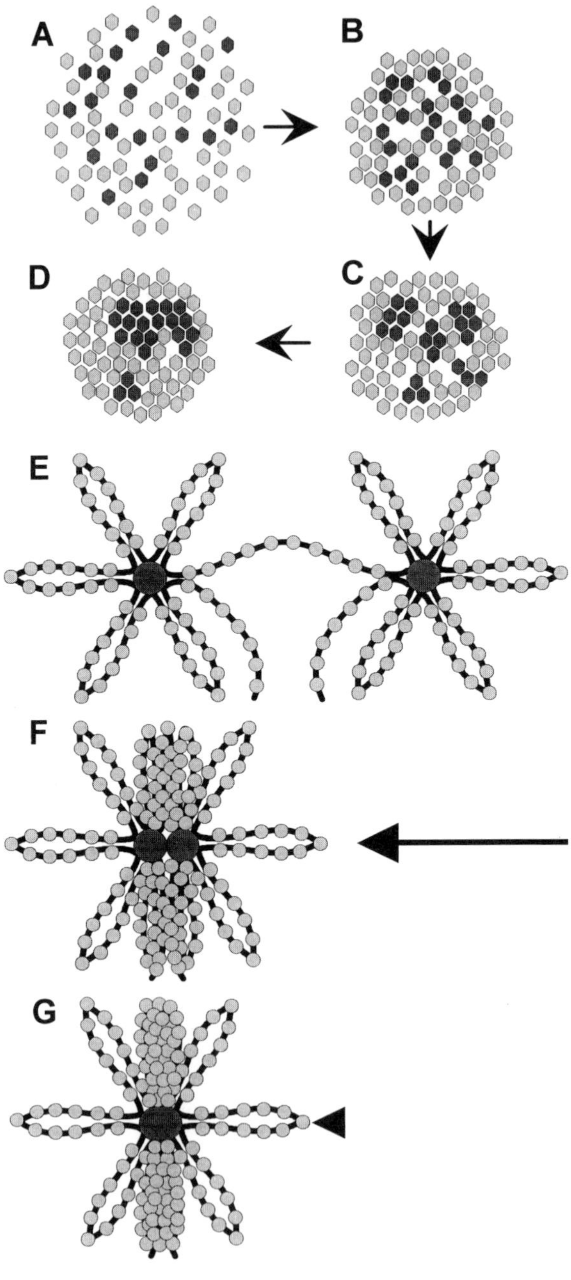
A
B
C
D
E
F
G

This model has been tested directly (Steinberg and Takeichi, 1994). L cells are not normally very adherent, but become more so if they express (after transfection of the appropriate cDNA) the adhesion molecule, cadherin, on their surface; mixtures expressing different amounts of cadherin do sort into low-expressers surrounding a core of high-expressers.

We can model interphase chromosomes as strings of less-adhesive nucleosomes running between, and looping from, more-adhesive factories (Fig. 10.2E—G). In mitosis, increased adhesiveness will drive 'sticky-end' aggregation to the most compact and stable structure — a cylinder of nucleosomes around a chromomeric core in which inter-nucleosomal and inter-chromomere interactions are maximized.

Fig. 10.2 (A-D) Cell sorting through differential adhesiveness and (E-G) an analogous 'sticky-end' aggregation and sorting during prophase. [From Cook (1995) with the permission of The Company of Biologists Ltd.] A-D: When two cell types (light and dark) are (A) mixed, (B) they aggregate into the most compact form, a sphere. As contacts created by random movement between the more-adhesive (dark) cells persist for longer, they clump internally. (D) Clumps of more-adhesive cells continue to aggregate until the less-adhesive cells surround a few large clumps, although sorting is rarely perfect. E-G: (E) A string of less-adhesive nucleosomes (small circles on loops) is attached to two more-adhesive factories (large circles). (F) As factories touch (arrow) during condensation, they stick together; large factories are probably the first to fuse. Displaced loops create a higher density in the plane of contact; for geometrical reasons this density is ~5× and ~10× higher if 86 kbp loops stretch 350 nm from factories of 50 and 25 nm diameter respectively. (G) As the touching halves of the two factories fuse into a cylinder (arrowhead), this central density increases further by one-third. The nucleosomal concentration is now much higher around the middle of the cylinder than at an end; this enhances nucleosomal aggregation and ensures that the next factory to fuse will do so at a more-accessible end. Moreover, the next factory is generally close by and tethered through a short inter-factory loop, as it split from its neighbour earlier during interphase when a larger loop attached to generate two smaller loops, one of which became the inter-factory loop. Additional factories now bind at the ends, elongating the cylinder. These geometrical considerations, and the principles illustrated in A—D, ensure that chromatids are cylindrical (and not spherical) and that loops initially incorporated into the 'wrong' cylinder (usually a sister chromatid) will sort into the 'right' one.

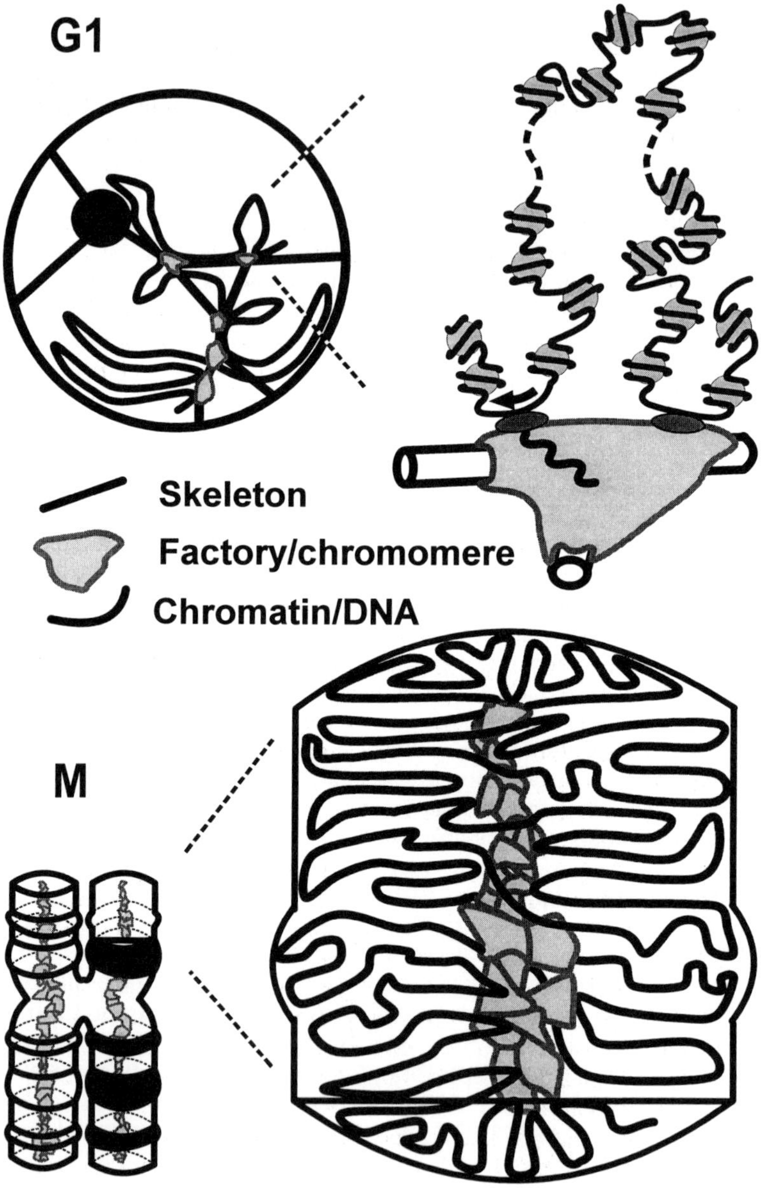
G1
Skeleton
Factory/chromomere
Chromatin/DNA
M

Fig. 10.3 A model for the structure of a human chromosome. [From Cook (1995) with the permission of The Company of Biologists Ltd.] Upper: Structure during G1 phase. Transcription factories are located at nodes on a lamin-containing endoskeleton and their number and size depend upon transcriptional activity. If there are 1250, each might be ~50 nm diameter, contain ~25 active RNA polymerases and be associated with ~56 chromatin loops (usually, but not invariably, derived from one chromatid) with a range of contour lengths (ie 5 sets of 11 loops centred around ~7.5, 50, 75, 100 and 175 kbp respectively; average 86 kbp). ~10% chromatin then lies within ~8.6 kbp of a factory, so genes within it are close enough to polymerases to attach and be transcribed; the remainder is too remote and condenses on to the lamina or nucleolus as heterochromatin. Increased transcription, generates more, smaller, factories each associated with fewer, shorter, loops. The enlargement (right) shows a loop attached through a transcription unit and a promoter/enhancer to a polymerizing complex and transcription factor (ovals). (In transcriptionally-inactive regions, transcription factors are the sole molecular ties.) A transcript (wavy line) is extruded as the template slides through the left-hand complex; this template movement 'opens' adjacent chromatin. The DNA duplex winds around nucleosomes (circles) in the loop; variations in linker length (±2 bp) and entry-exit angle (±15°) generate an irregular zig-zagging fibre (Horowitz *et al.*, 1994) that extends, on average, ~350 nm from a factory. Lower: Mitotic structure. Human chromosome 16 — which contains ~100 Mbp DNA (3% of genome) — is modelled, assuming each chromatid is 3,400 × 800 nm. (The dimensions of 'native' chromosomes vary significantly, depending on condensation.) On entry into mitosis, skeletons depolymerize, transcription ceases and nucleosomes plus ~38 factory remnants aggregate and then sort into a cylindrical nucleosomal 'cloud' surrounding a chromomeric axis, as described in Fig. 3E-G. Despite these rearrangements, loops retain their attachments and contour lengths. A typical factory/chromomere, plus an average loop extended on each side, fit within the width; longer loops fold back on themselves (enlargement, right). Factories associated with more shorter loops (and active genes) give narrower cylinders; examples include (I) Giemsa-light bands (5% thinner), (ii) several Mbp of yeast DNA translocated into a mouse genome (the closely-spaced yeast genes are transcribed during interphase and so will be in short loops; McManus *et al.*, 1994) and (iii) active rDNA loops which give the 'secondary constriction' of the nucleolar organizing region (Robert-Fortel *et al.*, 1993). The design principles are, almost certainly, further modified in other specialized chromosomal regions; for example, the several Mbp of non-transcribed α1-satellite probably condense into a spherical centromeric 'chromomere' through interactions involving CENP-B, rather than polymerases or transcription factors (Yoda *et al.*, 1992).

10.5 A MODEL

These considerations lead to a model involving three fundamental levels of organization: nucleosomes, loops and transcription factories/chromomeres (Fig. 10.3). During interphase, loops are tied through RNA polymerases or transcription factors to factories; increased transcription counteracts the tendency of factories to fuse and generates more, smaller, factories to which relatively more genes are attached in shorter loops. During mitosis, the skeleton depolymerizes, transcription ceases and proteins are phosphorylated, increasing adhesiveness between factories and between nucleosomes. Chromosome condensation would begin in heterochromatin and proceed bidirectionally to give cylindrical strings of chromomeres that first fuse into one large cylinder per (entangled) chromatid pair, which then splits into two as individual chromatids sort out. Longer strings would give longer cylinders of similar width. Each cylinder would contain nucleosomes arranged around a disordered chromomeric axis (Fig. 10.3B). In different metaphases, regions rich in transcribed genes (in short loops) associated with (small) factories will condense to form Giemsa-light bands in roughly the same places, but individual genes do not occupy absolutely fixed positions. On entry into G1, the appropriate intermediate filaments would repolymerize between decondensing chromomeres, around the nuclear periphery and throughout the cytoplasm to form the internal nucleoskeleton, lamina and cytoskeleton that together integrate cellular space.

10.6 CONCLUSIONS

This model is deliberately designed to be a minimalist one, and will obviously have to be embroidered as new details appear. It requires only three basic structural motifs in the organizational hierarchy — the nucleosome, chromatin loop and factory/chromomere — to explain why nuclei and chromosomes have the shapes they have. Its major assumption is that interphase transcription foci/factories are the precursors of mitotic chromomeres, which is supported by the established relationship of nucleolar foci/factories with the mitotic nucleolar organizing regions. The interactions involved are either numerous and nonspecific (i.e. between nucleosomes and between chromomeres) or fewer but more specific (ie between polymerases/transcription factors and transcription units). The structural role suggested for RNA polymerase as one loop tie is perhaps surprising, but the enzyme plays just such a role in looping bacterial DNA (Krawiec and Riley, 1990). The persistence of ties through mitosis inevitably means that transcriptional patterns will be inherited by daughter cells. Importantly, the best evidence for the model is derived from studies on intact or cryo-fixed cells, or those permeabilized in 'physiological' buffers.

ACKNOWLEDGEMENTS

We thank The Cancer Research Campaign, The Wellcome Trust, The Medical Research Council and the Junta Nacional de Investigação Científica & Tecnológica for support.

REFERENCES

Armstrong, P.B. (1989) Cell sorting out; the self-assembly of tissues *in vitro*. *Critical Reviews in Biochemistry and Molecular Biology*, **24**, 119—49.

Bak, A.L., Zeuthen, J. and Crick, F.H.C. (1977) Higher-order structure of the mitotic chromosomes. *Proceedings of the National Academy of Science USA*, **74**, 1595—9.

Boy de la Tour, E. and Laemmli, U.K. (1988) The metaphase scaffold is helically folded: sister chromatids have predominantly opposite helical handedness. *Cell*, **55**, 937—44.

Cook, P.R. (1988) The nucleoskeleton: artefact, passive framework or active site? *Journal of Cell Science*, **90**, 1—6.

Cook, P.R. (1995) A chromomeric model for nuclear and chromosome structure. *Journal of Cell Science*, **108**, 2927—35.

DuPraw, E.J. (1966) Evidence for a 'folded-fibre' organization in human chromosomes. *Nature*, **209**, 577—81.

Gerace, L. and Burke, B. (1988) Functional organization of the nuclear envelope. *Annual Review of Cell Biology*, **4**, 335—74.

Haaf, T., Hayman, D.L. and Schmid, M. (1991) Quantitative determination of rDNA transcription units in vertebrate cells. *Experimental Cell Research*, **193**, 78—86.

Hernandez-Verdun, D. and Gautier, T. (1994) The chromosome periphery during mitosis. *BioEssays,* **16**, 179—85.

Hochstrasser, M. and Sedat, J.W. (1987) Three-dimensional organization of *Drosophila melanogaster* interphase nuclei. I. Tissue-specific aspects of polytene nuclear architecture. *Journal of Cell Biology,* **104**, 1455—70.

Horowitz, R.A., Agard, D.A., Sedat, J.W. and Woodcock, C.L. (1994) The three-dimensional architecture of chromatin in situ: electron tomography reveals fibres composed of a continuously variable zig-zag nucleosomal ribbon. *Journal of Cell Biology,* **125**, 1—10.

Hozák, P., Cook, P.R., Schöfer, C. *et al.* (1994a) Site of transcription of ribosomal RNA and intra-nucleolar structure in HeLa cells. *Journal of Cell Science,* **107**, 639—48.

Hozák, P., Hassan, A.B., Jackson, D.A. and Cook, P.R. (1993) Visualization of replication factories attached to a nucleoskeleton. *Cell,* **73**, 361—73.

Hozák, P., Jackson, D.A. and Cook, P.R. (1994b) Replication factories and nuclear bodies: the ultrastructural characterization of replication sites during the cell cycle. *Journal of Cell Science,* **107**, 2191—202.

Hozák, P., Sasseville, A. M-J., Raymond, R. and Cook. P.R. (1995) Lamin proteins form an internal nucleoskeleton as well as a peripheral lamina in human cells. *Journal of Cell Science,* **108**, 635—44.

Jackson, D.A. and Cook, P.R. (1985) Transcription occurs at a nucleoskeleton. *EMBO Journal*, **4**, 919—25.

Jackson, D.A. and Cook, P.R. (1988) Visualization of a filamentous nucleoskeleton with a 23 nm axial repeat. *EMBO Journal,* **7**, 3667—77.

Jackson, D.A. and Cook, P.R. (1993) Transcriptionally-active minichromosomes are attached transiently in nuclei through transcription units. *Journal of Cell Science,* **105**, 1143—50.

Jackson, D.A., Dickinson, P. and Cook, P.R. (1990) The size of chromatin loops in HeLa cells. *EMBO Journal*, **9**, 567—71.

Jackson, D.A., Hassan, A.B., Errington, R.J. and Cook, P.R. (1993) Visualization of focal sites of transcription within human nuclei. *EMBO Journal*, **12**, 1059—65.

Jackson, D.A., Yuan, J. and Cook, P.R. (1988) A gentle method for preparing cyto- and nucleo-skeletons and associated chromatin. *Journal of Cell Science,* **90**, 365—78.

Jordan, E.G. and McGovern, J.H. (1981) The quantitative relationship of the fibrillar centres and other nucleolar components to changes in growth conditions, serum deprivation and low doses of actinomycin D in cultured diploid human fibroblasts (strain MRC-5). *Journal of Cell Science,* **52**, 373—89.

Krawiec, S. and Riley, M. (1990) Organization of the bacterial chromosome. *Microbiological Reviews,* **54**, 502—39.

Manton, I. (1950) The spiral structure of chromosomes. *Biological Reviews,* **25**, 486—508.

Manuelidis, L. (1990). A view of interphase chromosomes. *Science*, **250**, 1533—40.

McDowall, A.W., Smith, J.M. and Dubochet, J. (1986) Cryo-electron microscopy of vitrified chromosomes *in situ*. *EMBO Journal,* **5**, 1395—986.

McManus, J., Perry, P., Sumner, A.T. *et al.* (1994) Unusual chromosome structure of fission yeast DNA in mouse cells. *Journal of Cell Science,* **107**, 469—86.

Nakamura, H., Morita, T. and Sato, C. (1986) Structural organisation of replicon domains during DNA synthetic phase in the mammalian nucleus. *Experimental Cell Research,* **165**, 291—7.

Ohnuki, Y. (1968) Structure of chromosomes. I. Morphological studies of the spiral structure of human somatic chromosomes. *Chromosoma,* **25**, 402—28.

Paulson, J.R. and Laemmli, U.K. (1977) The structure of histone-depleted metaphase chromosomes. *Cell,* **12**, 817—28.

Rattner, J.B. (1992) Integrating chromosome structure with function. *Chromosoma,* **101**, 259—64.

Rattner, J.B. and Lin, C.C. (1985) Radial loops and helical coils coexist in metaphase chromosomes. *Cell,* **42**, 291—6.

Robert-Fortel, I., Junéra, H.R., Géraud, G. and Hernandez-Verdun, D. (1993) Three-dimensional organization of the ribosomal genes and Ag-NOR proteins during interphase and mitosis in PtK_1 cells studied by confocal microscopy. *Chromosoma,* **102**, 146—57.

Roussel, P., Andre, C., Masson, C. *et al.* (1993) Localization of the RNA polymerase I transcription factor hUBF during the cell cycle. *Journal of Cell Science,* **104**, 327—37.

Scheer, U. and Rose, K.M. (1984) Localization of RNA polymerase I in interphase cells and mitotic chromosomes by light and electron microscopic immunocytochemistry. *Proceedings of the National Academy of Science USA,* **81**, 1431—5.

Spector, D.L. (1993) Macromolecular domains within the cell nucleus. *Annual Review of Cell Biology,* **9**, 265—315.

Steinberg, M.S. (1964) The problem of adhesive selectivity in cellular interactions, in *Cellular membranes in development* (ed M. Locke) Academic Press, New York, pp. 321—66.

Steinberg, M.S. and Takeichi, M. (1994) Experimental specification of cell sorting, tissue spreading, and spatial patterning by quantitative differences in cadherin expression. *Proceedings of the National Academy of Science USA,* **91**, 206—9.

Townes, P.L. and Holtfreter, J. (1955) Directed movements and selective adhesion of embryonic amphibian cells. *Journal of Experimental Zoology,* **128**, 53—120.

Van Holde, K. and Zlatanova, J. (1995) Chromatin higher order structure: chasing a mirage? *Journal Biological Chemistry,* **270**, 8373—6.

Wansink, D.G., Schul, W., van der Kraan, I. *et al.* (1993). Fluorescent labelling of nascent RNA reveals transcription by RNA polymerase II in domains scattered throughout the nucleus. *Journal of Cell Biology,* **122**, 283—93.

Wilson, E.B. (1928) *The cell in development and heredity*, 3rd edition, pp. 906—11. Macmillan, Inc., New York.

Yoda, K., Kitagawa, K., Masumoto, H. *et al.* (1992) A human centromere protein, CENP-B, has a DNA binding domain containing four potential α helices at the NH_2 terminus, which is separable from dimerizing activity. *Journal of Cell Biology,* **119**, 1413—27.

Zatsepina, O.V., Polyakov, V.Y., Chentsov, Y.S. (1983) Chromonema and chromomere: structural units of mitotic and interphase chromosomes. *Chromosoma,* **88**, 91—7.

11

The replication of ribosomal RNA genes in eukaryotes

C. López-Estraño, J.B. Schvartzman and P. Hernández

11.1 INTRODUCTION

Replication of chromosomal DNA is a crucial process in the cell cycle of eukaryotes that is still poorly understood. Initiation of replication occurs when a cell is going to divide. However, entry into mitosis depends upon the successful termination of DNA replication. Initiation and termination of DNA replication are two relevant events for dividing cells.

Initiation of DNA replication in eukaryotic chromosomes occurs at numerous discrete sites along the parental DNA (Huberman and Riggs, 1968). In yeast, there is evidence that these sites are specific DNA sequences known as replication origins. Some DNA fragments of the budding yeast *Saccharomyces cerevisiae* confer to the plasmids containing them the ability to replicate autonomously once per S phase. Recently, it was demonstrated that many of these DNA fragments, named ARSs (for *a*utonomously *r*eplicating *s*equences), function as true replication origins not only when inserted into a plasmid but also in their chromosomal context. Sequence comparison and point mutation analysis led to the identification of an 11–bp ARS consensus sequence (ACS) (A/TTTTATA/GTTTA/T) that is essential for ARS function. This ACS is not sufficient, however, for efficient ARS activity. A less conserved AT-rich sequence located 3' to the T-rich strand of the ACS is also required. The ease of unwinding of this 3' region, named DUE (for *D*NA *u*nwinding *e*lement), seems to be important for ARS function (Natale *et al.*, 1993). In a detailed functional analysis of the ARS1, Marahrens and Stillman (1992) found that ACS is part of a 15-bp sequence called the A element. Linker substitution mutations across the 3' sequence detected three additional elements, referred to as B1, B2 and B3, that are collectively also essential for ARS function. More recently, a multiprotein complex was identified in yeast that specifically recognizes and binds to the ACS both *in vitro* (Bell and Stillman, 1992) and *in vivo* (Diffley and Cocker, 1992). Point mutations within the ACS that reduce DNA binding of this origin recognition complex (ORC) also result in a reduction of the ARS function. These results together with genetic evidence that ORC is required for DNA replication (reviewed in Huberman, 1994), suggest that ORC may be the initiator protein responsible for the activation of replication origins in yeast.

Contrary to the situation in yeast, the nature of DNA replication origins in higher eukaryotes is still unclear. Early experiments with DNA fibre autoradiography showed that initiation of replication takes place at discrete sites along the DNA. However, whether or not these sites are specific DNA sequences is still a matter of controversy (DePamphilis, 1993).

11.2 INITIATION OF DNA REPLICATION IN THE RIBOSOMAL RNA GENES

One experimental approach that simplifies the study of DNA replication in complex eukaryotic genomes is to focus the investigation on specific DNA loci. During the past decade, several techniques have been developed that permit *in vivo* physical mapping of DNA replication initiation sites by analysing either the newly synthesized DNA strands or the structure of the replication intermediates at a specific DNA locus (Vassilev and DePamphilis, 1992). Ribosomal DNA (rDNA) was chosen by a number of investigators as a model system to study chromosomal replication in eukaryotes.

The basic organization of the genes coding for the 18S, 5.8S and 25–28S rRNAs is conserved among most eukaryotic organisms. The rDNA consists of numerous repeats (from a few hundreds to several thousands) arranged in tandem (Fig. 11.1). Each repeat contains a transcription unit for the rRNA precursor and a non-transcribed spacer (NTS). Initiation of transcription starts upstream of the 5' end of the sequence coding for the 18S rRNA and, in most species, transcription terminates further downstream of the 3' end of the sequence coding

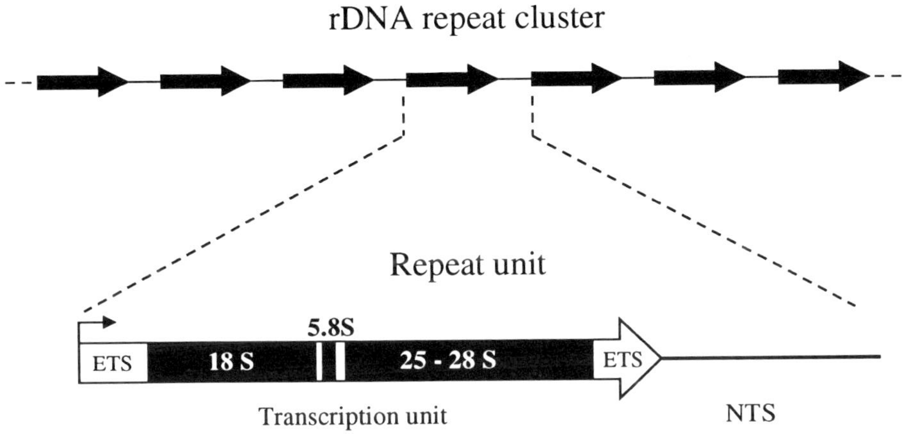

Fig 11.1 Organization and structure of eukaryotic rRNA genes. Above: Ribosomal DNA repeat cluster organization. Arrows represent transcription units and lines non-transcribed spacers. Below: Structure of the rDNA repeat unit. External (ETS) and internal transcribed spacers are represented by open boxes. Black boxes correspond to the sequences coding for the mature rRNAs. The small arrow indicates the transcription initiation site.

for the large rRNA. Thus, besides the coding sequences for the three rRNA species, the transcription unit contains two external and two internal transcribed spacers (ETS and ITS, respectively). The NTS is involved in the regulation of rRNA transcription and, in most species, it shows a complex primary structure containing a number of repeated sequences. In contrast with the high sequence conservation of the transcriptional unit among species, the sequence and organisation of the NTS show important divergence even between evolutionarily close species. Even within a given species, the NTS of the rDNA repeats frequently shows heterogeneity both in size and restriction enzyme cleavage sites.

Visualisation of replication bubbles in the rDNA by electron microscopy indicates that initiation of replication takes place at the NTS in *Physarum polycephalum* (Vogt and Braun, 1977), *Drosophila melanogaster* embryos (McKnight *et al.*, 1978), *Xenopus laevis* larvae (Bozzoni *et al.*, 1981), *Tetrahymena thermophila* (Cech and Brehm, 1981), sea urchin embryos (Botchan and Dayton, 1982) and the yeast

S. cerevisiae (Saffer and Miller, 1986). In agreement with these results, it was shown that the NTS is the earliest replicating rDNA region in G1/S synchronized cells from *Xenopus* (Bozzoni *et al.*, 1981) and *Pisum sativum* (Hernández *et al.*, 1988). In pea, the earliest replicated rDNA region is a 1.5 kb restriction fragment containing a cluster of four near matches to the yeast ACS. This cluster is embedded in an AT-rich domain flanked 3' by the transcription initiation site and 5' by an intrinsically bent DNA locus (Hernández *et al.*, 1993). Bent DNA is also present near other putative eukaryotic replication origins (Caddle *et al.*, 1990; Hernández *et al.*, 1993 and references therein). Moreover, several replication initiation factors are known to recognize or enhance DNA bending at some prokaryotic origins. The yeast ARS1 contains a stretch of bent DNA that has been proposed as a replication enhancer. Deletion of this locus impairs ARS activity which is restored when an unrelated synthetic bent DNA is inserted in its place (Williams *et al.*, 1988). The role of this bent locus in ARS1 function has been recently questioned by Marahrens and Stillman (1992). They show that the bent locus overlaps with the B3 element of ARS1. Their results strongly suggest that the importance of the B3 element in ARS1 function does not reside in its bent nature, but in the fact that B3 is the binding site for the ABF1 protein (ARS binding factor 1). ABF1 is a yeast transcriptional activator of a variety of genes that has been suggested to be also important for ARS activity.

It is known that bent DNA near transcription initiation sites can regulate promoter function. Thus, the DNA bending found in pea rDNA might be a transcriptional element directly involved in the initiation of rRNA transcription rather than in replication initiation. It is also known, however, that transcriptional elements can enhance the activation of replication origins located nearby (DePamphilis, 1988; Heintz *et al.*, 1992; Coverley and Laskey, 1994). Therefore, the DNA bent locus mapped about 200 bp upstream of the transcription initiation site in pea rDNA may indirectly influence initiation of DNA replication by affecting promoter function. Recent observations in the rDNA of *T. thermophila* and *P. polycephalum* support this hypothesis: both the replication origin region and DNA bent loci are located upstream of the transcription initiation site (Schroth *et al.*, 1992; Bénard *et al.*, 1995; Du *et al.*, 1995) In addition, intrinsically bent DNA has been recently mapped at the NTS in maize and *Cruciferae* species (Griess *et al.*, 1993; Da Rocha and Bertrand, 1995).

The two-dimensional agarose gel electrophoresis (2D gel) technique developed by Brewer and Fangman (1987) allows us to establish the replication mode of a specific DNA restriction fragment. Briefly, DNA is digested with the selected restriction enzyme(s) and the resulting fragments are separated predominantly according to their mass in the first dimension, which is run in a low percentage agarose gel at low voltage. The second dimension maximizes the retardation caused by the molecular topology of replication intermediates (RIs). This second dimension is run in a high percentage agarose gel at high voltage. Once electrophoresis is completed, the gel is blotted and then hybridized with a probe specific for the fragment to be analysed. After the first dimension, the RIs corresponding to any given DNA fragment (Fig. 11.2, RIs on top of each panels) are distributed between the positions where non-replicating linear forms (1×) and molecules of double its mass (2×) have moved. After the second dimension, due to the low mobility caused by their shape, RIs move above the diagonal corresponding to non-replicating linear fragments. Unique hybridization patterns are obtained depending on how the fragment is replicated: by a single fork moving from one end to the other (simple Y pattern) (Fig. 11.2A), by a bidirectionally growing bubble initiated at the centre of the fragment (bubble pattern) (Fig. 11.2C) or by two forks moving in opposite directions that meet at the centre of the fragment (double Y pattern) (Fig. 11.2D). If for any reason the progress of the replication fork along the fragment slows down or even stops at a specific site, a simple Y shaped RI would accumulate. This accumulation is observed in the 2D gel as a spot somewhere over the simple Y arc, depending on the location of the barrier within the fragment (Fig. 11.2B). Composite patterns also can be found if the fragment replicates from an origin asymmetrically located within the fragment (bubble-to-simple Y composite pattern) or by two forks entering the fragment asynchronously (simple Y-to-double Y composite pattern). This 2D gel technique has been also useful for analyzing DNA replication from unidirectional origins (Martín-Parras *et al.*, 1991). Huberman's group independently developed a different 2D gel technique to analyse DNA replication (Huberman *et al.*, 1987). In this case, the second dimension is run under alkaline denaturing conditions, which permits us to estimate the length of the nascent DNA strands. The direction of replication and the presence of origins can be determined by using different probes along the fragment of interest. During the last eight years these 2D gel techniques have been extensively used to analyse DNA replication in both prokaryotic and eukaryotic systems.

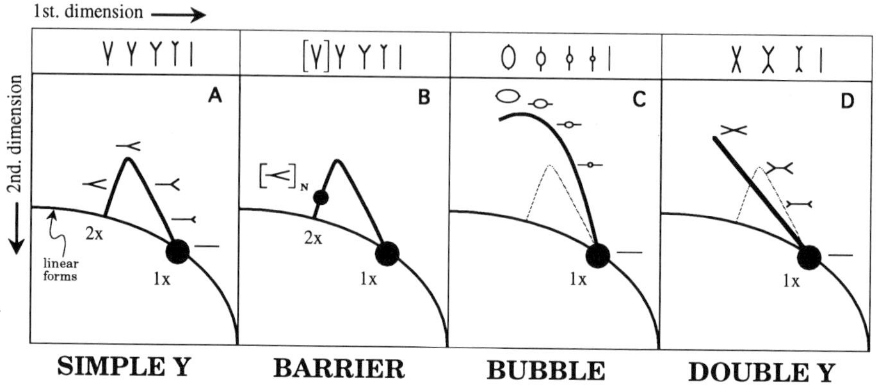

Fig. 11.2 Two-dimensional hybridization patterns generated by replication intermediates of a restriction fragment replicated by a single fork (panel A), by a single fork transiently arrested at a replication barrier (panel B), by a bidirectional origin located in the centre of the fragment (panel C) or by two forks that enter simultaneously into the fragment (panel D). Above each panel, the shape and distribution of replication intermediates after the first dimension are shown. The position of the linear unreplicated fragment (1x) and that one corresponding to a fragment twice its size (2x) are also indicated. In panels C and D, the simple Y pattern has been drawn for reference. See text for detail.

Analysis of *S. cerevisiae* rDNA replication by 2D gels showed a functional origin of replication that colocalized with an ARS and a DUE mapped to the NTS (Skryabin *et al.*, 1984; Linskens and Huberman, 1988; Brewer and Fangman, 1991; Miller and Kowalski, 1993). However, fragments containing the ARS generated not only a bubble pattern, but also a strong signal for a complete simple Y arc. This mixture of patterns is generated by DNA fragments containing an internal origin that is not activated in all replication rounds (Schvartzman *et al.*, 1990; Martín-Parras *et al.*, 1992). Therefore, although all yeast rDNA repeats seem to have an ARS, only in a fraction of them is that ARS used as an origin. This agrees with the observation that yeast rDNA replicons are large enough to accommodate at least five repeat units (Walmsley *et al.*, 1984). In the fission yeast *Schizosaccharomyces pombe* both a simple Y and a faint bubble pattern are seen in 2D gels (Fig. 3A), indicating again that not all potential origins are used in each S phase.

A low frequency of origin usage has been deduced also from 2D gel analysis of pea rDNA replication (Hernández *et al.*, 1993). It is known that only a fraction of the ribosomal genes are transcriptionally active showing an open chromatin configuration (Conconi *et al.*, 1992; Dammann *et al.*, 1993). Chromatin configuration might influence initiation of rDNA replication by changing the accessibility to the origin of required *trans*-acting factors. Therefore, one possible explanation for the low replication origin usage in the rDNA is that those origins flanked by active rRNA genes are preferentially used. This hypothesis is consistent with the observation that active genes are replicated early in the S phase whereas inactive genes are late replicated (reviewed in Fangman and Brewer, 1992).

Analysis of chromosomal rDNA replication in *Xenopus* early embryos by 2D gels gave different results compared to those obtained in yeast. A bubble pattern was found in each of a family of restriction fragments analysed covering the complete rDNA repeat, showing that replication initiation takes place randomly throughout the rDNA (Hyrien and Méchali, 1993). This conclusion extended to chromosomal rDNA that previously reached for the replication of plasmids containing a complete *Xenopus* rDNA repeat (Hyrien and Méchali, 1992; Mahbubani *et al.*, 1992). Moreover, the absence of sequence requirements for rDNA replication in *Xenopus* early embryos is in agreement with classic experiments showing that almost any molecule injected into *Xenopus* eggs replicates with an efficiency that depends only on the size of the DNA fragment and not on the presence of a particular DNA sequence (Harland and Laskey, 1980; Méchali and Kearsey, 1984). As mentioned above, however, initiation of rDNA replication takes place specifically at the NTS in *Xenopus* larvae and cultured cells (Bozzoni *et al.*, 1981). The apparent contradiction between these two groups of results can be understood in the light of a model recently proposed by Gilbert *et al.* (1995). This model suggests that in rapidly dividing early *Xenopus* embryos DNA is assembled into a relaxed nuclear structure that permits initiation of DNA replication at many sites. At the mid-blastula transition (MBT), changes in chromatin and nuclear structure and the onset of transcription would prevent initiation from many potential sites and facilitate it from others, resulting in site-specific initiation of DNA replication in differentiated cells. This model could be applied to rDNA replication. Coinciding with MBT, nucleolar structure changes, transcription of rRNA genes is turned on and, as a consequence, initiation of rDNA replication from sites at the coding regions would be repressed occurring only at the NTS. Due to sequence features and/or to promoter activation nearby, initiation could take place at a preferred site(s) within the NTS. This seems to be the case for rDNA replication in human cell cultures. Replication analysis by 2D gels showed a bubble pattern for all NTS segments tested, indicating initiation events occurring throughout the 31-kb NTS. On the other hand, no bubble patterns were seen for fragments composed solely of transcription unit sequences (Little *et al.*, 1993). More recently, these conclusions were confirmed using a different technique for mapping DNA replication origins. The latter study also suggests that preferred replication initiation sites are located a few kilobase pairs upstream of the transcription start site (Yoon *et al.*, 1995).

In summary, a universal feature for rDNA replication in eukaryotes, from unicellular microorganisms to mammals, is that initiation occurs at origins located in the NTS, outside the coding region. In some species, initiation is restricted to a relatively short DNA region within the NTS, whereas in others, like human, initiation seems to occur at any site throughout the NTS. The lack of detectable replication initiation events within the coding region of the rDNA could be the consequence of the absence of genetic elements required for initiation in the transcriptional unit. Alternatively, potential or actual transcriptional activity of rRNA genes could somehow repress the activation of origins located within the coding region. The observation that in the absence of rRNA transcription (before MBT), initiation of rDNA replication in *Xenopus* embryos can occur within the coding region favours the latter interpretation. It would be very interesting to address the prediction of specific initiation in the NTS for differentiated *Xenopus* cells.

11.3 TERMINATION OF DNA REPLICATION IN THE RIBOSOMAL RNA GENES

For a long time it have been assumed that once initiation of DNA replication takes place, nascent strands at a given replicating fork elongate at a constant speed until the fork meets another one moving in the opposite direction. This fork meeting ends up with completion of replication by the ligation of the growing DNA strands. However, recent analysis of replication fork movement showed that this is far from true, at least, in some particular regions of the genome of both prokaryotic and eukaryotic organisms. Eukaryotic rDNA is a clear example. Replication forks find it difficult to traverse the rDNA region located in the NTS close to the 3' end of the transcriptional unit. *S. cerevisiae* was the first species where this phenomenon was demonstrated (Brewer and Fangman, 1988; Linskens and Huberman, 1988). When restriction fragments spanning this region were analysed by the 2D gel technique, a conspicuous spot was found on the simple Y hybridisation signal. This spot corresponded to an accumulated replication intermediate containing a single fork positioned at a specific site (Fig. 11.2B). This site, named replication fork barrier (RFB), maps at the 3' end of the 35S rRNA transcription unit and arrests only those forks moving in the direction opposite to rRNA transcription. We demonstrated that this RFB is not a singular feature of the yeast rDNA replication, as a mechanistically identical barrier is also present at an equivalent site in the rDNA of the distant organism *P. sativum* (Hernández *et al.*, 1993). Both size classes of pea rDNA repeats contain an RFB at the 3' end of the transcriptional unit (Fig. 3B). As in yeast, this RFB is polar since it has no effect on the progress of those forks moving in the direction of transcription, but it terminates forks moving in the opposite direction. As a consequence, in these two species most of the rDNA cluster is replicated unidirectionally and cooriented with transcription. These striking coincidences led us to suggest that (i) the mechanism of fork blockage may be similar in these two distantly-related organisms, (ii) the RFB could be a feature of rDNA replication conserved in eukaryotes and, therefore, (iii) this barrier probably plays a relevant physiological role. The second of our predictions has been confirmed, which makes the third one more likely. RFBs have been detected at the 3' end of the transcriptional unit in the rDNA of the fission yeast *S. pombe* (Fig. 3A), *Lycopersicon esculentum* (Fig. 11.3C), *Xenopus* (Wiesendanger *et al.*, 1994), mouse (Fig. 11.3D) and human (Little *et al.*, 1993). However, the molecular mechanism that mediates fork blockage and the biological significance of this conserved RFB are still a mystery. Early electron microscopy studies of rDNA chromatin demonstrated that there is a temporal coincidence between replication and transcription of rRNA genes. Thus, the physiological role of the RFB might be to prevent head-on collision between replication and transcription, which would have deleterious consequences. In *Escherichia coli* and *Bacillus subtilis* this is specifically avoided by the appropriate orientation of actively transcribed genes so that their promoters are proximal to the bidirectional origin of replication (Brewer, 1988; Zeigler and Dean, 1990).

The coincidence of the map position of the polar RFB with the region where transcription terminates led Brewer and Fangman (1988) to suggest that transcription itself might be responsible for the arrest of replication forks in yeast. Fork arrest would be the consequence of collision between RNA polymerase I and the replication complex, when the fork reaches the 3' end of an active gene. However, these investigators (Brewer *et al.*, 1992) and others (Kobayashi *et al.*, 1992) soon demonstrated that arrest of replication forks occurs in the absence of transcription. The yeast RFB excised and cloned in an ARS plasmid is still active and in the same polar fashion as in the chromosome. However, a correlation between rRNA gene expression and RFB efficiency has been suggested, since stalled forks were found mostly at the 3' end of active rRNA gene copies (Lucchini and Sogo, 1994).

Fork arrest might be induced by the DNA structure at the barrier itself, without the requirement of *trans*-acting factors. This hypothesis is unlikely, since DNA fragments containing the pea or yeast RFB cloned in both orientation in pUC vectors and replicating in *E. coli* have no detectable effect on fork movement (Brewer *et al.*, 1992; Hernández *et al.*, 1993). Therefore, RFB must be mediated by *trans*-acting factors that somehow inhibit replication complex movement in a polar fashion. A similar mechanism of replication termination operates in the bidirectionally replicated chromosomes of *E. coli* and *B. subtilis* (reviewed by Baker, 1995). In both bacteria, the region where the two replication forks meet (about 180° from the origin) is flanked by polar fork barriers arranged in such a way that they form a replication trap. In *E. coli* six of such barriers, named *Ter* sites, have been identified so far. *Ter* sites are 22-bp binding sequences recognized by the Tus protein. The *Ter*-Tus complex seems to arrest replication forks by obstructing helicase action in an orientation-dependent manner. A similar mechanism could also function in the eukaryotic rDNA. The presence of these *Ter* sites in *E. coli* ensures that fork meeting occurs in a defined region of the chromosome and that each half of the circular chromosome is always replicated in the same direction and cooriented with transcription of active genes.

Figure 11.3 shows 2D gels of restriction fragments containing the 3' end of the rDNA transcription unit of four different eukaryotic species. The spots detected over the simple Y arcs demonstrate the presence of RFBs in this rDNA region of all these four species. The structure of the region where forks stop shows repeated sequences, that in some cases are suggested to play a role in transcription termination.

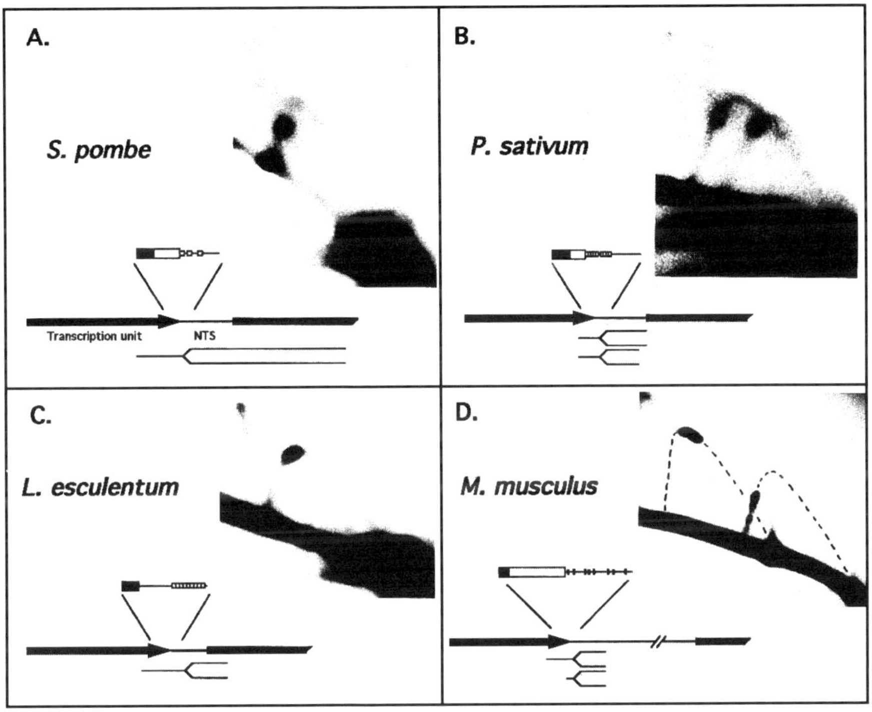

Fig. 11.3 2D gel analysis of replication intermediates corresponding to rDNA restriction fragments containing the 3' end of the transcription unit in four eukaryotic species. At the bottom of the panels, the accumulated Y-shaped intermediates of each fragment have been drawn at the corresponding position within the rDNA map. A detail of the rDNA structure at the region of fork arresting in each species is also shown, where the small boxes represent repeated DNA sequences. Notice the faint bubble signal in the 2D autoradiogram of panel A. The 2D autoradiogram in panel D has been underexposed for the discrete spots forming the long signal of accumulated Y-shaped intermediates to be visible. The simple Y arc is highlighted with a discontinuous line.

What *trans*-acting factors might be involved in the blockage of rDNA replication? Those factors implicated in transcription termination are good candidates since, as mentioned before, the RFBs colocalize with the region where rRNA transcription terminates. Among the ribosomal genes studied in Fig. 11.3, those of mouse are the best known in terms of transcription termination (reviewed in Evers *et al.*, 1995). Mouse rRNA transcription termination is mediated by mTTF-I (murine transcription termination factor-I) that specifically binds to an 18-bp sequence. At least eight of these terminator sequences, named Sal boxes, are present downstream of the mature 3' end of the rRNA. Sal boxes function in only one orientation, so that the Sal box/mTTF-I complex terminates rRNA transcription coming from the 5' end. Interestingly, the N-terminal half of the mTTF-I molecule can be truncated without affecting Sal box binding or transcription termination function (Evers *et al.*, 1995).

We used 2D electrophoresis to study the replication intermediates of mouse rDNA restriction fragments containing the Sal boxes. One representative 2D gel is shown in Fig. 11.3D. Due to restriction site heterogeneity, some mouse rDNA repeats lack the *Bam*HI site closer to the 3' end of the transcription unit. Thus, digestion of mouse rDNA with this enzyme generates two restriction fragments containing the region of interest. Both fragments show a simple Y pattern, indicating that they are replicated by a single fork in most of the rDNA repeats. The simple Y arc generated by the long *Bam*HI fragment (left-hand arc in Fig. 11.3D) clearly shows a prominent rod-shaped signal located near the inflexion of the arc. This signal corresponds to accumulated replication intermediates containing a fork stalled near the centre of the fragment, where Sal boxes are located (larger Y at the bottom of Fig. 11.3D). In the case of the small *Bam*HI fragment, the Sal boxes are found near the 5' end. The signal corresponding to accumulated replication intermediates of this fragment appears along the descending part of the simple Y arc, near the diagonal of linear molecules (right-hand arc in Fig. 11.3D). Therefore, these intermediates contain the stalled forks near the 5' end of the fragment (small Y at the bottom of Fig. 11.3D). These results demonstrate that mouse rDNA contains a polar RFB at or near the Sal boxes region, where forks moving in the direction opposite to transcription are arrested. The possible role of Sal boxes in mouse RFB is reinforced by the observation that forks are not stalled at a unique site. The long signal of accumulated Y-shaped replicative intermediates corresponding to the small *Bam*HI fragment is clearly composed of a series of discrete spots. Thus, mouse RFB consists of a series of alternative sites, close to each other, where forks can stop. If, as suggested from these results, Sal boxes are *cis*-acting signals for mouse rDNA RFB, mTTF-I would be a good *trans*-acting factor candidate.

The N-terminal half of the mTTF-I, which is dispensable for transcription termination activity (Evers *et al.*, 1995), might contain the domain required for replication arrest. Clearly, more work is needed to confirm or rule out this hypothesis.

The rDNA repeats of *S. pombe* and *L. esculentum* also contain a polar RFB at the 3' end of the transcription unit (Fig. 11.3A and C). In *S. pombe*, forks are arrested at or near a region where three alternative transcription termination sites have been recently mapped by S1 nuclease protection analysis. Each of these termination sites precedes a sequence, named the Sal box-like sequence because it shows analogy with the mouse Sal box sequence repeat (Melekhovets *et al.*, 1994). Unfortunately, termination of rRNA transcription has not yet been determined in *L. esculentum*. However, a repeated 53-bp element is present downstream of the 3' end of the mature 25S coding sequence, that may be involved in transcription termination (Perry and Palukaitis, 1990).

Transcription termination in human rDNA seems to occur by a mechanism very similar to that operating in mouse. Little et al. (1993) have also detected barriers to replication fork progression near the sequences involved in transcription termination in the human rDNA. This finding agrees with the hypothesis that transcription termination factors may play a role in rDNA replication termination. Contrary to what we observed for mouse, these authors found that human RFBs are bidirectional, although they seem to be more efficient on leftward moving forks. Our results show that mouse RFB mainly inhibits progression of forks moving in the direction opposite to transcription. They do not totally rule out, however, some effect on forks moving in the same direction as transcription. If this is the case, it must be very weak since even in overexposed 2D gel autorradiograms, we do not detect spots corresponding to accumulated replication intermediates for rightward moving forks.

The notion that prevention of replication forks from entering the 3' end of rRNA transcription unit is a universal feature of eukaryotic rDNA replication has been strengthened by the recent discovery of the same phenomenon in cultured *Xenopus* cells (Wiesendanger *et al.*, 1994). In *X. laevis*, the arrested replication forks map to a broad 450-bp region located at the 3' flanking spacer, suggesting the presence of a series of sites where forks can stop. In *X. borealis* rDNA, fork arrest takes place at a more define site near to two closely spaced transcription terminators. In both frog species, the RFB is polar arresting only those forks that approach the transcription unit from downstream (Wiesendanger *et al.*, 1994). Contrary to these observations, 2D gel electrophoresis failed to detect RFBs in the rDNA of early *X. laevis* embryos (before MBT) (Hyrien and Méchali, 1993). As mentioned before, transcription of rRNA genes are not yet active at this developmental stage. Therefore, either rRNA transcription is required for RFB activity in *Xenopus* rDNA or, as suggested by some authors, the expression or activation of RFB binding protein(s) takes place later during the embryo development coinciding with or after the time rRNA genes are turned on (Hyrien and Méchali, 1993; Wiesendanger *et al.*, 1994). Experiments focused to address this question may shed light in understanding the molecular mechanism of replication fork arrest in eukaryotic rDNA.

In summary, in the eukaryotic rDNA the NTS region close to the 3' end of the transcription unit functions as a barrier for replication forks. In most of the species studied so far this barrier is polar, arresting forks moving in the direction opposite to transcription. The molecular mechanism responsible for this fork arrest is still unknown. Since this phenomenon seems to be conserved in eukaryotes, it is likely that the molecular mechanism is similar in all species. The most likely hypothesis is that fork arrest is mediated by DNA binding protein(s) that somehow inhibit replication progress in an orientation dependent manner. These proteins may be unknown factors with this sole specific function or, at least some of them, protein factors also involved in other functions like regulation of rRNA transcription. Yeast RFB has been also related to mitotic recombination, since it colocalizes with one of the two sequences necessary for activity of the recombinational hot spot *HOT1* (Kobayashi *et al.*, 1992). Replication fork blockage might be required for *HOT1* activity and/or binding proteins involved in *HOT1* activity might act as barrier for replication forks. The former is consistent with the observation made in different systems that DNA regions where replication machinery is blocked or slowed down are susceptible to DNA rearrangements (Bierne and Michel, 1994).

The physiological role of this barrier is also unknown, although it has been hypothesized that it may be to prevent the putative deleterious effects of head-on collisions between transcriptional and replicational machineries (Brewer *et al.*, 1992; Hernández *et al.*, 1993). The consequences of *in vitro* head-on collision between the DNA replication apparatus of bacteriophage T4 and an *E. coli* RNA polymerase ternary transcription complex have been examined by Liu and Alberts (1995). Upon collision, the replication fork stalls for many minutes when the *in vitro* assay lacks a DNA helicase. In the presence

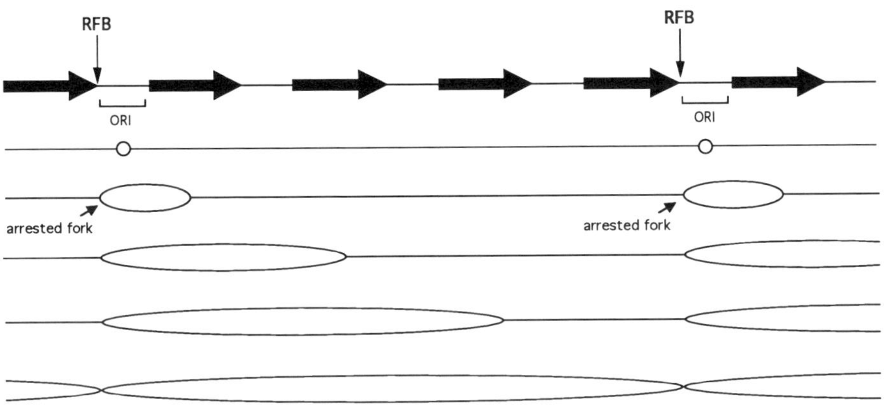

Fig 11.4 Model for replication of the rDNA repeat cluster. The location of the polar replication fork barrier (RFB) and the region where initiation can occur (ORI) are indicated. See text for detail. Re-drawn from Brewer and Fangman (1988) and Linskens and Huberman (1988).

of a helicase, the fork stalls for a few seconds and then it passes the RNA polymerase. Interestingly, this pause is longer than that produced by codirectional collision, suggesting an inherent disadvantage of head to head orientation of transcription and replication (Liu *et al.*, 1993; Liu and Alberts, 1995).

Taking together what is known about initiation and termination of rDNA replication, the mode of replication proposed for yeast rDNA (Brewer and Fangman, 1988; Linskens and Huberman, 1988) can probably apply to most eukaryotes with few modifications (Fig. 11.4). Initiation of bidirectional replication takes place within the NTS of only a subset of the rDNA repeat units, either at a specific site or at any site throughout most of this region. Forks moving toward the 3' end of the transcription unit are arrested at the RFB. On the other hand, each of the forks that move in the same direction as transcription replicate several rDNA repeats, since RFBs are inactive for forks progressing in this direction. Termination occurs when each moving fork encounters an arrested fork initiated in the next activated origin. This model predicts that rDNA replication is mostly unidirectional and cooriented with rRNA transcription. In *S. cerevisiae* it was confirmed that more than 90% of replication occurring in the transcription unit is carried out by rightward moving forks, suggesting a high efficiency of fork arrest. Due to technical difficulties, RFB efficiency has only been addressed in *Xenopus* rDNA. A significant fraction of the rDNA transcription units in cultured *Xenopus* cells is replicated in a 3'-to-5' direction (Wiesendanger *et al.*, 1994). This result is explained either as a consequence of transient fork arrest at the barrier or because only a subset of the potential barriers is active in *Xenopus* rDNA. Human rDNA replication do not totally fit the model shown in Fig. 11.4, since in this case RFBs do not appear to be polar. If initiation of human rDNA replication takes place, as in other species, only in a subset of the repeats, forks must be only transiently arrested at the RFBs for the complete rDNA replication to be achieved. Consistently, convergence of replication forks in human rDNA occurs specifically at the region of the barrier in some repeats and at apparently random sites in others (Little *et al.*, 1993).

ACKNOWLEDGEMENTS

We are grateful to M. Martínez-Robles and P. Robles for technical assistance. We thank Drs. P.R. Gaillardin (Institut National Agronomique Paris–Grignon, France), P. Palukaitis (Cornell Univ., College of Agriculture and Life Sciences, Ithaca, USA) and I. Grummt (German Cancer Research Center) for providing rDNA probes. This work was partially supported by grant 93/0161 from the Spanish Fondo de Investigaciones Sanitarias, grant AE00118/94 from the Comunidad Autónoma de Madrid (Spain) and a fellowship of the Consejo Nacional de Investigaciones Científicas y Tecnológicas (Venezuela) to C. L.-E.

REFERENCES

Baker, T.A. (1995) Replication arrest. *Cell*, **80**, 521–4.

Bell, S.P. and Stillman, B. (1992) ATP-dependent recognition of eukaryotic origins of DNA replication by a multiprotein complex. *Nature*, **357**, 128–34.

Bénard, M., Lagnel, C. and Pierron, G. (1995) Site-specific initiation of DNA replication within the non-transcribed spacer of *Physarum* rDNA. *Nucleic Acids Research*, **23**, 1447–53.

Bierne, H. and Michel, B. (1994) When replication forks stop. *Molecular Microbiology*, **13**, 17–23.

Botchan, P.M. and Dayton, A.I. (1982) A specific replication origin in the chromosomal rDNA of *Lytechinus variegatus*. *Nature*, **299**, 453–6.

Bozzoni, I., Baldari, C.T., Amaldi, F. and Buongiorno-Nardelli, M. (1981) Replication of ribosomal DNA in *Xenopus laevis*. *European Journal of Biochemistry*, **118**, 585–90.

Brewer, B.J. (1988) When polymerases collide: replication and the transcriptional organization of the *E. coli* chromosome. *Cell*, **53**, 679–86.

Brewer, B.J. and Fangman, W.L. (1987) The localization of replication origins on ARS plasmids in *S. cerevisiae*. *Cell*, **51**, 463–71.

Brewer, B.J. and Fangman, W.L. (1988) A replication fork barrier at the 3' end of yeast ribosomal RNA genes. *Cell*, **55**, 637–43.

Brewer, B.J. and Fangman, W.L. (1991) Mapping replication origins in yeast chromosomes. *BioEssays*, **13**, 317–22.

Brewer, B.J., Lockshon, D. and Fangman, W.L. (1992) The arrest of replication forks in the rDNA of yeast occurs independently of transcription. *Cell*, **71**, 267–76.

Caddle, M.S., Dailey, L. and Heintz, N.H. (1990) RIP60, a mammalian origin-binding protein, enhances DNA bending near the dihydrofolate reductase origin of replication. *Molecular and Cellular Biology*, **10**, 6236–43.

Cech, T.R. and Brehm, S.L. (1981) Replication of the extrachromosomal ribosomal RNA genes of *Tetrahymena thermophila*. *Nucleic Acids Research*, **9**, 3531–43.

Conconi, A., Sogo, J.M. and Ryan, C.A. (1992) Ribosomal gene clusters are uniquely proportioned between open and closed chromatin structures in both tomato leaf cells and exponentially growing suspension cultures. *Proceedings of the National Academy of Sciences USA*, **89**, 5256–60.

Coverley, D. and Laskey, R.A. (1994) Regulation of eukaryotic DNA replication. *Annual Review of Biochemistry*, **63**, 745–76.

Da Rocha, P.S.C.F. and Bertrand, H. (1995) Structure and comparative analysis of the rDNA intergenic spacer of *Brassica rapa*: implications for the function and evolution of the *Cruciferae* spacer. *European Journal of Biochemistry*, **229**, 550–7.

Dammann, R., Lucchini, R., Koller, T. and Sogo, J.M. (1993) Chromatin structures and transcription of rDNA in yeast *Saccharomyces cerevisiae*. *Nucleic Acids Research*, **21**, 2331–8.

DePamphilis, M.L. (1988) Transcriptional elements as components of eukaryotic origins of DNA replication. *Cell*, **52**, 635–8.

DePamphilis, M.L. (1993) Origins of DNA replication in metazoan chromosomes. *Journal of Biological Chemistry*, **268**, 1–4.

Diffley, J.F.X. and Cocker, J.H. (1992) Protein DNA interactions at a yeast replication origin. *Nature*, **357**, 169–72.

Du, C., Sanzgiri, R.P., Shaiu, W.-L. *et al.* (1995) Modular structural elements in the replication origin region of *Tetrahymena* rDNA. *Nucleic Acids Research*, **23**, 1766–74.

Evers, R., Smid, A., Rudloff, U. *et al.* (1995) Different domains of the murine RNA polymerase I-specific termination factor mTTF-I serve distinct functions in transcription termination. *EMBO Journal*, **14**, 1248–56.

Fangman, W.L. and Brewer, B.J. (1992) A question of time – replication origins of eukaryotic chromosomes. *Cell*, **71**, 363–366.

Gilbert, D.M., Miyazawa, H. and DePamphilis, M.L. (1995) Site-specific initiation of DNA replication in *Xenopus* egg extract requires nuclear structure. *Molecular and Cellular Biology*, **15**, 2942–54.

Griess, E.A., Grasser, K.D. and Feix, G. (1993) Repeat units from a maize rDNA external spacer region exhibit DNA curvature and interact with high-mobility-group proteins. *Planta*, **191**, 524–31.

Harland, R.M. and Laskey, L.A. (1980) Regulated replication of DNA microinjected into eggs of *Xenopus laevis*. *Cell*, **21**, 761–771.

Heintz, N.H., Dailey, L., Held, P. and Heintz, N. (1992) Eukaryotic replication origins as promoters of bidirectional DNA synthesis. *Trends in Genetics*, **8**, 376–81.

Hernández, P., Bjerknes, C.A., Lamm, S.S. and Hof, J.V.t. (1988) Proximity of an ARS consensus sequence to a replication origin of pea (*Pisum sativum*). *Plant Molecular Biology*, **10**, 413–22.

Hernández, P., Martín-Parras, L., Martínez-Robles, M.L. and Schvartzman, J.B. (1993) Conserved features in the mode of replication of eukaryotic ribosomal RNA genes. *EMBO Journal*, **12**, 1475–85.

Huberman, J.A. (1994) A tale of two functions. *Nature*, **367**, 20–1.

Huberman, J.A. and Riggs, A.D. (1968) On the mechanism of DNA replication in mammalian chromosomes. *Journal of Molecular Biology*, **32**, 327–41.

Huberman, J.A., Spotila, L.D., Nawotka, K.A. *et al.* (1987) The *in vivo* replication origin of the yeast 2 mm plasmid. *Cell*, **51**, 473–81.

Hyrien, O. and Méchali, M. (1992) Plasmid replication in *Xenopus* eggs and egg extracts: a 2D gel electrophoretic analysis. *Nucleic Acids Research*, **20**, 1463–9.

Hyrien, O. and Méchali, M. (1993) Chromosomal replication initiates and terminates at random sequences but at regular intervals in the ribosomal DNA of *Xenopus* early embryos. *EMBO Journal*, **12**, 4511–20.

Kobayashi, T., Hidaka, M., Nishizawa, M. and Horiuchi, T. (1992) Identification of a site required for DNA replication fork blocking activity in the rRNA gene cluster in *Saccharomyces cerevisiae*. *Molecular and General Genetics*, **233**, 355–62.

Linskens, M.H.K. and Huberman, J.A. (1988) Organization of replication of ribosomal DNA in *Saccharomyces cerevisiae*. *Molecular and Cellular Biology*, **8**, 4927–35.

Little, R.D., Platt, T.H.K. and Schildkraut, C.L. (1993) Initiation and termination of DNA replication in human rRNA genes. *Molecular and Cellular Biology*, **13**, 6600–13.

Liu, B. and Alberts, B.M. (1995) Head-on collision between a DNA replication apparatus and RNA polymerase transcription complex. *Science*, **267**, 1131–1137.

Liu, B., Wong, M.L., Tinker, R.L. *et al.* (1993) The DNA replication fork can pass RNA polymerase without displacing the nascent transcript. *Nature*, **366**, 33–9.

Lucchini, R. and Sogo, J.M. (1994) Chromatin structure and transcriptional activity around the replication forks arrested at the 3' end of the yeast rRNA genes. *Molecular and Cellular Biology*, **14**, 318–26.

Mahbubani, H.M., Paull, T., Elder, J.K. and Blow, J.J. (1992) DNA replication initiates at multiple sites on plasmid DNA in *Xenopus* egg extracts. *Nucleic Acids Research*, **20**, 1457–62.

Marahrens, Y. and Stillman, B. (1992) A yeast chromosomal origin of DNA replication defined by multiple functional elements. *Science*, **255**, 817–23.

Martín-Parras, L., Hernández, P., Martínez-Robles, M.L. and Schvartzman, J.B. (1991) Unidirectional replication as visualized by two-dimensional agarose gel electrophoresis. *Journal of Molecular Biology*, **220**, 843–53.

Martín-Parras, L., Hernández, P., Martínez-Robles, M.L. and Schvartzman, J.B. (1992) Initiation of DNA replication in ColE1 plasmids containing multiple potential origins of replication. *Journal of Biological Chemistry*, **267**, 22496–505.

McKnight, S.L., Bustin, M. and Miller, O.L. (1978) Electron microscopic analysis of chromosome metabolism in the *Drosophila melanogaster* embryo. *Cold Spring Harbour Symposium on Quantitative Biology*, **42**, 741–54.

Méchali, M. and Kearsey, S. (1984) Lack of specific sequence requirement for DNA replication in *Xenopus* eggs compared with high sequence specificity in yeast. *Cell*, **38**, 55–64.

Melekhovets, Y.F., Good, L., Elela, S.A. and Nazar, R.N. (1994) Intragenic processing in yeast rRNA is dependent on the 3' external transcribed spacer. *Journal of Molecular Biology*, **239**, 170–80.

Miller, C.A. and Kowalski, D. (1993) *cis*-acting components in the replication origin from ribosomal DNA of *Saccharomyces cerevisiae*. *Molecular and Cellular Biology*, **13**, 5360–9.

Natale, D.A., Umek, R.M. and Kowalski, D. (1993) Ease of DNA unwinding is a conserved property of yeast replication origins. *Nucleic Acids Research*, **21**, 555–60.

Perry, K.L. and Palukaitis, P. (1990) Transcription of tomato ribosomal DNA and the organization of the intergenic spacer. *Molecular and General Genetics*, **221**, 103–12.

Saffer, L.D. and Miller, O.L. (1986) Electron microscopic study of *Saccharomyces cerevisiae* rDNA chromatin replication. *Molecular and Cellular Biology*, **6**, 1148–57.

Schroth, G.P., Siino, J.S., Cooney, C.A. *et al.* (1992) Intrinsically bent DNA flanks both sides of an RNA polymerase-I transcription start site — both regions display novel electrophoretic mobility. *Journal of Biological Chemistry*, **267**, 9958–64.

Schvartzman, J.B., Adolph, S., Martín-Parras, L. and Schildkraut, C.L. (1990) Evidence that replication initiates at only some of the potential origins in each oligomeric form of bovine papillomavirus type 1 DNA. *Molecular and Cellular Biology*, **10**, 3078–86.

Skryabin, K.G., Eldarov, M.A., Larionov, V.L. *et al.* (1984) Structure and function of the nontranscribed spacer regions of yeast rDNA. *Nucleic Acids Research*, **12**, 2955–68.

Vassilev, L.T. and DePamphilis, M.L. (1992) Guide to identification of origins of DNA replication in eukaryotic cell chromosomes. *Critical Reviews in Biochemistry and Molecular Biology*, **27**, 445—72.

Vogt, V.M. and Braun, R. (1977) The replication of ribosomal DNA in *Physarum polycephalum*. *European Journal of Biochemistry*, **80**, 557–66.

Walmsley, R.M., Johnston, L.H., Williamson, D.H. and Oliver, S.G. (1984) Replicon size of yeast ribosomal DNA. *Molecular and General Genetics*, **195**, 260–6.

Wiesendanger, B., Lucchini, R., Koller, T. and Sogo, J.M. (1994) Replication fork barriers in the *Xenopus* rDNA. *Nucleic Acids Research*, **22**, 5038–46.

Williams, J.S., Eckdahl, T.T. and Anderson, J.N. (1988) Bent DNA functions as a replication enhancer in *Saccharomyces cerevisiae*. *Molecular and Cellular Biology*, **8**, 2763–9.

Yoon, Y., Sanchez, J.A., Brun, C. and Huberman, J.A. (1995) Mapping of replication initiation sites in human ribosomal DNA by nascent-strand abundance analysis. *Molecular and Cellular Biology*, **15**, 2482–9.

Zeigler, D.R. and Dean, D.H. (1990) Orientation of genes in the *Bacillus subtilis* chromosome. *Genetics*, **125**, 703–8.

12

Genomic interactions: gene expression, DNA methylation and nuclear architecture

N. Neves, A. Castilho, M. Silva, J.S. Heslop-Harrison and W. Viegas

12.1 INTRODUCTION

New varieties of plants are obtained by breeders producing new combinations of characters. With this goal, some strategies aim to introduce alien chromatin from related species into cultivated crops. Genetic engineering and hybridization techniques permit assembly in a single organism of the advantages of different organisms; however, in many cases the regular expression of the inserted genes cannot be ensured. Thus, integrated in a context of gene, chromosome and genomic manipulation, further studies of genomic interactions are vital for fundamental studies of gene control and for plant breeding.

12.2 GENOMIC INTERACTIONS

The production of new plants by introducing new characteristics from different species can be achieved by somatic hybridization or transformation, or through sexual manipulation based on the production of new hybrids. Chromosome engineering allows plant breeders to produce chromosome or chromosome arm addition, substitution or translocation lines. When this new chromatin is introduced into cultivated crops it is hoped that this new information is stably transmitted in every cell cycle and from one generation to another; furthermore, the regular expression of the inserted genes is also expected. However, the simultaneous presence in the same nucleus of two partial or entire genomes may cause complex gene, chromosome and genomic interactions. Hence, it is often observed that 'one plus one does not always equal two', meaning that new chromatin combinations may not obey arithmetic or Mendelian rules. Thus, only by understanding the mechanisms underlying genomic interactions it is possible to produce new and useful phenotypic variability.

Genomic interactions may be expressed in two different ways: the first concerns the preferential elimination of chromosomes from one of the parental genomes in a hybrid; the other deals with the preferential inactivation of gene(s) from one parent. In both plants and animals the elimination of whole genomes or parental sets of chromosomes at cell division has been extensively reported, with the eliminated chromosomes failing to congress at metaphase and migrate at anaphase (Heslop-Harrison and Schwarzacher, 1993). Karyotypically unstable crosses which eliminate one of the parental genomes prevent the introduction of variability and are useful for the production of haploids (Kasha and Kao, 1971). It should be noted that elimination events can occur as regular developmental processes in species at meiosis as in *Sciara* (Abbott *et al.*, 1981) or at mitosis as in *Metasciulus* (Nelson-Rees *et al.*, 1980); others take place in hybrids as in *Hordeum* (Lange, 1971; Subrahmanyam and Kasha, 1973), *Poeciliopsis* (Cimino, 1972) and *Rana* (Heppich *et al.*, 1982), not always at defined developmental stages.

One of the most studied elimination events resulting from intrageneric genomic interactions occurs in *Hordeum* hybrids (Davies, 1958; Lange, 1971; Finch and Bennett, 1983). In this case, further complexity of the process was observed in some hybrid combinations with the elimination of alternative parental genomes being tissue specific (embryo/endosperm) depending on the hybrids initially produced (Finch, 1983; Finch and Bennett, 1983). Despite this feature, genome elimination is always a uniparental event, i.e. the same genome is excluded in reciprocal crosses (Finch and Bennett, 1983). Centromere dysfunction and consequent loss of centric activity (Finch, 1983), peripheral position in the nucleus (Norum and Migeon, 1974; Finch, 1983; Finch and Bennett, 1983) and chromosome elimination progressing in a non-random way (Vig, 1981; Finch and Bennett, 1983) are some of the special features that have been associated with the genome loss. Why only particular chromosomes usually belonging to the same parent tend to be lost in certain hybrids is as yet unknown.

Interactions between genomes can also be detected where the genes from one parent have different expression properties from those of the other parent (Heslop-Harrison, 1990). This results solely as a consequence of the origin of those genes and is due to the presence of the other genome(s), and may be the reason why karyotypically stable hybrids resemble one of the parents in many morphological and physiological aspects. For example, the hybrid *Hordeum vulgare* × *Secale africanum*, produced by Finch *et al.* (1981) does not present an intermediate phenotype, but resembles the *Secale* parent in many morphological characters, namely the ligule, auricule, leaf length, lemma length, leaf surface features, etc. (Bennett, 1984; Heslop-Harrison and Bennett, 1984; Fujigaji and Tozu, 1987). Also, the perennial growth habit and sensitivity to low temperatures observed in the hybrid indicates a degree of dominance of the *Secale* parent over the *Hordeum* parent (Finch and Bennett, 1983; Fujigaki and Tozu, 1987). The same is observed in triticale (×*Triticosecale* Witt.), an artificial amphiploid obtained by doubling the chromosome number after crossing *Triticum* (wheat) × *Secale* (rye). The triticale plant does not truly resembles an intermediate form between the parental species, but has many features of rye rather than wheat (Heslop-Harrison, 1990).

Genomic interactions between *Triticeae* genera can also modify the expression patterns of genetic resistance to mildew and rust fungi, as detected in *Tritordeum* (*Triticum* sp.× *Hordeum* sp.) and *Hordeum* sp. × *Secale* sp. (review in Rubiales *et al.*, 1993). Wheat heat shock proteins (HSP) expression is also suppressed in the presence of the rye genome (Somers *et al.*, 1992); the triticale HSP profile resembles more the rye one indicating that the rye genome has a suppressive effect on wheat HSP expression, located on rye chromosome 5 (Somers and Filion, 1994).

12.3 NUCLEOLAR DOMINANCE

One of the most studied genomic interaction systems is the one concerning the dominance of one progenitor in the expression of ribosomal loci - nucleolar dominance or the amphiplasty phenomenon (Navashin 1928, cited in Lacadena et al. 1984a).

Systems involving preferential inactivation of rRNA loci according to their parental origin are found widely in plant and animal hybrids

(Reeder, 1985) and hence are of wide interest and general significance. The interaction system is a valuable model for studying gene expression because the regulation of individual known loci can be examined directly by cytological methods.

In cereals, nucleolar dominance is a common feature in both intra and intergeneric hybrids. In crosses between *Agropyrum elongatum* and *Triticum aestivum* (hexaploid wheat), a partial inactivation of rDNA from *A. elongatum* origin is observed, stronger in the addition lines of *A.elongatum* nucleolar organizer chromosomes 6E and 7E to the wheat genome than in the amphiploid (Lacadena *et al.*, 1984b). Conversely, a preferential inactivation of wheat rDNA occurs in the presence of *Aegilops umbellulata* NOR bearing chromosomes 1U and 5U (Martini *et al.*, 1982).

Wheat-rye nucleolar dominance is an important research topic in our group and the basis of the work described here. In wheat × rye F1 hybrids and in triticale, nucleolar dominance of wheat is observed, resulting in an almost total inactivation of rye origin rRNA genes (Thomas and Kaltsikes, 1983; Cermeño *et al.*, 1984; Lacadena *et al.*, 1984; Neves *et al.*, 1995; Silva *et al.*, 1995). Consequently, in the large majority of cells, rye NOR activity is undetectable by the silver staining procedure, a method that reveals interphase nucleoli and stains metaphase NORs that were transcribed during the previous interphase (Goodpasture and Bloom, 1975; Schwarzacher *et al.*, 1978; Hubbell, 1985; Jiménez *et al.*, 1988).

12.4 DNA METHYLATION AND rRNA GENE EXPRESSION IN WHEAT-RYE HYBRIDS

The control of gene expression and epigenetic effects by DNA methylation has been and is still widely discussed. However, since Jones and Taylor (1980) proposed a correlation between cytosine residue hypomethylation and gene transcription, relatively few cases have been reported where modifications of methylation levels were shown to directly alter expression (Barlow, 1993).

In plants, DNA methylation and rDNA expression were first related from studies made in pea and flax (Reeder, 1985). In wheat, Flavell and collaborators (Flavell and O'Dell, 1979; Flavell *et al.*, 1988) found that the extent of CpG methylation within rDNA units is particularly high and as the number of ribosomal genes increases, the proportion of methylated units also increases. This suggests the maintenance of a certain level of rDNA transcription between different genotypes. The distribution of hypomethylated sites between different chromosomal sites correlates with their relative activities based upon nucleolar volumes, a fact that was also established by the finding that the proportion of unmethylated sites at a locus decreases when this locus became more active, due to deletion of another rRNA locus (Martini and Flavell, 1985; Flavell *et al.*, 1988; Sardana *et al.*, 1993).

12.4.1. DNA methylation and wheat-rye nucleolar dominance

In wheat × rye hybrids and triticale, rye rRNA gene expression patterns are correlated with, and arguably ultimately modulated by, DNA methylation. 5-azacytidine (5-AC) is a cytidine analogue which is incorporated in the DNA during replication but cannot be methylated, causing the newly replicated strands to be undermethylated. By using 5-AC in the germination of F1 hybrids and triticale seeds, where rRNA genes of rye origin are usually silent, it is possible to observe the activity of rye rDNA (Neves *et al.*, 1995). More than 92% of untreated metaphases show three or four silver stained satellite NORs, located on wheat chromosome pairs 1B and 6B. After 5-AC treatment however more than 90% of metaphase cells present three pairs of subterminal NORs, identified as 1B and 6B from wheat origin and 1R from rye. These three chromosome pairs are the only ones with satellite NORs present in triticale.

After Southern blot hybridization using wheat rDNA as probe on DNA extracted from control and 5-AC treated triticale plants, an extra fragment is found in the DNA from the treated plants, after double digestion with DraI and ApaI restriction enzymes (ApaI is methylation sensitive). Reprobing of the same blot with the rye rDNA inter-25S-18S-spacer showed that the extra fragment is of rye origin (Neves *et al.*, 1995). Silver staining of triticale root-tip cells shows that 5-AC treatment causes the activation of rRNA genes from rye which are normally inactive in the presence of wheat genomes. The Southern blot analysis indicates that the effect correlates with reduction of methylation in a cytosine residue within the rye rDNA intergenic spacer. Hence, external application of 5-AC causes reduction of methylation at defined sites in the rye rRNA gene unit. Such methylation reduction, at least for those particular cytosine residues, seems to directly enable its expression.

Two main questions emerge:

(i) If rye meiotic cells carry an active NOR which is no longer active in the ensuing wheat × rye F1 hybrid root-tip cells, when is the 1R NOR methylation imprint set in place?
(ii) If DNA methylation is involved in determining rye origin rDNA activity, are these genes reprogrammed during meiosis with consequent changes in methylation and expression patterns?

12.4.2 Developmental stage of 1R NOR imprinting establishment

If we consider the natural activity of the 1R NOR in rye meiotic cells which is suppressed in root-tip cells of wheat × rye hybrid plants, there must be a time for establishing the methylation imprint responsible for modifying rye rRNA gene expression. This imprinting phenomenon probably occurs during embryogenesis as in animals where developmental patterning happens at this stage (Ferguson-Smith and Surani, 1993). Silver staining was performed in F1 embryos and endosperms of developing seeds collected at different times after controlled pollination. After evaluation of nucleolar activity it was possible to conclude that the inactivation of rDNA from rye occurs simultaneously in both tissues (diploid embryo and triploid endosperm) between the 4th and the 5th days after fertilization (Castilho *et al.*, 1995). Three and five major NORs were silver positive in embryos (ABDR genome designation) and endosperms (AABBDDR genome designation) respectively until 4 days after fertilization; however, silver stained ovules rescued 5 days post-fertilization revealed only two and four active subterminal NORs in embryos and endosperms. As the imprinting event takes place simultaneously in the embryo and endosperm, this indicates that the process is independent of the number of cell cycles, i.e., DNA replication rounds. Genomic interactions leading to the inactivation of rye rDNA are time-dependent, related to a specific developmental stage.

The effects of DNA hypomethylation induced during embryo development were also examined in wheat × rye hybrids and triticale. Hexaploid wheat and triticale plants were emasculated and artificially pollinated with rye and triticale pollen respectively. The volume enclosing the developing embryos was injected with 5-AC over a period of three consecutive days centred 3, 6 and 9 days after pollination in the three different treatments. After normal seed

Fig. 12.1 ***Colour Plate 6*** Root tip metaphases from cells of triticale after silver staining of NORs active during the previous interphase. (a) Untreated cell showing activity of the four major NORs (arrows) originating from wheat chromosomes 1B and 6B. (b) Cell from a plant of 5-AC treatment 9, showing six active NORs (arrows), four from wheat origin and two from rye origin.

Fig. 12.2 ***Colour Plate 6*** Ontogenic map of rye rRNA gene expression in triticale. Both wheat and rye NORs are active at fertilization, but rye-origin genes are suppressed 5 days after fertilization (point 1). The suppression is not reversible by 5-AC treatment up to 6 days after fertilization (point 2), after which it becomes stably reversible (treatment 9, figure 1b). Rye NORs become active immediately after meiosis (point 3).

Fig. 12.3 ***Colour Plate 6*** Genomic in situ hybridization to a root-tip metaphase cell from triticale. The 14 rye chromosomes are labelled red, while the 28 wheat chromosomes are visible by weak cross-hybridization of the rye genomic DNA probe.

Fig. 12.4 ***Colour Plate 6*** Examples of confocal sections of root tip interphase nuclei labelled with genomic rye DNA (green) and rDNA (red). (a) Nuclei from an untreated root-tip showing non-intermixing of rye (light green) and wheat (dark green) chromatin, with the chromatin of both genomes occurring in discrete domains, often as interspread slices. (b) Nuclei from 5-AC treated root tips show much less structure, with few discrete domains of rye or wheat chromatin.

maturation and germination, nucleolar activity was analysed by silver staining in root-tip cells (Fig. 12.1). Triticale seedlings from treatment 9 regularly had five or six active and subterminal NORs (1B, 6B and 1R), contrasting with the control and earlier treatments where most cells showed only four major NORs of wheat origin. In the F1 hybrids no cells with more than two subterminal NORs were found from controls and treatment 3, while 80% of cells from treatment 9 had three subterminal NOR loci. These results show that the inactivation is established early in embryogenesis, but is reversible by 5-AC incorporation at a later stage. Hence there is an important change during embryo development concerning the effect of an external stimulus or factors which may affect the establishment of new epigenetic patterns. Before 7 days post fertilization, modifications of rye rDNA methylation patterns with 5-AC are not maintained as no effect on rRNA gene expression is detected by silver staining in the ensuing seedling. The data imply that *de novo* methylases are active until that moment establishing the methylation imprint on rye rRNA genes. Later, the methylation imprint can be erased by 5-AC treatment. Thus we suggest that *de novo* methylases are active until the 7th day after fertilization when the imprinting is completed and the activity of maintenance methylases keeps the rye rDNA methylation patterns largely constant.

12.4.3 Meiotic reprogramming of rye rRNA genes

It is known that in both animals and plants some DNA methylation patterns are modified during meiosis (Monk, 1987; Holliday, 1987; Neves *et al.*, 1992). To evaluate the existence of a possible mechanism of meiotic reprogramming of rye rDNA in triticale, through changes in DNA methylation patterns established early in seed development, nucleolar activity was analysed and compared in meristematic root-tip cells, pollen mother cells at metaphase I and young pollen grains at first mitotic division. The results show that, in contrast to meristematic root-tip cells and pollen mother cells where wheat nucleolar dominance is observed, in young pollen grains at first mitosis the 1R NOR is expressed in the large majority of cells indicating its full activity at this stage (Silva *et al.*, 1995). Thus, we conclude that nucleolar dominance is erased between pollen mother cells and young pollen grains, the 1R NOR becoming active in the later stage. As initially proposed by Holliday (1984), we suggest that meiotic reprogramming of rye rDNA is mediated by the resetting of methylation patterns, allowing the efficient transcription of rye rRNA genes in young pollen grains.

Thus, an integrated, ontogenic map of rye rDNA expression in triticale can be proposed (Fig. 12.2): rye NORs are active throughout fertilization; shortly after this event (between 4 and 5 days post-fertilization) these genes became inactive by a methylation imprint process, and this state is maintained during most vegetative growth. During meiosis rye rRNA genes are reprogrammed, probably due to the resetting of the rDNA methylation patterns established earlier, allowing their expression immediately after meiosis.

12.5 CHROMATIN ORGANIZATION AND NUCLEOLAR DOMINANCE IN TRITICALE

The study of patterns of nuclear architecture has been of interest for more than a century, but recent advances in microscopical methods and *in situ* hybridization have enabled chromatin organization to be evaluated in new ways - particularly in three dimensions at interphase. Comings (1968) pointed out that the high precision of nuclear functions and mitotic and meiotic structures suggests a non-random disposition of chromatin within the nucleus.

The analysis of interphase disposition of polytene chromosomes in *Drosophila* (Hochstrasser and Sedat, 1987) and of telomeric and centromeric sequences in humans (Manuelidis and Borden, 1988) revealed different patterns of interphase chromatin organization in functionally distinct cell types. In plants, Cecarelli and Cionini (1993) found different patterns of heterochromatin association in meristematic and differentiated cells. Also, the maintenance of specific nuclear domains occupied by repetitive, non-transcribed sequences was observed in the same cell types between different mammalian species (Manuelidis and Borden, 1988). The functional compartmentalization of the nucleus with respect to gene transcription, DNA replication or RNA processing (splicing) has been reported for different types of animal cells (Spector, 1990; Carmo-Fonseca *et al.*, 1991; Carter *et al.*, 1993; Cremer *et al.*, 1993; Xing and Lawrence, 1993; Hassan *et al.*, 1994). Thus, the positioning of specific regions of the genome in particular nuclear areas with different transcriptional features allows physical and functional nuclear integration (Jackson, 1991).

Sexual reproduction requires nuclear organization, since the two parental chromosome sets physically pair and recombine. Parental genome organization has been for the last two decades one of the most exciting cytogenetic research areas, looking at possible implications of intergenomic interactions and the resulting effects of parental dominance. Evidence for genomic nuclear compartmentalization emerged from studies by Odartchenko and Keneklis (1973) using *in vivo* radioactively labelled rat sperm DNA following fertilization of non-labelled ova; in the early developing egg, parental genomes tended to stay apart. However, the differentiation of the two haploid complements within a species is very difficult; intergeneric and interspecific hybrids have proved to be valuable tools for studying the disposition of parental genomes.

In plant hybrids, non-random intergenomic distribution was

extensively studied by Finch *et al.* (1981) in somatic cells of *Hordeum vulgare* × *Secale africanum* hybrids. After reconstruction of sectioned metaphase cells it was possible to observe a concentric distribution of the two genomes, where *H. vulgare* chromosomes tended to occupy a central position in the metaphase plate. These results lead Bennett (1982) and Heslop-Harrison and Bennett (1983a, b) to describe a model of organization of haploid chromosome sets, based on the 'chain linkage' of non-homologous chromosomes according to arm lengths and ratios. Despite the criticism of Dorninger and Timisch (1987) and Callow (1985), regarding the results obtained for odd and even numbers of chromosomes and the difficulty in evaluating the natural variability associated with biological processes, the model has been helpful in predicting the spatial disposition of chromosomes at metaphase.

During interphase, genomes also tend to occupy specific domains in animal hybrid nuclei, as revealed by *in situ* hybridization using total genomic DNA (Manuelidis, 1985, 1990; Schardin *et al.*, 1985; Lichter *et al.*, 1988). In plants, *in situ* hybridization allowed visualization of discrete domains occupied both by entire chromosomes (Heslop-Harrison and Bennett, 1990; Schwarzacher *et al.*, 1992) and translocated arms (Schwarzacher *et al.*, 1992). Nuclear three-dimensional analysis also showed that *Hordeum chilense* and *Secale africanum* genomes tend to lie in discreet nuclear domains in the hybrid (Schwarzacher-Robinson *et al.*, 1987); in this and other hybrids genome separation occurs not only in interphase but throughout the cell cycle and also throughout the plant life cycle (Leitch *et al.*, 1990, 1991; Schwarzacher *et al.*, 1992).

Genomic fluorescent *in situ* hybridization allows discrimination between wheat and rye genomes in triticale nuclei (Fig. 12.3). Through *in situ* hybridization followed by confocal microscopy analysis of the nucleus, we have recently made preliminary investigations of the spatial organization of wheat and rye genomes in root-tip optical sections of wheat × rye hybrids and triticale. The relative distribution of the two genomes was significantly different between squashed and intact nuclei. In squashed triticale nuclei, *in situ* hybridization allowed us to distinguish one or two rye domains in more than 50% of interphase cells; however, in root-tip sections analysed by confocal microscopy only 15% of cells presented the rye genome clustered in one or two domains, with the majority of cells showing the rye and wheat genomes as interspread slices of chromatin (Fig. 12.4a). This reveals that this combined method of *in situ* hybridization and confocal microscopy, at least at this level of resolution, is more suitable for analysing genomic organization than examination of two-dimensional spread preparations, probably due to the maintenance of nuclear integrity. The universal model of somatic association between homologous chromosomes (review in Avivi and Feldman, 1981) was shown to be incorrect after three-dimensional analysis of nuclei. It seems that relative chromosome positioning tends to be modified from prophase until anaphase in humans (Mosgoller *et al.*, 1991). Heslop-Harrison *et al.* (1988) found no evidence for homologous chromosome association in reconstructed nuclei of four diploid or pseudodiploid cereal species.

The natural, relative disposition of wheat and rye genomes in hexaploid triticale revealed here contrasts with intergenomic organization in diploid intergeneric hybrids, where parental genomes segregate into distinct domains (Leitch *et al.*, 1991). In order to evaluate if F1 hybrids and amphiploids could reveal significant differences in intergenomic organization, relative genome disposition was also studied in nuclei of root-tip sections of hexaploid wheat × rye F1 hybrids. The results obtained were not significantly different from those observed in triticale nuclei, indicating that the disomic condition does not affect intergenomic nuclear organization and showing that the differences observed between triticale as an amphiploid and other intergeneric hybrids can not be ascribed to this factor, although the differences between 2x and 3x/6x hybrids may well be important.

Effects of 5-AC incorporation on triticale nuclear organization were also evaluated. After germination of triticale seeds in 5-AC the parental genomes were found usually to be organized quite differently when compared with normal, non treated nuclei. Whereas the latter show defined slices of wheat and rye chromatin, the 5-AC treatment induces general interspersion of chromatin from both genomes throughout the entire nuclear volume in 87% of the cells analysed (Fig. 12.4b).

Thus, as previously observed, 5-AC incorporation leads to DNA hypomethylation at specific rye rDNA intergenic sites causing the activation of rye rRNA genes usually silent in triticale. Moreover, after 5-AC treatment the triticale nuclear topology concerning the relative distribution of the parental genomes also changes from a series of defined and alternate wheat/rye slices to a mixed and interspersed genomic organization.

The spatial organization of chromosomes and genomes within the nucleus seems to correlate with several characteristics such as

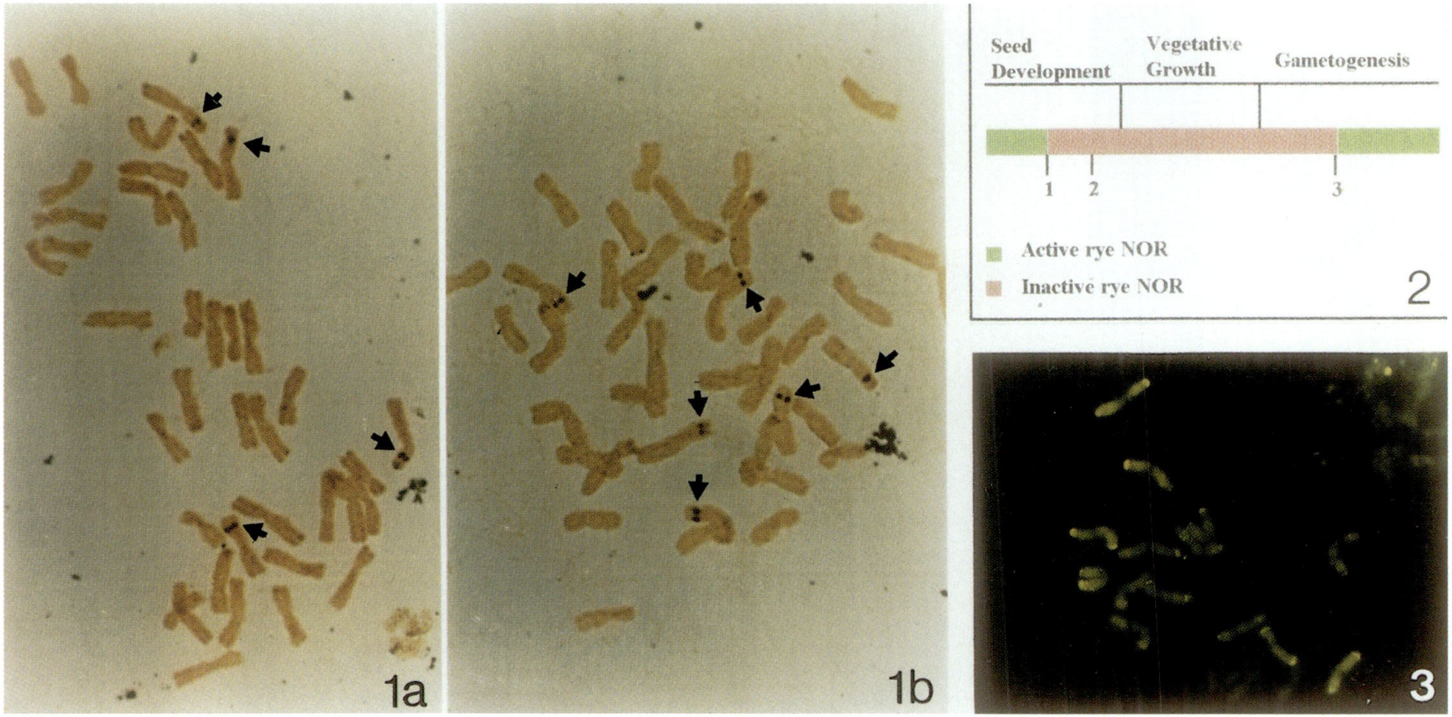

Plate 6 **Fig. 12.1** a–b; **Fig. 12.2**; **Fig. 12.3** (see p.188)

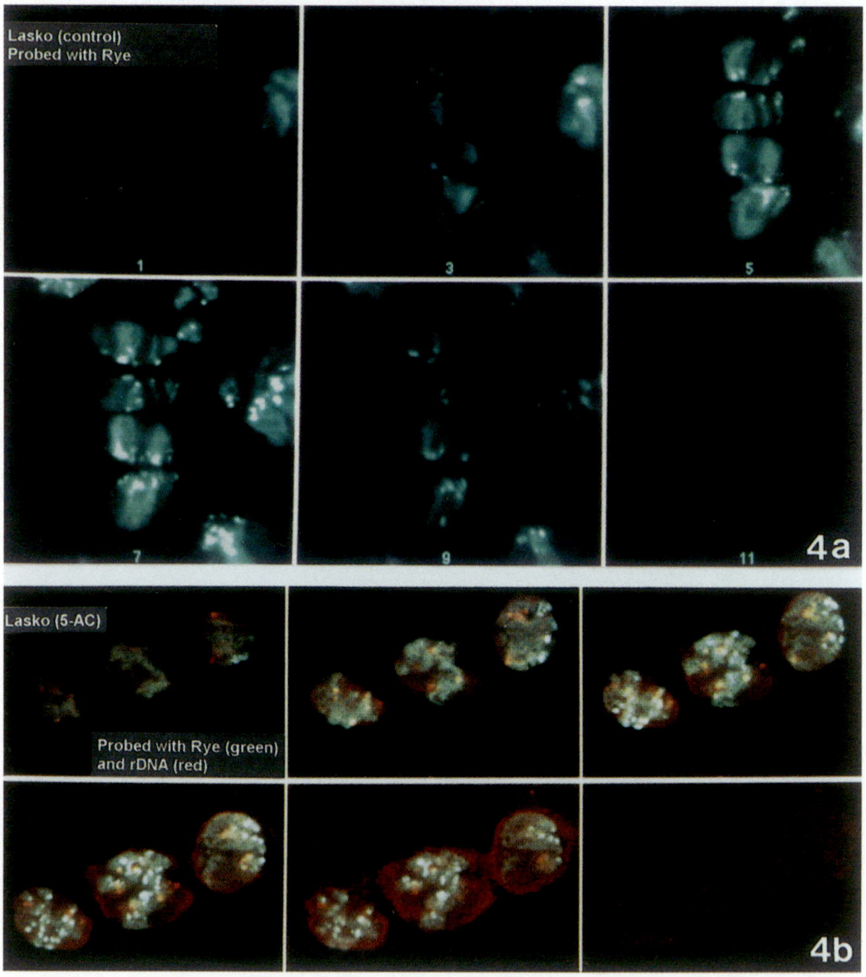

Plate 6 **Fig. 12.4** a–b (see p.188)

differential gene expression (e.g. nucleolar dominance), differential chromosome behaviour and phenotype (Heslop-Harrison and Bennett, 1984; Bennett, 1988; Heslop-Harrison, 1990). In the interspecific hybrid *Hordeum vulgare* × *H. bulbosum*, the NORs of the *H. bulbosum* genome are strongly suppressed (Bennett, 1984); the *H. vulgare* genome, carrying the active NORs in the hybrid, tends to occupy a central domain in the nucleus (Anamthawat-Jónsson *et al.*, 1993), suggesting a possible correlation between chromosome positioning and preferential gene expression.

The reorganization of intergenomic relationships observed in triticale nuclei after 5-AC incorporation can be related to the activation of rye rRNA genes as a consequence of specific DNA hypomethylation induced by this treatment. If DNA hypomethylation leads to the expression of the ribosomal genes of rye, associated with changes in nuclear topology, it seems likely that new genome, chromosome and gene positioning is essential for rye rDNA transcription. This repositioning may be ascribed to the need for proximity of the newly expressing genes to specific protein attachment points, essential for efficient transcription. The fact that not only the rye NOR bearing chromosomes change position may be associated with expression of other unlinked loci, some of which optimize rye NOR expression. The need of species-specific transcription factors, implying the expression of other NOR expression related genes which optimize rRNA gene expression (Reeder, 1985; Neves *et al.*, 1996), may account for the general reorganization of parental genomes observed in triticale nuclei after 5-AC treatment.

12.6 CONCLUSIONS

Interspecific plant hybrids provide a useful model for understanding genomic organization and gene expression because chromosome painting methods (Schwarzacher *et al.*, 1989, 1992; Heslop-Harrison, 1990) enable chromosomes originating from each genome to be identified and followed not only at metaphase but also in interphase when chromosomes are active in gene expression. The parental origin of specific gene activity can also be assessed relatively easily because allelic differences are also usually considerable. The results presented here show that nuclear architecture in such hybrids has many non-random elements, and that the organization can be modulated by changing DNA methylation patterns with 5-AC. Such treatments are shown to have relatively specific effects on expression of rDNA genes, which may lead to dominance of rRNA gene expression from one parent. The generality of such effects is not yet proven, but it seems likely that genomic interactions may play an important role in the control of gene expression in the nucleus and hence have an impact on the success of gene manipulation strategies in plants and animals.

REFERENCES

Abbott, A.G.; Hess, J.E.; and Gerbi, S.A. (1981). Spermatogenesis in *Sciara coprophilis*. I. Chromosome orintation in the monopolar spindle of meiosis. *Chromosoma,* **83**, 1—18.

Anamthawat-Jónsson, K., Schwarzacher, T., and Heslop-Harrison, J.S. (1993) Behaviour of parental genomes in the hybrid *Hordeum vulgare* × *H. bulbosum*. *Journal of Heredity,* **84**, 78—82.

Avivi, L. and Feldman, M. (1981) Arrangement of chromosomes in the interphase nucleus of plants. *Human Genetics,* **55**, 281—95.

Barlow, D.P. (1993) Methylation and imprinting: from host defense to gene regulation. *Science,* **260**, 309—10.

Bennett, M.D. (1982) Nucleotype basis of the spatial ordering of chromosomes in eukariotes and the implications of the order for genome evolution and phenotypic variation. In Genome evolution (eds. Dover G.A. and Flavell, R.B.), Academic Press London, pp. 239—61.

Bennett, M.D. (1984) Nuclear architecture and its manipulation. In Gene manipulation in plant improvement (ed. Gustafson, J.P.), Plenum Publish. Corp., London, pp. 469—502.

Bennett, M.D. (1988) Parental genome separation in F1 hybrids between grass species. In Proc. Kew Chromosome Conference III, (ed. Brandham P.B.), HMSO London, pp. 195—208.

Callow, R.S. (1985) Comments on Bennett's model of somatic chromosome disposition. *Heredity,* **54**, 171—7.

Carmo-Fonseca, M., Tollervey, D., Pepperkok, R. *et al.* (1991) Mammalian nuclei contain foci which are highly enriched in components of the pre-mRNA splicing machinery. *EMBO Journal,* **10**, 195—204.

Carter, K.C., Bowman, D., Carrington, W. *et al.* (1993) A three-dimensional view of precursor messenger RNA metabolism within the mammalian nucleus. *Science,* **259**, 1330—5.

Castilho, A., Queiroz, A., Neves, N. *et al.* (1995) The developmental stage of inactivation of rye origin rRNA genes in the embryo and endosperm of wheat × rye F1 hybrids. *Chromosome. Research*, **3**(3), 169—74.

Ceccarelli, M. and Cionini, P.G. (1993) Tissue-specific nuclear repatterning in plant cells. *Genome,* **36**, 1092—8.

Cermeño, M.C., Orellana, J., Santos, J.L., and Lacadena, J.R. (1984) Nucleolar activity in wheat, rye and derivatives analyzed by a silver staining procedure. *Chromosoma,* **89**,370—6.

Cimino, M.C. (1972) Egg production, polyploidization and evolution in a diploid all-female fish of the genus *Poeciliopsis*. *Evolution,* **26**, 294—306.

Comings, D.E. (1968) The rationale for an ordered arrangement of chromatin in the interphase nucleus. *American Journal of Human Genetics,* **29**, 440—60.

Cremer, T., Kurz, A., Zirbel, R. *et al.* (1993) Role of chromosome territories in the functional compartmentalization of the cell nucleus. *Cold Spring Harbour Symposium of Quantitative Biology,* **58**, 777—92.

Davies, D.R. (1958) Male parthenogenesis in barley. *Heredity,* **12**, 493—8.

Dorninger, D. and Timisch, W. (1987) Geometrical constraints on Bennett's prediction of chromosome order. *Heredity,* **58**, 321—5.

Ferguson-Smith, A.C. and Surani, M.A. (1993) Parental imprinting in mouse development. In *John Innes Review — The Chromosome* (eds. J.S. Heslop-Harrison, R.B. Flavell), Bios Scient. Publish., Oxford, pp 207—20.

Finch, R.A. (1983) Tissue specific elimination of alternative whole parental genomes in one barley hybrid. *Chromosoma,* **88**, 386—93.

Finch R.A. and Bennett, M.D. (1983) The mechanism of somatic chromosome elimination in *Hordeum*. In Proc. Kew Chromosome Conference II (eds. Brandham, P.E., Bennett, M.D.), G. Allen and Unwin Ltd. London, pp. 147—54.

Finch, R.A., Smith, J.B. and Bennett, D. (1981) *Hordeum* and *Secale* mitotic genomes lie apart in a hybrid. *Journal of Cell Science,* **52**, 391—403.

Flavell, R.B. and O'Dell, M. (1979) The genetic control of the nucleolus formation in wheat. *Chromosoma,* **71**, 135—52.

Flavell, R.B. and O'Dell, M. (1990) Variation and inheritance of cytosine methylation patterns in wheat at the high molecular weight glutenin and ribosomal RNA gene loci. *Development* (Supplement), 15—20.

Flavell, R.B., O'Dell, M. and Thompson, W.F. (1988) Regulation of cytosine methylation in ribosomal DNA and nucleolus organizer excpression in wheat. *Journal of Molecular Biology,* **204**, 523—34.

Fujigaki, J. and Tozu, T. (1987) Morphological and cytological studies of barley-rye hybrids *Hordeum vulgare* L. × *Secale africanum* Stapf. *Barley Genetics V,* (eds. Yasuda S. Konishi T.), pp. 325—35.

Goodpasture, C. and Bloom, S.E. (1975) Visualization of nucleolar organiser regions in mammalian chromosomes using silver staining. *Chromosoma* **53**, 37—50.

Hassan, A.B., Errington, R.J., White, N.S. *et al.* (1994) Replication and transcription sites are colocalized in human cells. *Journal of Cell Science,* **107**, 425—34.

Heppich, S., Tunner, H.G. and Greilhuber, J. (1982) Premeiotic chromosome doubling after genome elimination during spermatogenesis in the species hybrid *Rana esculenta*. *Theoretical and Applied Genetics,* **61**, 101—4.

Heslop-Harrison, J.S. (1990) Gene expression and parental dominance in hybrid plants. *Development* (Supplement), 21—8.

Heslop-Harrison, J.S. and Bennett, M.D. (1983a) Prediction and analysis of spatial order in haploid chromosome complements. *Proceedings of the Royal Society of London Series B.* **218**, 211—23.

Heslop-Harrison, J.S. and Bennett, M.D. (1983b) The spatial order of chromosomes in root-tip metaphases of *Aegilops umbellulata*. *Proceedings of the Royal Society of London Series B,* **218**, 225—39.

Heslop-Harrison, J.S. and Bennett, M.D. (1984) Chromosome order — possible implications for development. *Journal of Embryology and Experimental Morphology,* **83**, 51—73.

Heslop-Harrison, J.S. and Bennett, M.D. (1990) Nuclear architecture in plants. *Trends in Genetics,* **6**, 401—5.

Heslop-Harrison, J.S. and Schwarzacher, T. (1993) Molecular cytogenetics — biology and applications in plant breeding. *Chromosomes Today,* **11**, 191—8.

Heslop-Harrison, J.S., Smith, J.B. and Bennett, M.D. (1988) The absence of the somatic association of centromeres of homologous chromosomes in grass mitotic metaphases. *Chromosoma,* **96**, 119—31.

Hubbell, H.R. (1985) Silver staining as an indicator of active ribosomal genes. *Stain Technology,* **60**, 285—94.

Hochstrasser, M. and Sedat, J.W. (1987) Three-dimensional organization of *Drosophila melanogaster* nuclei. II. Chromosome spatial organization and gene regulation. *Journal of Cell Biology,* **104**, 1471—83.

Holliday, R. (1984) The biological significance of meiosis. In Controlling events in meiosis (eds. Evans C. and Dickenson, H.), Society for Experimental Biology, London, pp. 381—94.

Holliday, R. (1987) Ageing: X-chromosome inactivation. *Nature,* **327**, 661—2.

Jackson, D.A. (1991) Structure — function relationships in eukaryotic nuclei. *BioEssays,* **13**, 1—10.

Jiménez, R., Burgos, M. and Diaz de la Guardia, R. (1988) A study of the Ag-staining significance in mitotic NOR's. *Heredity,* **60**, 125—7.

Jones, P.A. and Taylor, S.M. (1980) Cellular differentiation, cytidine analogs and DNA methylation. *Cell,* **20**, 85—93.

Kasha, K.J. and Kao, K.W. (1970) High frequency of haploid production in barley (*Hordeum vulgare* L.). *Nature,* **225**. 874—6.

Lacadena, J.R., Cermeño, M.C., Orellana, J. and Santos, J.L. (1984a) Evidence for wheat-rye nucleolar competition (amphiplasty) in triticale by silver staining procedure. *Theoretical and Applied Genetics,* **67**, 207—13.

Lacadena, J.R., Cermeño, M.C., Orellana, J. and Santos, J.L. (1984b) Analysis of nucleolar activity in *Agropyron elongatum*, its amphiploid with *Triticum aestivum* and the chromosome addition lines. *Theoretical and Applied Genetics,* **68**, 75—80.

Lange, W. (1971) Crosses between *Hordeum vulgare* and *H. bulbosum*. II. Elimination of chromosomes in hybrid tissues. *Euphytica,* **20**, 181—94.

Leitch, A.R., Mosgoller, W., Schwarzarcher, T. *et al.* (1990) Genomic *in situ* hybridization to sectioned nuclei shows chromosome domains in grass hybrids. *Journal of Cell Science,* **95**, 335—41.

Leitch, A.R., Schwarzarcher, T., Mosgoller, W. *et al.* (1991) Parental genomes are separated throughout the cell cycle in a plant hybrid. *Chromosoma* **101**, 206—13.

Lichter, P., Cremer, T., Borden, J. *et al.* (1988) Delineation of individual human chromosomes in metaphase and interphase cells by in situ suppression hybridization using recombinant DNA libraries. *Human Genetics,* **80**, 224—34.

Manuelidis, L. (1985) Individual interphase chromosome domains revealed by in situ hybridization. *Human Genetics,* **71**, 288—93.

Manuelidis, L. (1990) A view of interphase chromosomes. *Science* **250**, 1533—9.

Manuelidis, L. and Borden, J. (1988) Reproducible compartmentalization of individual chromosome domains in human CNS cells revealed by in situ hybridization and three-dimensional reconstruction. *Chromosoma* **96**, 397—410.

Martini, G. and Flavell, R. 1985. The control of the nucleolus volume in wheat, a genetic study of three developmental stages. *Heredity,* **54**, 111—120.

Martini, G., O'Dell, M. and Flavell, R.B. (1982) Partial inactivation of wheat nucleolus organizers by the nucleolus organizer chromosomes from *Aegilops umbellulata*. *Chromosoma,* **84**, 687—700.

Monk, M. (1987) Genomic imprinting: memories of the mother and the father. *Nature,* **328**, 203—4.

Mosgoller, W., Leitch, A.R., Brown, J.K.M. *et al.* (1991) Chromosome arrangements in human fibroblasts at mitosis. *Human Genetics,* **88**, 27—33.

Nelson-Rees, W.A., Hoy, M.A. and Roush, R.T. (1980) Heterochromatinization, chromatin elimination and haploidization in the parahaploid mite *Metasciulus occidentalis*. *Chromosoma,* **77**, 263—76.

Neves, N., Barão, A., Castilho, A. *et al.* (1992) Influence of DNA methylation on rye B-chromosome nondisjunction. *Genome,* **35**, 650—2.

Neves, N., Heslop-Harrison, J.S. and Viegas, W. (1995) rRNA genes activity and control of expression mediated by methylation and imprinting during embryo development in wheat × rye hybrids. *Theoretical and Applied Genetics,* **91**, 529—33.

Neves, N., Silva, M., Heslop-Harrison, J.S. and Viegas, W. (1996) Nucleolar dominance in triticale: control by unlinked genes. *Genes and Development* (in press).

Norum, R.A. and Migeon, B.R. (1974) Non-random loss of human markers from man-mouse somatic cell hybrids. *Nature,* **251**, 72—4.

Odartchenko, N. and Keneklis,T. (1973) Localization of parental DNA in interphase nuclei of mouse eggs during early cleavage. *Nature,* **241**, 193—4.

Reeder, H.R. (1985) Mechanisms of nucleolar dominance in animals and plants. *Journal of Cell Biology,* **101**, 2013—6.

Rubiales, D., Ballesteros, J. and Martin, A. (1992) Resistance to *Septoria tritici* in *Hordeum chilense* × *Triticum* amphiploids. *Plant Breeding,* **109**, 281—6.

Rubiales, D., Niks, R.E. and Martin, A. (1993) Genomic interactions in the resistance to mildew and rust fungi in hybrids and amphiploids involving the genera *Triticum, Hordeum* and *Secale*. *Cereal Research Communications,* **21**, 187—94.

Sardana, R., O'Dell, M. and Flavell, R.B. (1993) Correlation between the size of the intergenic regulatory region, the status of cytosine methylation of rRNA genes and nucleolar expression in wheat. *Molecular and General Genetics,* **236**, 155—62.

Schardin, M., Cremer,T., Hager, H.D. and Lang, M. (1985) Specific staining of human chromosomes in Chinese hamster × man hybrid cell lines demonstrates interphase chromosome territories. *Human Genetics,* **71**, 281—7.

Schwarzacher, H.G., Mikelsaar, A.V. and Schnedl, W. (1978) The nature of Ag-staining of nucleolus organizer regions. Electron- and light-microscopic studies on human cells in interphase, mitosis and meiosis. *Cytogenetics and Cell Genetics,* **20**, 24—39.

Schwarzacher, T., Leitch, A.R., Bennett, M.D., and Heslop-Harrison, J.S. (1989) *In situ* localization of parental genomes in a wide hybrid. *Annals of Botany,* **64**, 315—24.

Schwarzacher, T., Heslop-Harrison, J.P., Anamthawat-Jonsson, K. *et al.* (1992) Parental genome separation in reconstructions of somatic and premeiotic metaphases of *Hordeum vulgare* × *H. bulbosum*. *Journal of Cell Science,* **101**, 13—24.

Schwarzacher-Robinson, T., Finch, R.A., Smith, J.B. and Bennett, M.D. (1987) Genotypic control of centromere positions of parental genomes in *Hordeum* × *Secale* hybrid metaphases. *Journal of Cell Science,* **87**, 291—304.

Silva, M., Queiroz, A., Neves, N. *et al.* (1995) Reprogramming of rye rDNA in triticale during microsporogenesis. *Chromosome Research,* **3**(8), 492—7.

Somers, D.J. and Filion, W.G. (1994) The influence of the rye genome on the accumulation of HSP18 and HSP70 transcripts in a wheat genetic background. *Theoretical and Applied Genetics,* **88**, 298—304.

Somers, D.J., Gustafson, J.P. and Filion, W.G. (1992) The influence of the rye genome on expression of heat-shock proteins in triticale. *Theoretical and Applied Genetics* **83**, 987—93.

Spector, D.L. (1990) Higher ordered nuclear organization — 3-Dimensional distribution of small nuclear ribonucleoproteins particles. *Proceedings of the National Academy of Science USA.,* **87**, 147—51.

Subrahmanyam, N.C. and Kasha, K.J. (1973) Selective chromosome elimination during haploid formation in barley following interspecific hybridization. *Chromosoma,* **42**, 111—25.

Thomas, J. and Kaltsikes P.J. (1983) Effect of chromosomes 1A and 6B on nucleolus formation in hexaploid triticales. *Canadian Journal of Genetics and Cytology,* **25**, 292—7.

Vig, B.K. (1981) Sequence of centromere separation: analysis of mitotic chromosomes in man. *Human Genetics,* **57**, 247—52.

Xing, Y. and Lawrence, J.B. (1993) Nuclear RNA tracks: Structural basis for transcription and splicing. *Trends in Cell Biology,* **3**, 346—53.

13

Gametic imprinting at the mouse and human IGF2R/MPR300 locus

O.W. Smrzka and D.P. Barlow

13.1 INTRODUCTION

Mammals inherit a maternal and a paternal homologue of every autosomal gene. Remarkably, a small number of autosomal loci are not equally expressed from the maternal and paternal allele. Instead, a phenomenon known as gametic (or genomic) imprinting, acts to restrict expression to one parental allele. Since the possession of two functional gene copies confers an obvious advantage in protecting the organism from genetic mistakes, the existence of imprinted genes has stirred considerable interest. The most burning question of 'why' gametic imprinting exists in mammals is difficult to address directly. However, supplementary questions, such as the type of gene that is imprinted and the molecular mechanisms involved in allele-specific expression, can be directly investigated. For the purposes of this chapter, gametic imprinting is defined as 'a reversible process whereby a gamete-imposed modification in the parental generation can lead to functional differences between the maternal and paternal allele in the offspring' (Barlow, 1994). Imprinting is thus a phenomenon that can be traced back to differential activities of the parental gametes which result in autosomal genes showing parent-of-origin effects. The first imprinted genes were isolated in 1991 but since then, a total of 14 have been isolated and of these 4 are maternally, and 10 paternally, expressed (see Table 15.1). Already sufficient analyses have been done to outline some features of the imprinting mechanism and support the hypothesis, suggested from an analysis of androgenetic and gynogenetic embryos (McGrath and Solter, 1984, Surani *et al.*, 1984), that

Table 13.1 Mammalian imprinted genes.

Gene	*Expressed allele*	*Phenotype by gene knockout*	*Reference*
IGF2R/MPR300	maternal	embryonic growth/lysosomes	1
IGF2	paternal	embryonic growth	2
H19	maternal	embryonic growth	3
MASH2	maternal	placental development	4
WT1***	maternal	organ development/ lethal E15.5	5, 6
INS***	paternal	ND*	7, 8
XIST	maternal/paternal	ND**	9
MAS****	paternal	ND	10
U2afbprs (SP2)	paternal	ND	11, 12
SNRPN	paternal	ND	13
ZNF127	paternal	ND	14
PAR1	paternal	ND	15
PAR5	paternal	ND	15
IPW	paternal	ND	16

ND, not done, *mutations in the human insulin receptor suggest a function in fetal growth. **associated with X chromosome inactivation. *** limited monoallelic expression, 1: Barlow *et al.*, 1991, 2: DeChiara *et al.*, 1991, 3: Leighton *et al.*, 1995, 4: Guillemont *et al.*, 1994, 5: Jinno *et al.*, 1994, 6: Kreidberg *et al.*, 1993, 7: Giddings *et al.*, 1994, 8: Wertheimer *et al.*, 1993, 9: Kay *et al.*, 1994, 10: Villar and Pederson 1994, 11: Hatada *et al.*, 1993, 12: Hayashizaki *et al.*, 1993, 13: Cattanach *et al.*, 1992, 14: Nicholls *et al.*, 1994, 15: Sutcliffe *et al.*, 1994, 16: Wevrick *et al.*, 1994.

imprinted genes could share a common function in regulating embryonic and placental growth. While most attention has centered on imprinted genes, other processes such as sister-chromatid asynchrony, trinucleotide repeat amplification and DNA methylation, also show parent-of-origin effects. However, it will not be clear until the molecular mechanism of imprinting is better understood, whether these latter processes are 'imprinted' or merely reflect a difference in mutability between male and female gametes. This chapter concentrates on an analysis of the IGF2R/MPR300 gene and describes data that suggest that results from IGF2R/MPR300 can be used as a model for all imprinted genes.

13.2 IGF2R/MPR300

The mouse Insulin-like growth factor type 2 receptor (IGF2R) was first identified as a 'cation-independent mannose 6-phosphate receptor' (MPR300), a receptor involved in transporting mannose 6-phosphate tagged proteins to lysosomes which is now referred to as either IGF2R or MPR300 (Morgan *et al.*, 1987, Oshima *et al.*, 1988). The MPR300 receptor binds IGF2 in mammals but not in other vertebrates, and, after binding, IGF2 is transported to lysosomes as part of a mechanism to reduce plasma concentration (von Figura, 1991). The function of the IGF2R/MPR300 gene has been the subject of intense study for many years because of the interest in human diseases involving abnormalities of lysosome function (Nolan and Sly, 1989). Thus it has been clearly shown to act in the Golgi apparatus and at the cell surface where it plays a role in intra- and extracellular transport of mannose 6-phosphate tagged proteins into lysosomes. However, transport of IGF2 does not depend on mannose 6-phosphate (von Figura, 1991). The function of the IGF2R/MPR300 gene has recently been analyzed in mouse knockout mutants that specifically lack a functional gene (Wang *et al.*, 1994 and Lau *et al.*, 1994). Interestingly, these experiments show that the primary role of IGF2R/MPR300 may be different for embryos and adults — acting in the embryo to inhibit growth by reducing plasma IGF2 levels and in the adult to maintain lysosome function by transporting lysosomal enzymes. The mouse gene maps to the proximal part of chromosome 17 and the human gene to the distal end of chromosome 6q (Laureys *et al.*, 1988) and both are highly conserved in sequence and in genomic organization (Ludwig *et al.*, 1995, Oshima *et al.*, 1988, Stöger *et al.*, 1993, Szebenyi and Rotwein, 1994, Smrzka *et al.*, 1995).

13.3 PARENT-OF-ORIGIN EFFECTS AT THE IGF2R/MPR300 LOCUS

13.3.1 Allele-specific expression

Several mouse mutants, including chromosome deletions, such as $T^{hairpin}$ (T^{hp}) and IGF2R/MPR300 knockout mice, have been used to demonstrate that the mouse IGF2R/MPR300 gene is imprinted and maternally expressed. The T^{hp} deletion removes IGF2R/MPR300, plus approximately 5 Mbp of flanking DNA. This deletion is lethal when maternally inherited and was initially used to demonstrate the imprinted status of IGF2R/MPR300 (Barlow *et al.*, 1991). More recently IGF2R/MPR300 knockout mutant mice, that show a similar phenotype to the T^{hp} mutant, have been used to demonstrate that this locus does not always show maternal-specific expression. In many situations biallelic expression (i.e., expression of both parental alleles) can be seen. For example, expression of the paternal allele is seen in embryonic stem cells and in some rare adult mice that survive maternal inheritance of the gene knockout (Wang *et al.*, 1994). Thus repression of the paternal allele is developmentally regulated in mice.

The human IGF2R/MPR300 gene differs from the mouse because it appears to be normally expressed from both parental alleles. To date, biallelic expression has been identified in a variety of human embryonic and adult tissues (Kalscheuer *et al.*, 1993, Ogawa *et al.*, 1993), and in 69/70 informative lymphoblastoid cell lines (Smrzka *et al.*, in press). However, it has also been reported that maternal-specific (monoallelic) expression was seen in 2/14 foetal samples (Xu *et al.*, 1993), suggesting that imprinting of IGF2R/MPR300 could be a polymorphic trait in the human population. The allelic-expression profiles of the mouse and human IGF2R/MPR300 genes is shown in Table 13.2.

Analysis of other imprinted genes supports the argument that allelic repression of imprinted genes is plastic in development and in disease. For example, biallelic expression of IGF2 has been demonstrated in preimplantation mouse embryos and in some differentiated mouse tissues (Latham *et al.*, 1994, DeChiara *et al.*, 1991) and also in human tumours as well as in normal human adult liver (Rainier and Feinberg, 1994, Vu *et al.*, 1994). Re-expression of the silent allele of an imprinted gene could arise either from changes to the imprint itself (Rainier and Feinberg, 1994) or changes in the transcription machinery

Table 13.2 Allele-specific expression of the IGF2R/MPR300 gene.

	maternal	*paternal*	*mono- or biallelic*
Preimplantation mouse embryos*	+	+	bi
Mouse embryonic stem cells	+	+	bi
Post implantation mouse embryos	+	—	mono
Adult mouse	ND	ND	ND
Adult mouse (maternal knock out)*	KO	++	bi
Foetal human tissues	+	12/14	sporadic mono
Adult human tissues and LB cell lines	+	+	bi

* the paternal allele is expressed in preimplantation embryos and re-expressed in rare mice that survive a maternally-inherited IGF2R/MPR300 knockout (Wang *et al.*, 1994), LB; lymphoblastoid, ND; not done, KO; mouse knock out.

which overcome the imprint. The latter case has been shown to occur when the silent IGF2 allele is re-expressed from an alternative promoter in normal adult liver (Vu *et al.*, 1994).

13.3.2 Allele-specific methylation

The action of the only known mammalian methyltransferase, plus the well-known link between hypermethylation and gene repression, has long suggested that methylation may be part of the cis-acting molecular mechanism that produces allele-specific expression of imprinted genes (Monk 1987). Examination of the IGF2R/MPR300 locus in the human and mouse shows that each possesses two CpG islands (Fig. 13.1), one island spans the promoter and first exon and the second lies in the second intron of the gene.

CpG 'islands' are genomic regions up to 1000 bp long, commonly associated with genes and rich in CpG dinucleotides (Cross and Bird 1995, Larsen *et al.*, 1992). Normally CpG islands are unmethylated, which has been attributed to the presence of Sp1 sites (Macleod *et al.*, 1994, Brandeis *et al.*, 1994). Both the promoter and intron CpG islands

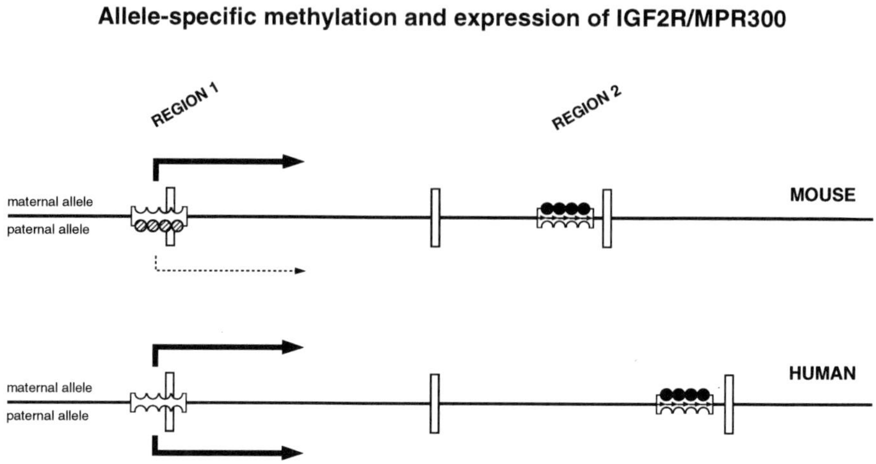

Fig. 13.1 CpG islands in the human and mouse IGF2R/MPR300 genes. The promoter and intron CpG islands each contain a CG:GC ratio of > 0.6 for approximately 1 kb. Human region 1 contains 3 potential SpI sites, mouse region 1 contains 8 potential SpI sites, human region 2 contains 4 potential SpI sites and mouse region 2 contains 2 potential SpI sites. Open boxes denote the first 3 gene exons, which span 29 kb in the mouse and 41 kb in the human. Arrowheads within region 2 indicate the presence of direct repeats, filled circles — maternal methylation inherited from the gamete and stable in development, shaded circles — paternal methylation acquired after implantation. Transcription is indicated by the large arrows originating at region 1. The dashed arrow at the paternal mouse allele indicates a low level of expression.

in the IGF2R/MPR300 gene fit the parameters for standard CpG islands and also contain several SpI sites (Fig. 13.1). However, in contrast to typical autosomal CpG islands, those in the IGF2R/MPR300 gene are subject to allele-specific methylation. The mouse promoter CpG island is methylated on the paternal chromosome, but only late in post implantation development (Stöger *et al.*, 1993) and not in all adult tissues (unpublished data). Thus methylation of the promoter island is considered at this time, to be secondary to the silencing of the paternal allele, however this has not yet been directly demonstrated. The promoter CpG island of the human IGF2R/MPR300 gene, that does not normally show allele-specific expression (Table 13.2), is free of methylation. The intronic

CpG island (region 2), is methylated on the maternal chromosome and free of methylation on the paternal chromosome in both the mouse and human gene (Stöger *et al.*, 1993, Smrzka *et al.*, 1995). Part of the methylation on M-region 2 is inherited from the female gamete (Brandeis *et al.*, 1993) and the prediction that methylation of region 2 could serve as a gametic imprint is currently being tested in transgenic mice. The intronic CpG island (M-region 2 and H-region 2) share little sequence homology (Smrzka *et al.*, 1995) but each contain a large number of direct repeats varying from 20—75 bp in size (arrowheads in Fig. 13.1).

DNA methylation was suggested to modify an imprinted gene during male or female gametogenesis in a manner that could act later in development to silence the locus. This hypothesis was supported by the observation that all imprinted genes and transgenes so far examined, were associated with allele-specific methylation (Razin and Cedar, 1994). The strongest support for a role for DNA methylation comes from experiments using mice that have an inactive methyltransferase gene (Li *et al.*, 1993). This work showed that DNA methylation is necessary for the allele-specific expression of imprinted genes. In its absence, the paternally repressed H19 gene becomes active. Thus for this gene, as expected methylation is necessary for repression. However, the involvement of methylation in imprinting is more complicated than predicted because these experiments also showed that the maternally expressed IGF2R/MPR300 gene and the paternally expressed IGF2 gene become completely silent in the absence of methylation. Thus for these latter two genes, in contrast to expectation, methylation appears necessary for their expression. At first sight these results may appear contradictory. However, such opposite behaviour can be explained by assuming that in both scenarios methylation is performing one function — that of changing the access of DNA binding proteins. In the case of H19, methylation could be presumed to inhibit the access of an enhancer and in the case of IGF2 and IGF2R, a repressor would be inhibited. The allele-specific methylation pattern at the IGF2R/MPR300 gene could cause maternal expression in two ways (Fig. 13.2), however, the fact that this gene is initially biallelically expressed and then maternally expressed supports the suggestion that methylation acts to block the repression of the active maternal allele.

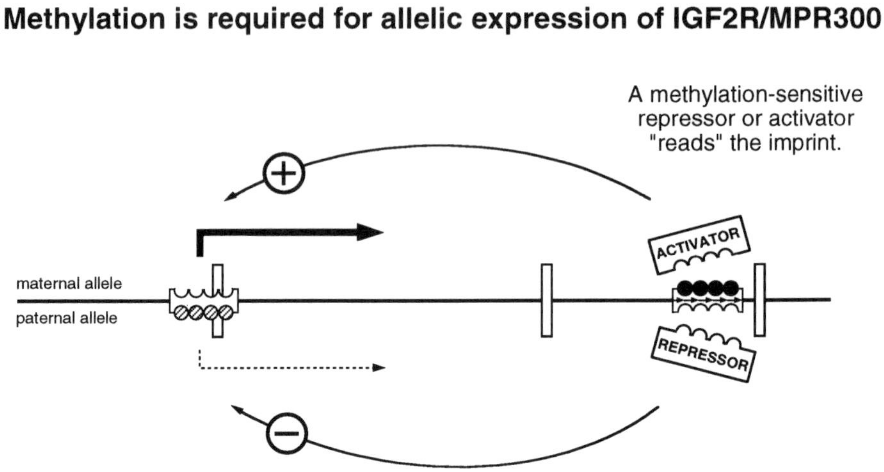

Fig. 13.2 The maternal allele is shown above the line and the paternal allele is shown below. Maternal-specific methylation could either allow the binding of a transcriptional activator leading to maternal-specific transcription or inhibit the binding of a repressor. The unmethylated paternal allele would then bind the repressor at region 2 and transcription would be inhibited from region 1. Region 1 is then subject to secondary methylation following the silencing of the allele. Symbols as Fig. 13.1.

13.3.3 Allele-specific sister-chromatid asynchrony

In addition to the link between methylation and gene activity, there also exists a link between chromosome replication timing in the cell cycle and gene activity (Goldman *et al.*, 1984). This feature has recently been examined for imprinted gene loci using FISH (Fluorescent *in situ* hybridization), and as expected, all imprinted loci so far examined show asynchrony in sister-chromatid behavior in S-phase (Kitsberg *et al.*, 1993, Gunaratne *et al.*, 1995). A similar FISH analysis of autosomal loci showing allelic exclusion, such as the immunoglobulin or olfactory receptor gene complexes, also revealed similar asynchrony (Efstratiadis, 1994). However, imprinted genes differ from those showing allelic exclusion because there is no correlation between allele asynchrony and expression — at imprinted loci the paternal allele shows early duplication of the hybridisation signal regardless of expression. Thus the expressed maternal IGF2R/MPR300 locus is late duplicating (has a single FISH hybridisation signal) in the mouse. The human IGF2R/MPR300 locus also shows similar asynchrony (Kitsberg *et al.*, 1993, Smrzka *et al.*, 1995) despite the finding that this gene only rarely shows allele-specific expression. An additional feature of the sister-chromatid asynchrony of imprinted genes is that the region involved is quite extensive, for example, up to 3 cM for the mouse IGF2R/MPR300 locus, and includes non-imprinted genes (Kitsberg *et al.*, 1993, Barlow *et al.*, 1991).

13.4 IGF2R/MPR300 — A MODEL IMPRINTED GENE?

The analysis of the human and mouse IGF2R/MPR300 genes has led to two concepts concerning the molecular basis of imprinting. Firstly, the identification of allelic methylation on the maternal allele in mice demonstrated that expressed alleles can carry imprints and allowed the prediction that for some imprinted genes methylation was necessary for expression (Stöger *et al.*, 1993). The analysis of mouse mutants deficient in methylation (Li *et al.*, 1993) confirmed this prediction. These experiments showed that for IGF2R/MPR300 and IGF2 methylation was necessary for allelic-expression, but in the case of H19 methylation was necessary for allelic-repression. Secondly, the identification of an intronic CpG island rich in direct repeats as carrying the methylation imprint in IGF2R/MPR300, led to the proposal that this element could act as the 'imprinting box' for this gene. The observation that direct repeats do not occur in the IGF2R/MPR300 promoter island, nor in the islands of non-imprinted genes, allowed the prediction that repeat-induced secondary structure could play a role in attracting maternal-specific methylation (Smrzka *et al.*, 1995). While the relevance of region 2 in imprinting the mouse gene is currently being tested in transgenic mice, the significance of the repeats has been tested by scanning the sequence of known imprinted genes for the presence of direct repeats associated with allele-specific methylation. Surprisingly, many other imprinted genes and transgenes share this feature, albeit to different degrees (Neumann *et al.*, 1995). In addition, a recent analysis of an imprinted transgenes (Chaillet *et al.*, 1995) has demonstrated that repeats play a significant role in imprinting this gene. The association of direct repeats with allelic methylation has suggested possible strategies to clone novel imprinted genes (Neumann *et al.*, 1995).

13.5 SUMMARY

The mouse and human IGF2R/MPR300 genes show several parent-of-origin effects. Both genes show sister-chromatid asynchrony and allele-specific methylation patterns. However, while mice show developmentally-regulated allele-specific expression, humans may only show sporadic monoallelic expression in a minority of the population. Thus, the question of whether the human IGF2R/MPR300 gene can be described as 'imprinted', remains open, since the true extent of replication asynchrony and allele-specific methylation in the remainder of the genome are unknown. A recent paper describing the linkage of a disease phenotype (Transient Neonatal Diabetes Mellitus) with uniparental paternal disomy of chromosome 6q (Temple *et al.*, 1995), does support the existence of an imprinted gene on this chromosome. However, this disease linkage could be explained as arising from other imprinted genes on 6q. In support of this, the MAS gene that flanks the mouse IGF2R/MPR300 locus, has recently been described as imprinted and paternally expressed in a small number of tissues in mice (Villar and Pederson, 1994). Despite the possibility of additional imprinted genes, the existence in the human IGF2R/MPR300 gene of allele-specific methylation of an element homologous to the proposed imprinting box for the mouse gene (Smrzka *et al.*, 1995, Stöger *et al.*, 1993) clearly shows that human gene is imprinted by methylation. This finding, combined with the report of sporadic allele-specific expression (Xu *et al.*, 1993) suggests, if the proposed role for the mouse imprinting box is confirmed experimentally, that the human IGF2R/MPR300 has the potential for monoallelic expression.

ACKNOWLEDGMENTS

We thank Heinz Himmelbauer, Walter Lerchner, Bjorn Neumann, Norbert Schweifer and Anton Wutz for insights and arguments about the imprinting process and Bjorn Neumann and Koichi Matsuo for reading the manuscript.

REFERENCES

Barlow, D.P., Stöger R., Herrmann, B.G. *et al.* (1991) The mouse insulin-like growth factor type 2 receptor is imprinted and closely linked to the Tme locus. *Nature,* **349**, 84—7.

Barlow, D.P., (1994) Imprinting: a gamete point of view. (Perspective) *Trends in Genetics,* **10,** 194—8.

Brandeis, M., Kafri, T., Ariel, M. *et al.* (1993) The ontogeny of allele-specific methylation associated with imprinted genes in the mouse. *EMBO Journal,* **12,** 3669—77.

Brandeis, M., Frank, D., Keshet, I. *et al.* (1994) Sp1 sites protect a CpG island from de novo methylation. *Nature,* **371**, 435—8.

Cattanach, B.N., Barr, J.A., Evans, E.P. *et al.* (1992) A candidate mouse model for Prader—Willi syndrome which shows an absence of Snrpn expression. *Nature Genetics,* **2**, 270—4.

Chaillet, J.R., Bader, D.S. and Leder, P. (1995) Regulation of genomic imprinting by gametic and embryonic processes. *Genes and Development,* **9**, 1177—87.

Cross, S.H. and Bird, A.P. (1995) CpG islands and genes. *Current Opinion in Genetics and Development,* **5**, 309—14.

DeChiara, T.M., Robertson, E.J. and Efstratiadis, A. (1991) Paternal imprinting of the mouse insulin-like growth factor II gene. *Cell,* **64**, 849—59.

Efstratiadis, A. (1994) Parental imprinting of autosomal mammalian genes. *Current Opinion in Genetics and Development,* **4**, 265—80.

Giddings, S.J., King, C.D., Harman, K.W. *et al.* (1994) Allele-specific inactivation of insulin 1 and 2, in the mouse yolk sac indicates imprinting. *Nature Genetics,* **6**, 310—3.

Goldman, M.A., Holmquist, G.P., Gray, M.C. *et al.* (1984) Replication timing of mammalian genes and middle repetitivie sequences. *Science,* **224**, 686—92.

Guillemont, F., Caspary, T., Tilghman, S.M. *et al.* (1994) Genomic imprinting of Mash 2, a mouse gene required for trophoblast development. *Nature Genetics,* **9**, 235—41.

Gunaratne, P.H., Nakeo, M., Ledbetter, D.H. *et al.* (1995) Tissue-specific and allele-specific replication timing control in the imprinted human Prader—Willi syndrome region. *Genes and Development,* **9**, 808—20.

Hatada, I., Sugama, T. and Mukai, T. (1993) A new imprinted gene cloned by a methylation sensitive genome scanning method. *Nucleic Acid Research* **21**, 5577—82.

Hayashizaki, Y., Shibata, H., Hirotsune, S. *et al.* (1993) Identification of an imprinted U2af binding protein related sequence on mouse chromosome 11 using the RLGS method. *Nature Genetics,* **6**, 33—9.

Jinno, Y., Yun, K., Nishiwaki, K., Kuboto, T., Ogawa, O., Reeve, A.E., Niikawa, N. (1994) Mosaic and polymorphic imprinting of the WT1 gene in humans. *Nature Genet.,* **6**, 305—9.

Kalscheuer, V.M., Mariman, E.C., Schepens, M.T. *et al.* (1993) The insulin like growth factor type-2 receptor gene is imprinted in the mouse but not in humans. *Nature Genetics,* **5**, 74—8.

Kay, G.F., Barton, S.C., Surani, M.A. and Rastan, S. (1994) Imprinting and X chromosome counting mechanisms determine Xist expression in early mouse development. *Cell,* **77**, 639—50.

Kitsberg, D., Selig, S., Brandeis, M. *et al.* (1993) Allele-specific replication timing of imprinted gene regions. *Nature,* **364**, 459—63.

Kreidberg, J.A., Sariola, H., Loring, J.M. *et al.* (1993) WT-1 is required for early kidney development. *Cell,* **74**, 679—91.

Larsen, F., Gunderson, G., Lopez, R., Prydz, H. (1992) CpG islands as markers in the human genome. *Genomics,* **13**, 1095—107.

Latham, K.E., Doherty, A.S., Scott, C.D. and Schultz, R.M. (1994) IGF2R and IGF2 gene expression in androgenetic gynogenetic and parthenogenetic preimplantation mouse embryos: absence of regulation by genomic imprinting. *Genes and Development,* **8**, 290—9.

Lau, M.M.H., Stewart, C.E.H., Liu, Z. *et al.* (1994) Loss of the imprinted IGF2/mannose 6-phosphate receptor results in fetal overgrowth and peri-natal lethality. *Genes and Development,* **8**, 2953—63.

Laureys, G., Barton, D.E., Ullrich, A. and Francke, U. (1988) Chromosomal mapping of the gene for the type II insulin like growth factor/cation independent mannose 6-phosphate receptor in man and mouse. *Genomics*, **3**, 224—9.

Leighton, P.A., Ingram, R.S., Eggenschwiler, J. *et al.* (1995) Disruption of imprinting caused by deletion of the H19 gene region in mice. *Nature*, **375**, 34—9.

Li, E., Beard, C. and Jaenisch, R. (1993) The role of DNA methylation in genomic imprinting. *Nature*, 366, 362—5.

Ludwig, T., Tenscher, K., Remmler, J. *et al.* (1995) cloning and sequencing of cDNAs encoding the full length mouse mannose 6-phosphate/insulin like growth factor type 2 receptor. *Gene,* **142**, 311—2.

Macleod, D., Charlton, J., Mullins, J. and Bird, A.P. (1994) Sp1 sites in the mouse APRT gene promoter are required to prevent methylation of the CpG island. *Genes and Development*, **8**, 2282—92.

McGrath, J. and Solter, D. (1984) Completion of mouse embryogenesis requires both the maternal and paternal genomes. *Cell,* **37**, 179—83.

Monk, M. (1987) Memories of mother and father. *Nature,* **328**, 203—4.

Morgan D.O., Edman J.C., Standring D.N. *et al.* (1987) Insulin like growth factor II receptor as a multifunctional binding protein. *Nature,* **329**, 301—7.

Neumann, B., Kubicka, P. and Barlow, D.P. (1995) Characteristics of imprinted genes. *Nature Genetics,* **9**, 12—3.

Nicholls, R.D. (1994) New insights reveal complex mechansims involved in genomic imprinting. *American Journal of Human Genetics,* **54**, 733—40.

Nolan, C.M. and Sly, W.S. (1989) I Cell disease and pseudo-hurler polydystrophy: disorders of lysosomal enzyme phosphorylation and localisation. pp 1589—601 in the Molecular Basis of Inherited disease 6th edit. McGraw—Hill, New York.

Ogawa, O., McNoe, L.A., Eccles, R. *et al.* (1993) Human insulin-like growth factor type I and type II receptors are not imprinted. *Human Molecular Genetics,* **2**, 2163—5.

Oshima, A., Nolan, C.M., Kyle, J.W. *et al.* (1988) The human cation-independent mannose-6-phosphate receptor. *Journal of Biological Chemistry,* **263**, 2553—62.

Rainier, S. and Feinberg, A.P. (1994) Genomic imprinting, DNA methylation and cancer. *Journal of the National Cancer Institute,* **86**, 753—9.

Razin, A. and Cedar, H. (1994) DNA Methylation and Genomic Imprinting. *Cell* **77**, 473—6.

Smrzka, O.W., Faé, I., Stöger, R. *et al.* (1995) Conservation of a maternal-specific methylation signal at the human IGF2R locus. *Human Molecular Genetics*, **4**, 1945—52.

Stöger R, Kubicka P, Liu C-G *et al.* (1993) Maternal-specific methylation of the imprinted mouse Igf2r locus identifies the expressed locus as carrying the imprinting signal. *Cell,* **73**, 61—71.

Surani, M.A., Barton, S.C. and Norris, M.L. (1984) Development of reconstituted mouse eggs suggests imprinting of the mouse genome during gametogenesis. *Nature,* **308**, 548—50.

Sutcliffe, J.S., Nakao, M., Christian, S. *et al.* (1994) Deletions of a differentially methylated CG island at the SNRPN gene define a putative imprinting control region. *Nature Genetics,* **8**, 52—8.

Szebenyi, G. and Rotwein, P. (1994) The mouse insulin-like growth factor II/Cation-independent mannose 6-phosphate (IGF-II/MPR) receptor gene: molecular cloning and genomic organization. *Genomics,* **19**, 120—9.

Temple, I.K., James, R.S., Crolla, J.A. *et al.* (1995) An imprinted gene for diabetes? *Nature Genetics,* **9**, 110—2.

Villar, A.J. and Pederson, R.A. (1994) Parental imprinting of the MAS protooncogene in the mouse. *Nature Genetics,* **8**, 373—9.

von Figura, K. (1991) Molecular recognition and targeting of lysosomal proteins. *Current Opinion in Cell Biology,* **3**, 642—6.

Vu, T. and Hoffman, A.R. (1994) Promoter-specific imprinting of the human insulin-like growth factor-II gene. *Nature,* **371**, 714—7.

Wang, Z.-Q., Fung, M., Barlow, D.P. and Wagner, E.F. (1994) The imprinted Igf2/Mpr gene regulates embryonic growth and lysosomal targeting. *Nature,* **372**, 464—7.

Wertheimer, E., Lu, S.-P., Backeljauw, P.F., Davenport, M.L. and Taylor, S.I. (1993) Homozygous deletion of the human insulin receptor gene results in leprechaunism. *Nature Genetics,* **5**, 71—3.

Wevrick, R., Kerns, J.A. and Francke, U. (1994) Identification of a novel paternally expressed gene in the Prader—Willi syndrome region. *Human Molecular Genetics,* **3**, 1877—82.

Xu, Y., Goodyer, C.G., Deal, C. and Polychronakos, C. (1993) Functional polymorphism in the parental imprinting of the human IGF2R gene. *Biochemical and Biophysical Research Communications,* **197**, 747—54.

PART FOUR

MEIOTIC MOBILITY AND RECOMBINATION

Harry Scherthan

Terry Orr-Weaver

Pilar Arana

Bikram Gill

14

Chromosome behaviour in earliest meiotic prophase

H. Scherthan

14.1 INTRODUCTION

The formation of haploid gametes in sexually reproducing organisms is dependent on the pairing and recombination of homologous chromosomes (homologues) during meiotic prophase I. This ensures the faithful segregation of homologues at Metaphase I. Chromatids are separated at metaphase II (Hawley, 1988). At the onset of prophase I, sister chromatids of meiotic chromosomes become connected by axial elements (cores) which run along their entire length. Axial elements are called lateral elements, when they become interconnected by transverse filaments to result in the well known tripartite synaptonemal complex (SC) structure (reviewed by Moses, 1968; von Wettstein *et al.*,, 1984, Giroux, 1988). While chromosome pairing and meiotic recombination can apparently occur without SC formation (Roeder, 1990; Padmore *et al.*, 1991, Hawley and Arbel, 1993; Loidl *et al.*, 1994, Scherthan *et al.*, 1994; Weiner and Kleckner, 1994, Nag *et al.*, 1995), tripartite SC assembly seems to be a prerequisite for proper chiasma distribution, i.e. chiasma interference (Sym and Roeder, 1994).

Chromosome pairing at meiotic prophase involves distinct events like homologue search, alignment pairing (homologue alignment prior to, or in the absence of synapsis), DNA sequence matching, recombination and synapsis.

This article will discuss the most recent findings on chromosome and telomere behaviour during earliest meiotic prophase, made possible by the delineation of homologous chromosomes and subregions in meiotic and premeiotic cells by fluorescence in situ hybridization (FISH). For further aspects of meiotic chromosome behaviour such as the initiation of recombination, its relation to synapsis, chiasma formation and segregation at metaphase I the reader may refer to other relevant reviews (e.g., Gillies, 1975; Smithies and Powers, 1986; Maguire, 1988, Hawley, 1988, Roeder, 1990, Loidl, 1990, 1994, John 1990, Kleckner *et al.*, 1991, Moens, 1994).

14.2 CHROMOSOME PAIRING AT MEIOSIS IS ASSOCIATED WITH DRASTIC CHANGE OF THEIR SHAPE

The original cytological investigations from the turn of the century, performed on formaldehyde/heavy metal fixed, sectioned material, demonstrated a drastic reorganization of meiotic chromosomes associated with their side by side pairing at pachytene (see Fig. 14.1, and Belar, 1928; Wilson, 1925; Darlington, 1931 for further lit. and illustrations). In general, chromosomes decondense after the last premeiotic division and perform DNA synthesis. In some organisms, chromosomes contract during premeiotic interphase (see Walters, 1970). At leptotene, chromosomes develop into long thin threads, which in some organisms appear first unpaired and then approach each other and align (distance pairing or presynaptic alignment, cf. Loidl, 1990). Leptotene chromosomes have their chromatin organized in loops which are attached at their basis to the proteinaceous axial core. The size of the loops is species specific (Moens and Pearlman, 1988) and their attachment sites may involve specialized repeated sequences (Pearlman *et al.*, 1992, Heng *et al.*, 1995). Where segments of homologous axial cores (elements) have approached each other to within 300nm, synapsis is initiated (zygotene). The initiation of synapsis has been predominantly observed near the chromosome ends but may also occur interstitially (reviewed by von Wettstein *et al.*, 1984, John, 1990). After synapsis has been completed, chromosomes are seen as thick threads (pachytene) in the meiocyte.

Chromatin reorganization at meiotic prophase sees the expression of SC and associated proteins (Heyting *et al.*, 1989; Chen *et al.*, 1992; Dietrich *et al.*, 1992; Dobson *et al.*, 1994; Sym *et al.*, 1993; Bishop 1994; Lammers *et al.*, 1994; Ross-Macdonald and Roeder 1994; Terasawa *et al.*, 1995), meiosis specific lamins (Moss *et al.*, 1993; Venster *et al.*, 1993), histones (Meistrich *et al.*, 1985; Cole *et al.*, 1986; Poccia *et al.*, 1987) and related proteins (e.g. Fuchikawa *et al.*, 1994; Riggs 1994).

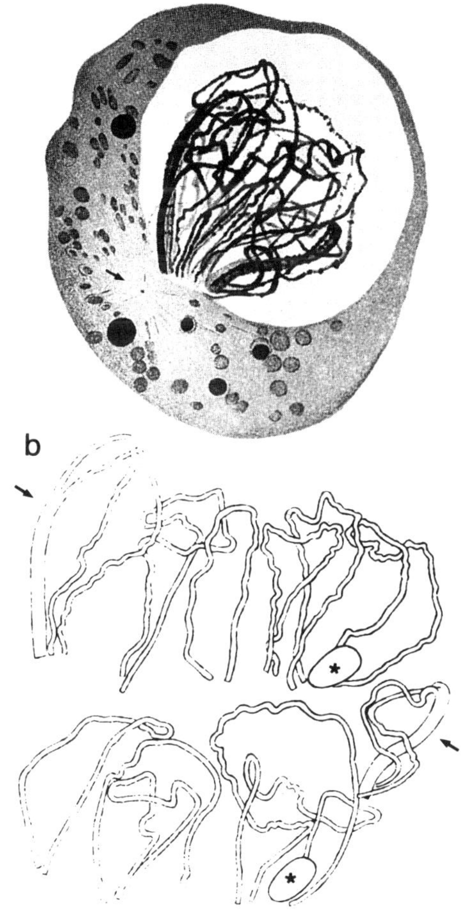

Fig. 14.1 Classical drawing in an early zygotene bouquet of *Dendrocoelum lacteum*. Ends of bent, elongated chromosomes have accumulated in parallel at the cluster site. Heavy metal staining. (b), Expanded reconstruction of the conformation and position of all individual chromosomes in the bouquet. Synapsis has initiated between two pairs of homologs (arrows). The single nucleolus has been drawn twice (marked by asterisks) to indicate positions of homologs which have aligned near the bouquet base. Note the chromomere patterns of meiotic chromosomes and the weakly indicated position of the centrosome near the bouquet base (black arrow). Reproduced from Gelei 1921, with modifications.

14.3 FISH: A NOVEL TOOL TO REVISIT CLASSICAL PROBLEMS OF MEIOTIC CHROMOSOME BEHAVIOUR

Over the past decades, chromosome behaviour during meiotic prophase in a variety of organisms has been studied using conventional light and electron microscopy on intact or spread meiocytes (for reviews see, Belar, 1928; Darlington, 1931; Rhoades, 1961; John, 1969; von Wettstein *et al.*, 1984; Loidl, 1990). Progress towards the understanding of the molecular basics of the meiotic process has been achieved due to the recent application of yeast genetics to meiosis (for reviews see Hawley and Arbel, 1993; Roeder, 1995).

Until recently, behaviour of meiotic chromosomes at early prophase, when homologue search and recognition is believed to take place, was not amenable to direct cytological observation. This drawback has been overcome by the adaptation of FISH techniques with whole chromosome-specific DNA library probes (chromosome painting; Lichter *et al.*, 1988; Pinkel *et al.*, 1988) or subregional probes to meiotic prophase chromosomes (Moens and Pearlman, 1989; Scherthan *et al.*, 1992; Goldman and Hultén, 1992; Albini and Schwarzacher, 1992). FISH allows for the decoration of entire chromosomes as well as subregions by fluorescent reporter molecules and thus provides for the investigation of chromosome morphology and position throughout the entire cell cycle at the molecular level (for reviews see Lichter *et al.*, 1991; Cremer *et al.*, 1993).

Up to now this approach has been fruitfully applied in the investigation of numerous aspects of chromosome organization at meiosis (e.g., Moens and Pearlman, 1990; Goldman and Hultén, 1992; Scherthan *et al.*, 1992, 1994; Goldman and Hultén, 1993; Ashley *et al.*, 1993; Speed *et al.*, 1993; Heng *et al.*, 1994; Loidl *et al.*, 1994, 1995; Armstrong *et al.*, 1994; Cheng and Gartler, 1994; Chevret *et al.*, 1995 and refs. therein). By painting meiotic chromosomes or subregions thereof in spread meiotic nuclei of *S. cerevisiae* (Scherthan *et al.*, 1992, 1993) it has become possible to assess the impact of the molecular phenotype of a mutation on chromosome pairing at the cytological level (Bähler *et al.*, 1993; Loidl *et al.*, 1994; Weiner and Kleckner 1994; Scherthan *et al.*, 1994; Nag *et al.*, 1995; Rockmill *et al.*, 1995).

Finally, protocols have become available which allow the study of topological aspects of chromosome behaviour in three dimensionally well preserved premeiotic and meiotic nuclei (e.g., Dawe *et al.*, 1994; Scherthan and Cremer, 1994; Dernburg *et al.*, 1995; Scherthan, 1995).

Fig. 14.2 ***Colour Plate 7*** (**a**) Painting of human chromosome 1 territories in a spermatogonium reveals two separate, compacted territories (yellow, arrows). The basement membrane of the testis tubule is seen to the left. (**b**) Nucleus showing elongated, bent chromosome 1 territories in the process of pairing. Ends of territories seem to be associated with the nuclear envelope. (**c**) A single signal tract meandering throughout the nucleus indicates tightly paired homologs at pachytene. Bar in c: 10μm; it applies to all details.

14.4 CHROMOSOME ORGANIZATION IN MEIOTIC STEM CELLS

In somatic interphase nuclei chromosomes occupy distinctly compacted, often mutually exclusive territories (Lichter *et al.*, 1988; Pinkel *et al.*, 1988; reviewed by Haaf and Schmi; 1991, Cremer *et al.*, 1993). Recently, chromosome painting in structurally well preserved paraffin-embedded human testis tissue section nuclei (Scherthan and Cremer, 1994) revealed that in meiotic stem cells (spermatogonia, as identified by their close association with the tubule membrane; see de Rooij 1988) chromosomes occupy compacted, predominantly separate chromosome territories (Fig. 14.2a; Scherthan *et al.*, submitted). At leptotene/zygotene, chromosome territories are seen as thread like structures (Fig. 14.2b), while a single signal tract indicates synapsis at pachytene (Fig. 14.2c; Goldman and Hultén, 1992; Armstrong *et al.*, 1994; Cheng and Gartler, 1994). The question to what extent pairing of homologues during meiotic prophase benefits from their premeiotic arrangement has been under debate ever since (e.g. Gillies, 1975; Rassmusen and Holm, 1980; John 199; Maguire, 1988; Loidl, 1990; Kleckner and Weiner, 1993). Vegetative homologue association has been confirmed by FISH in Diptera (Hiraoka *et al.*, 1993). It was also detected in diploid strains of fission yeast (Scherthan *et al.*, 1994) and in a substantial fraction of vegetatively growing budding yeast cells (Loidl *et al.*,, 1994; Weiner and Kleckner, 1994).

FISH analysis has also revealed the absence of a pronounced homologue association for: (i) X homologues in the premeiotic mouse ovary (O'Keeffe *et al.*, 1995) and at leptotene in human ovaries (Cheng and Gartler, 1994), (ii) chromosome 8 and 12 subsatellite regions in mouse spermatogonia (S. Weich and H. Scherthan, unpublished results), Y homologues in human prepuberal XYY gonocytes (Ragg *et al.*, 1995), human XY chromosomes at leptotene (Armstrong *et al.*,

1994), and chromosome 1 territories and their centromeric regions in spermatogonia of human testis (Scherthan *et al.*, submitted). In man and mouse high levels of pairing were first detected at zygotene (Scherthan *et al.*, submitted), indicating that the mammalian pairing process acts during early meiotic prophase.

These findings are consistent with earlier observations that in the majority of organisms a precondition of synapsis is the *de novo* prealignment of homologues, i.e. a process which establishes their close proximity and proper orientation during early meiotic prophase (e.g. Zickler, 1977; Therman and Sarto, 1977; Rassmusen and Holm 1978; Boiko, 1983 Albini and Jones, 1987; Scherthan *et al.*, 1992; Armstrong *et al.*, 1994; Cheng and Gartler, 1994; Dawe *et al.*, 1994). Knowledge about the nature of the chromosome pairing process and its general preconditions is still fragmentary and thus heavily under debate (see e.g., Fussell, 1987; Maguire, 1988; Loidl, 1990; John, 1990; Kleckner and Weiner, 1993; Klein, 1995).

Several models of the topological prerequisites of the pairing process have been proposed. These imply for example, the benefit from a Rabl orientation (Fussell, 1987), an increasing degree of homologue associations during a series of mitotic cycles of premeiotic cells before these enter meiotic prophase (Brown and Stack, 1968; Stack and Brown, 1969), or that homologues are connected by contractile elements which draw them together at meiotic prophase (Maguire, 1984). The possibility of a preordering of chromosomes in haploid chromosome sets (Ashley and Pocock, 1981; Bennett, 1983, 1984) and the arrangement of the chromosomes in chains (Wagenaar, 1969; Ashley 1979) have also been considered. On the basis of the observations that rearranged chromosomes seem to pair efficiently during meiotic prophase, most of the available models have been questioned (see Maguire 1988).

Molecular (Padmore *et al.,* 1991) and chromosome painting analysis (Scherthan *et al.*, 1992) of *S. cerevisiae* meiosis has recently suggested that synapsis is preceded by homologue search and prealignment and that homologues condense during this process. It has been assumed that prophase chromatin condensation along with growing unstable interactions between homologues may provide the forces which drive homologue alignment (Kleckner and Weiner, 1993). Once aligned, homologues may be linked by the assembly of recombination nodules, transverse filament proteins or other protein complexes involving components of the double strand/mismatch repair machinery (Meuwissen *et al.*, 1992; Sym *et al.*, 1993; Bishop, 1994; Dobson *et al.*,

1994; Ross-McDonald and Roeder, 1994; Rockmill *et al.*, 1995). A double strand break independent homology search and alignment process may account for the residual pairing capacity observed in yeast meiotic mutants (see Loidl, 1994; Roeder, 1995 for review). It has been speculated that vegetative homologue associations facilitate the presynaptic alignment process (Weiner and Kleckner, 1994). However, these associations disappear when the cells become committed to meiosis, presumably during premeiotic S (Weiner and Kleckner, 1994), resulting in separated homologues at the onset of the pairing process (Scherthan *et al.*, 1992). The nature of the process establishing meiotic levels of chromosome pairing is not known at present. In this context, one has to recall that chromosome behaviour during meiotic prophase in yeasts and other organisms has often been deduced from the analyses of spread nuclei. This implies that important topological interphase chromosome arrangements (like e.g., chromosomal bouquets) associated with the pairing process might have been disrupted by the spreading procedure.

14.5 RENEWED INTEREST IN THE BEHAVIOUR OF MEIOTIC TELOMERES

Telomeres have been implicated as key structures of meiotic chromosomes (reviewed by Fussell, 1987; Gilson *et al.*, 1993; Blackburn, 1994; Ashley, 1995). A solid body of cytological investigations indicates that they play an important role in the pairing process (e.g., Belar, 1928; Darlington, 1937; Rhoades, 1961; Gillies, 1975; von Wettstein *et al.*, 1984; Loidl, 1990; Dernburg *et al.*, 1995).

Most recently, Chigashige *et al.*, (1994) combined the live analysis of meiotic prophase cells with FISH of telomere and centromere sequences on fixed meiotic cells of *Schizosaccaromyces pombe*. They demonstrated that at induction of zygotic meiosis, telomeres attach to the spindle pole body (the equivalent of the centrosome of higher eukaryotes) of the parental nuclei and lead chromosome movements, which persist throughout karyogamy and the entire meiotic prophase (Chigashige *et al.*, 1994). In vegetative cells of diploid *S. pombe* strains homologues occupy a joint territory (vegetative pairing), while telomere clustering is the first meiosis specific event at the onset of azygotic meiosis (Scherthan *et al.*, 1994). The resulting bouquet configuration is maintained throughout the entire prophase and is thought to be a prerequisite for meiotic chromosome pairing and the high levels of meiotic recombination in this asynaptic organism (for review see Kohli, 1994).

Like in diploid fission yeast, mosquito (Diptera) cells also display alignment of homologues in joint territories prior to the onset of meiosis (Wandall and Svendsen, 1985). During prepachytene, the transition from somatic to meiotic pairing is characterized by changes of intranuclear chromatin morphology, while telomeres aggregate opposite to the centrosome at the cytoplasmic nuclear envelope (cf. Fig. 14.1a, Wandall and Svendsen, 1985). In *Drosophila* meiocytes, a remodeling of chromosomal chromatin during prepachytene and the non-random peripheral distribution of chromosome ends have also been observed (Carpenter, 1975). However, further analysis is required before final conclusions about telomere behaviour in *Drosophila* meiosis may be reached.

Clustering of telomeres at a nuclear sector in the leptotene/zygotene meiocyte of the majority of synaptic organisms leads to the unique polarized organization of the elongated prophase chromosomes - the *bouquet* stage. This meiotic chromosome polarization has been consistently observed during synapsis initiation at the leptotene/zygotene transition (Dernburg *et al.*, 1995). It was first described nearly a century ago (Eisen, 1900) and since then has been detected in the meiotic prophase of most organisms analyzed (see von Wettstein *et al.*, 1984; Dernburg *et al.*, 1995 for review). At this stage the telomeres of the meiotic prophase chromosomes, which are attached to the inner nuclear membrane (von Wettstein *et al.*, 1984), have accumulated at a nuclear sector opposite to the yet unseparated centrioles at the outer nuclear membrane (e.g., Schreiner and Schreiner, 1905; Gelei, 1921; Hughes-Schrader, 1943; Moens, 1969, 1974; Church, 1976; Virkki, 1977). In plant meiosis the aggregation of ends at a nuclear envelope (NE) sector is observed (Hiraoka, 1952; Therman and Sarto, 1977; Gillies, 1975; Dernburg *et al.*, 1995). In *S. cerevisiae* bouquet cells (Dresser and Giroux, 1988) and the accumulation of end attachments at an intrusion of the nuclear envelope has been observed (Zickler and Olson, 1975).

In summary, the data available in the literature suggest telomere clustering being a conserved motif of the pairing process, irrespective of the premeiotic chromosome arrangement.

Due to the transient nature of chromosome polarization (Belar, 1928, Moens, 1973; Rasmussen and Holm, 1980; Scherthan, 1995) and its disruption by spreading techniques its significance for chromosome pairing has remained enigmatic. It has been noted that the bouquet

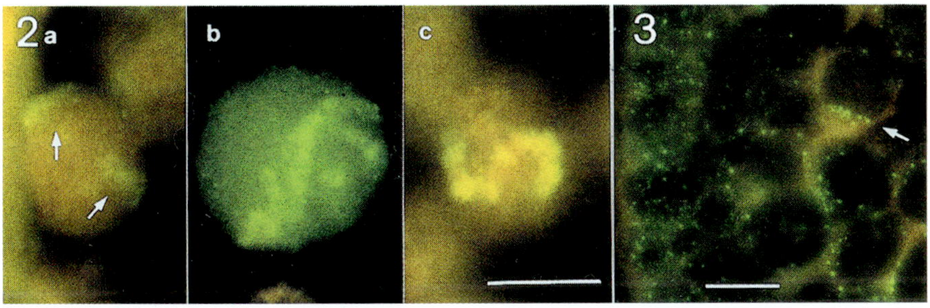

Plate 7 **Fig. 14.2** a–c (see p.221); **Fig. 14.3** (see p.226)

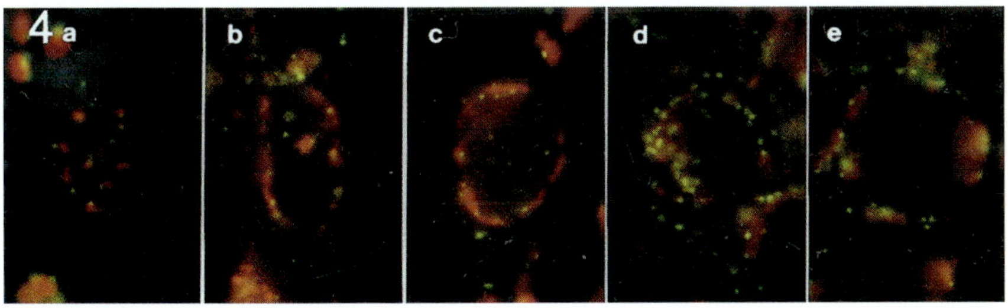

Plate 7 **Fig. 14.4** a–e (see p.226)

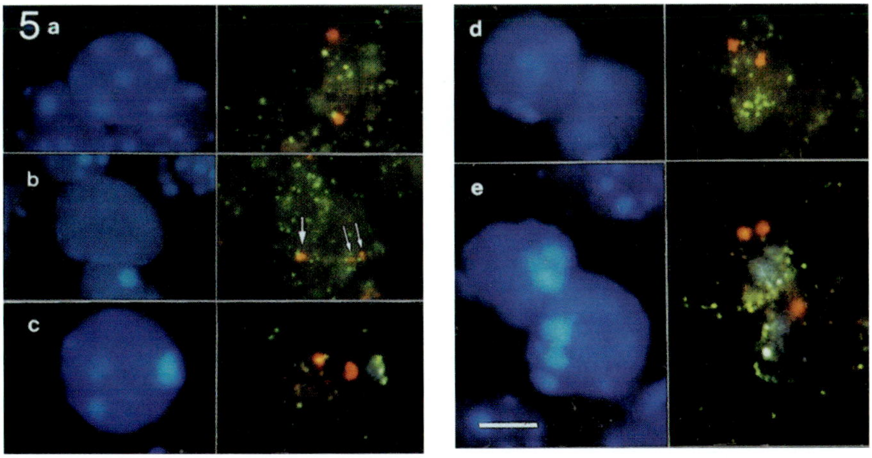

Plate 7 **Fig. 14.5** a–e (see p.226)

arrangement should 'help in some way the sort out process of homologues prior to synapsis' (see Therman and Sarto, 1977) and that NE attachment of chromosome ends may reduce a three dimensional pairing problem to a two dimensional one (Rhoades, 1961). However, it has been questioned whether the short ranged and saltatory nuclear prophase movements observed (Salonen *et al.*, 1976; Parvinen and Söderström, 1982) could substantially contribute to efficient pairing of homologues (Maguire, 1988; Loidl, 1990). While the movement of elongated chromosomes throughout the chromatin filled nucleus may be associated with intra nuclear motor protein action, the movement of telomeres to the cluster site may lend itself to bringing about a favourable topological conformation to the elongated prophase chromosomes (below).

14.6 CENTROMERE AND TELOMERE MOVEMENTS DURING MAMMALIAN MEIOSIS

Three dimensional serial section EM studies revealed that at leptotene homologous telomeres attach widely separated to the NE. Subsequent movements of telomeres along the meiotic NE bring about chromosome polarization (for review see Rasmussen and Holm 1980). These movements are distinct from the ones observed during the mitotic cell cycle of mouse (Vourc'h *et al.*, 1993) and human (Ferguson and Ward 1992, Weimer *et al.*, 1992).

Recently, we studied the pairing process in paraffin embedded testis tissue sections of prepuberal and adult mice by FISH with telomeric $(TTAGGG)_7$ repeat (Moyzis *et al.*, 1988), centromeric major satellite (Pardue and Gall, 1970) and chromosome 8 specific subsatellite (Boyle and Ward, 1992) DNA probes (Scherthan *et al.*,, submitted). The onset of spermatogenesis in the prepuberal mouse is fairly synchronous (Bellvé *et al.*, 1977, Kluin and de Rooij, 1981, Goetz *et al.*, 1984, Vergowen *et al.*, 1991) and allows for investigation of meiotic prophase without drug synchronization protocols. From this we could deduce a sequence which revealed sequential changes of centromere and telomere positions during the onset and course of meiotic prophase. It was observed that centromeric satellite DNA and telomeres were dispersed throughout spermatogonial nuclei, while they transited and attached to the NE during mid preleptotene (Fig. 14.3 and 14.4).

Fig. 14.3 ***Colour Plate 7*** A detail of a prepuberal mouse testis tubule (day 12p.p.) hybridized to telomere repeat probes (FITC, green) reveals drastic telomere redistribution associated with the onset of meiotic prophase. In meiotic stem cells at the basement membrane of the tubule (to the left) numerous telomere signals are dispersed throughout the nuclear lumen. In spermatocytes I, which are remote from the basement membrane and display an increased nuclear diameter, these have become peripheral. Aspects of telomere clustering are apparent in two cells near the center of the tubule (to the right). Arrow: bouquet spermatocyte. Bar: 10μm.

Fig. 14.4 ***Colour Plate 7*** Distribution of centromeric major satellite DNA (red) and telomere DNA sequences (green) in spermatogonia and spermatocytes nuclei from mouse testis sections. The order of the stages was determined by the sequential appearance of the nuclei in the prepuberal testis (Scherthan *et al.*,, submitted). (a), A-type spermatogonium showing numerous satellite clusters and telomeres distributed throughout the nucleus. (b,c), Mid preleptotene nuclei showing centromeric satellite DNA compressed against the NE. Some distal telomeres and a single satellite cluster (in b) are still seen within the nuclear lumen. (d), Satellite DNA forming peripheral clusters in an early zygotene nucleus. Satellite DNA and telomeres are confined to at a limited region of the nuclear periphery. (e), Pachytene nucleus showing distinctly peripheral localization of telomeres and satellite clusters. Positions of centromeres and telomeres have also been observed by confocal laser scanning microscopy (H. Scherthan and T. Cremer, unpublished). Scale bar in Fig. 14.5e applies.

Fig. 14.5 ***Colour Plate 7*** Sequential telomere clustering and homologous chromosome pairing during mouse meiotic prophase as revealed by two color FISH of chromosome 8 specific subsatellite (red) and telomere (green) DNA probes to testis tissue sections from mice days 6 (**a,b**) and 10 (**c-e**) p.p.. DAPI counterstain, indicating the position of DAPI bright centromeric satellite DNA, is shown to the left of the hybridization image. (**a**) Prepuberal germ cell with chromosome 8 specific signals (hereafter referred to as #8 signals) separate and telomeres dispersed throughout the nuclear lumen. (**b**) Mid preleptotene nucleus with #8 signals separate (twin arrows + arrow). Major satellite DNA does not shown in the DAPI image cause it forms a layer at the NE (see Fig. 14.4b,c). Proximal telomeres are associated with the NE while numerous distal telomeres are still in the nuclear lumen. One #8 signal appears as signal doublet (twin arrows) which indicates replicated target DNA (Selig *et al.*, 1992). (**c**) Late preleptotene nucleus showing telomeres exclusively located at the NE (only few telos are seen in this focal plain). Satellite DNA forms few, DAPI bright, large satellite clusters at the periphery. #8 signals separate. (**d**) Telomeres captured during cluster formation, #8 signals separate. (**e**) Two adjacent nuclei of a symplast exhibit tightly clustered telomeres. #8 signals appear aligned in the upper nucleus and fused in the lower nucleus. In the latter the telomere cluster is in the process of dissolution. Bar: 5μm.

Centromeres preceded the attachment of telomeres to the NE (Fig. 14.4; Scherthan *et al.*, submitted). Experiments on human bouquet meiocytes showed centromeres remote from the nuclear periphery (Scherthan *et al.*,, submitted), suggesting an exchange of peripheral centromere and telomere positions. Exchange of centromere/telomere positions in the vicinity of the spindle pole body has been observed in parental nuclei, prior to karyogamy, in fission yeast meiosis (Chigashige *et al.*, 1994). This exchange seems to be a highly conserved motif of the pairing process and may represent the earliest detectable event at the chromosomal level.

After telomeres have attached to the NE they move towards an envelope sector where they cluster. Mouse chromosome 8-specific repeat signals, which were predominantly separate in spermatogonia and preleptotene spermatocytes (>87% of nuclei), paired at the tightest telomere clustering (Fig. 14.5) during the preleptotene/zygotene transition (Scherthan *et al.*, submitted). It has been estimated that the time from premeiotic DNA replication to the onset of zygotene in the male mouse takes only a few hours (Oud and Reutlinger, 1981; Dietrich and de Boer, 1983). Thus, the pairing process, which presumably initiates after premeiotic DNA replication (Oud and Reutlinger, 1981; Dietrich and de Boer, 1983; Guitart *et al.*, 1985; Scherthan *et al.*, 1995, submitted), has to be very rapid and efficient.

Axial element protein (AEP) immunostaining (Lammers *et al.*, 1994) combined with telomere FISH showed that these proteins appear as large aggregates within the late preleptotene nucleus, which has its telomeres attached to the NE (Scherthan *et al.*,, submitted). Such aggregates have also been observed throughout prophase in meiocytes of female rats, but seemed to be sensitive to spreading procedures (Dietrich *et al.*, 1992). AEPs dispersed onto the chromosomal chromatin during telomere clustering, indicating that mouse chromosomes adopt an elongated chromosome conformation before the axial/lateral elements are assembled. The appearance of transverse filament proteins and paired chromosome 8 signals, generally observed in polarized mouse spermatocytes (Scherthan *et al.*, submitted), indicates that chromosome alignment is transformed into synapsis during this stage. According to these observations synapsis initiation seems to be associated with a leptotene/zygotene bouquet in most organisms (reviewed by Dernburg *et al.*, 1995; e.g. Fig.14.1).

14.7 TELOMERE CYTOSKELETON INTERACTIONS

It has been shown that clustering of NE attached telomeres is associated with the aggregation of nuclear pores at the polar region of the meiocyte (Chemes *et al.*, 1978; Fawcett and Chemes, 1979; Church, 1976), a phenomenon which is not known from somatic cells (see de Boni 1994). Redistribution of organelles (Virkki, 1971; Hiraoka, 1952; Church ,1976; Hiraoka and Fuchikawa, 1993), the sensitivity of nuclear motility to microtubule disrupting drugs observed in live meiocytes of fission yeast (Chigashige *et al.*, 1994), rat and house cricket (Salonen *et al.*, 1982; Rickards, 1975) and the pairing failure induced at metaphase I by colchicine (e.g., Driscoll and Darvey, 1970; Bennett *et al.*, 1979; Puertas *et al.*, 1984; Loidl, 1989) indicate that chromosome pairing may be associated with cytoskeletal components. In somatic nuclei the outer nuclear membrane is associated with microtubules and (micro)filaments (Franke, 1971; Penman, 1995). In mouse spermatocytes I, prominent bundles of microtubuli were observed to encase the nuclear surface and to converge at the edge of the nucleus (the centrosome?) (Cherry and Hsu, 1984). This would provide a polarized microtubule network along which attached ends could move.

In human female and male meiocytes, telomere attachment plaques at the inner nuclear membrane are linked by filament bundles to the spherical dense structures at the cytoplasmic nuclear membrane (see Boiko, 1983). Telomere spindle pole body connections were also observed in *Neurospora* (Boiko, 1990). The spheres at the cytoplasmic NE could be a means to link telomere attachments to the rails of tubulin surrounding the spermatocyte nucleus. Chromosome/ microtubuli interactions through the intact NE are possible and have been observed to mediate chromosome segregation in dinoflagellate division (Kubai and Ris, 1969).

In the mouse *in situ* hybridization combined with electron microscopy revealed that telomere repeat sequences may be involved in the attachment of chromosome cores to the NE (Scherthan *et al.*, submitted). This is supported by the observation that in human somatic nuclei telomeric DNA sequences copurify with the nuclear matrix fraction (de Lange 1992). In *S. cerevisae,* telomeres interact with a nuclear matrix protein, RAP1, both in vegetative and meiotic cells (Klein *et al.*, 1992; Palladino *et al.*, 1993). FISH to human somatic nuclei revealed that telomeres locate preferentially at the surface of interphase chromosome territories (T. Cremer and S. Dietzel, unpublished data). Interactions of nuclear matrix and telomere sequences could provide for their efficient transition to the NE at earliest meiotic prophase.

Most of the nuclear movements observed in live meiocytes are exceedingly complicated and of a saltatory nature (Parvinen and Söderström, 1976; Rickards, 1975). It can be assumed that the concerted movements of groups of ends or movements of individual ends towards the cluster site could induce rotations and movements of the entire nucleus with respect to a cytoskeleton anchored centrosome. Aggregation of groups of proximal telomeres, as identified by their associated satellite DNA, was observed during bouquet formation of the mouse (Scherthan *et al.*, submitted; see DAPI images in Fig. 14.2). Increasingly unstable interactions during the pairing of homologues (Kleckner and Weiner, 1993), could lead to movement of NE attached telomeres towards each other. However, the convergence of most, if not all ends at a confined region of the NE requires the action of an additional force producing machinery, which directs the ends towards the focal point of movements and by this bends elongated chromosome territories and leads to bouquet formation.

The observed number of fully polarized nuclei in mouse testis sections was <1% at zygotene in male mice (Scherthan *et al.*,, submitted) and about 0.5% at pachytene in air dried preparations of mouse ovaries (Speed, 1982), indicating full bouquet formation to be a short, transitional stage, while the possibility remains that some meiocytes might never reach clustering of all telomeres. It has been observed, that small chromosomes were the first to initiate (and finish) synapsis (e.g., Zickler, 1977; Rasmussen and Holm, 1980; Jones and Croft, 1986; Santos *et al.*, 1993). Asynchronous synapsis initiation was observed in *Dendrocoelum lacteum* bouquet cells (Gelei, 1921; Fig. 14.1). It can be assumed that alignment of homologues occurs in a bent fashion in the polarized nucleus (Fig. 14.6). Upon recognition, matched partners initiate stable interactions and relocate from the cluster site, either by centriole movements (e.g., Hughes-Schrader, 1943; Moens, 1974) or condensational forces, leaving chromosome polarization in some species nearly undetectable at pachytene (e.g., Glamann, 1986).

14.8 THE SIGNIFICANCE OF THE POLARIZED NUCLEUS FOR HOMOLOGUE SEARCH

The aggregation of the terminal parts (in terms of Mbp) of elongated chromosome territories within a limited area of the nucleus could stimulate surface interactions among chromosome territories and thus facilitate their search. Homologue search and recognition may involve paranemic DNA/DNA interactions between intact DNA duplexes (Kleckner and Weiner, 1993; Weiner and Kleckner, 1994), or DNA/DNA interactions dependent on a double strand break repair mechanism (Smithies and Powers, 1986; Carpenter, 1987; Moens, 1994; Loidl, 1994). Multiple interactions during homologue search (Weiner and Kleckner, 1994; Scherthan *et al.*, 1994) may occur at the surface of chromosome territories at sites of specialized chromatin conformation. Pairing sites may be found in the early replicating, GC and gene rich fraction of a genome (Chandley, 1986), located at the surface of chromosome territories. Such an organization has been proposed for somatic chromosome territories (Cremer *et al.*, 1993). The efficient interactions of homologous DNA sequences at non-homologous (ectopic) positions in a complement has been observed (Jinks-Robertson and Petes, 1985; Lichten *et al.*, 1987). Homologue search throughout the looped chromatin of elongated prophase chromosomes would be consistent with this observation.

A homologue search would be most efficient at telomeric regions (in terms of megabases), as potential testing sites (regions) become aligned most accurately near the bouquet base. This may explain why in many organisms synapsis starts near the telomeres. In support of this view, the extra length of one chromosome core has been observed to prevent synapsis with its normal homologue near the NE attached telomeres in *Chloealtis* (Moens *et al.*, 1989).

Chromosome-specific variation of DNA sequence composition (GC content) and gene densities along chromosomes have been observed (see Bernardi, 1989). This variation is expressed as banding patterns (in vertebrates) along mitotic chromosomes (Holmquist, 1992; Craig and Bickmore, 1994) and patterns of chromomeres along meiotic chromosomes (Fig. 14.1; see Lima-de-Faria, 1983 for lit.). Specific variations in GC content and gene densities along *S. cerevisiae* chromosomes have recently emerged from the complete nucleotide sequences of chromosomes I, II, III, VIII and XI (Bussey *et al.*, 1995, Feldman *et al.*, 1994; Oliver *et al.*, 1992; Johnston *et al.*, 1994; Dujon *et al.*, 1994, respectively).

Thus I propose that the chromosome specific variations of DNA sequence (gene) composition confer a molecular address (bar code) to each individual homologue pair (for illustration see banding ideograms of e.g., human chromosomes). The elongated chromosomes at meiotic prophase expose the more subtle variations of sequence composition (bands) along the leptotene chromosomes and thus facilitate testing for homology at multiple points along chromosomes. It has been assumed that 1 kb of perfectly matched heteroduplex DNA is required for promotion of stable recombination intermediates and synapsis (Radman and Wagner, 1995). The close spacing of unique (coding) sequences at telomeric regions of warm blooded vertebrates (Saccone *et al.*, 1993; Craig and Bickmore, 1994) may in these species increase the efficacy of homologue search.

In organisms with large genomes, like certain plants, a Rabl orientation might facilitate meiotic chromosome alignment and pairing (Fussell, 1987). In these cases and in cases of anti Rabl orientation (see Dernburg *et al.*, 1995) chromosomes are organized in a bent configuration prior to telomere attachment and clustering at the NE. By this orientation, prealignment along a substantial amount of their entire length could be facilitated, which is indicated by almost simultaneous use of multiple synapsis initiation points in species showing Rabl orientation (e.g., Hasenkampf, 1984; Albini and Jones, 1987).

In organisms with somatic association of homologues, telomere clustering at the nuclear periphery has been observed at the onset of meiotic prophase (Wandall and Svendsen, 1985; Scherthan *et al.*, 1994). This would be required to pull pairing sites of territories into register, facilitating specific side-by-side pairing. As long range chromosome movements are not required, the entire pairing process is more efficient. Prophase stages like leptotene/zygotene can therefore be omitted, as has been observed in Diptera (Wandall and Svendsen, 1985; Carpenter, 1975).

Clustering of chromosome ends may *per se* not lead to pairing, but rather confers close proximity and proper spacing of homologous regions to elongated chromosome territories. Prealignment in a bent fashion and telomere movements would allow for resolution of entangled chromosome ends prior to the initiation of stable interactions between homologues. The adapted bent configuration furthermore facilitates the alignment and pairing of rearranged chromosomes such as ring chromosomes, inversion heterozygotes or centric fusion heterozygotes (see Fig. 14.6 and model below).

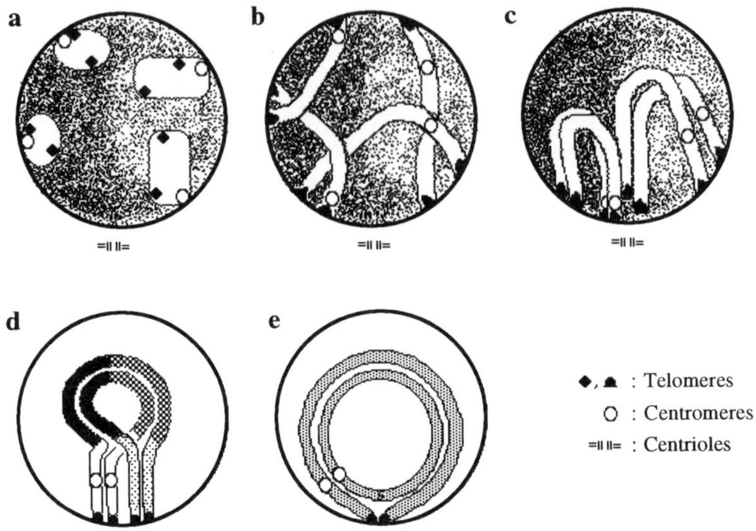

Fig. 14.6 A generalized, highly schematic illustration of the sequential changes of chromosome morphology and telomere motility events during the meiotic pairing process. It is based on the observations from the literature and in our lab (see text). Chromosome territories in premeiotic cells of an imaginary organism occupy compacted, in this case separate territories. Telomeres (black dots) locate predominantly to the surface of territories. (**a**), Mid preleptotene: all centromeres (open circles) have moved to the nuclear periphery, while telomeres lag behind. Chromosome territories are about to elongate. (**b**), Telomeres have attached at various points to the NE (black thickenings) and chromosomes have adopted an elongated conformation (late preleptotene, leptotene). Centromeres of sub-metacentric chromosomes (longer chromosome pair, to the right) are remote from the NE, while centromeres of acrocentric chromosomes (shorter chromosome pair, to the left) remain close to it. Stretching throughout the nucleus and telomere movements towards the cluster site (at the position of the cytoplasmic centrioles) bring about encounters among territories, which contribute to homology search. (**c**), Telomeres have accumulated at the cluster site and territories have partially aligned at leptotene/zygotene transition, still allowing for entanglement resolution. Presynaptic alignment is transformed into synaptic pairing (presumably starting at small homologs). Connected homologs will be removed from the cluster site by progression into pachytene (not shown). (**d,e**), Alignment pairing of rearranged chromosomes with their normal homologs at the bouquet stage. (**d**) Loop formation during bent-alignment in an inversion heterozygote. Differential patterns along twisted oblong indicates the orientation of the inverted segments in relation to the normal chromosome (outer homolog). (**e**) Pairing of a ring chromosome with its normal homolog. Bent conformation and telomere clustering of the normal homolog facilitates homology search and pairing. Size and positions arbitrary.

14.9 MODEL CONSIDERATIONS

Summarizing the present knowledge of the pairing process, the following model can be suggested. Initially separate and compacted homologous chromosome territories become oriented in a proper way by the transition of centromeres to the NE (Fig. 14.6,a). Telomeres follow and move along the surface of territories towards, and finally attach to, the inner nuclear membrane. When telomeres have attached to the NE (late preleptotene), chromosome territories develop into long, thin cords which stretch through the nuclear lumen (Fig. 14.6,b). Telomeres rapidly move to the cluster site, which produces numerous encounters among territories. These allow for homology testing at exposed pairing sites. An increasing number of interactions between homologues may provide for their prealignment (see Kleckner and Weiner, 1993). Subsequently, tight clustering of ends near the centrosome or its equivalent lead to bent alignment of homologues (Fig. 14.6,c). This polarization of territories permits tightest surface interactions near the attachment points and facilitates resolution of entangled chromosomes by further movements of ends at the cluster site. At this stage stable interactions between properly aligned homologues and recombination intermediates are initiated.

Rapid distribution of SC proteins along the bent-aligned chromosomes facilitates synapsis initiation at various sites and the rapid distribution of proteins along chromosomes. According to this view, breakage and reunion of lateral elements for interlock resolution during zygotene (von Wettstein *et al.*, 1984) probably represents a mechanism for the removal of a few persistent interlocks. Once connected, paired homologues will rapidly relocate from the cluster site, either by condensational forces and/or NE associated movements of centrioles. As yet unsynapsed partners remain at the cluster site, where a reduced number of encounters facilitates the pairing of the remaining ones.

One attractive feature of the present model is that it allows for the efficient pairing of rearranged chromosomes. Inversion heterozygotes would lead to the formation of inversion loops (Fig. 14.1,d), reciprocal translocation heterozygotes would form quadrivalents and even a ring chromosome could efficiently pair with its intact partner (Fig 14.6,e). The only prerequisite for the efficient pairing under these circumstances is that chromosomes maintain their bent configuration at the cluster site to allow the exposure of enough well-oriented homologous chromosome segments for homology testing and initiation of synapsis. A twist in one chromosome of, e.g., a pair of bent-aligned inversion heterozygotes, would facilitate inversion loop formation (Fig. 14.6,d). When this spatial conformation is stabilized by crossovers, it will be seen as an inversion loop after spreading (e.g., Moses *et al.*, 1984; Maguire and Riess, 1994). As this process requires additional movements as compared to the pairing of normal homologues, a rearranged chromosome and its unaltered partner should show a pronounced tendency to be the last to pair.

Efficient pairing of multiple Robertsonian translocation heterozygotes (Johannisson and Winking, 1994) is obviously facilitated by bent alignment. Prolonged prophase has been observed in mice with more severely rearranged chromosomes (for review see, de Boer and de Jong, 1989). Furthermore, formation of a membrane associated ring bivalent (Glamann, 1986) occurred presumably during bent alignment. Foldback pairing (Gillies, 1974; Santos *et al.*, 1994) and end associations (Loidl *et al.*, 1991) observed in haploid meiosis most likely initiate during bent alignment of non-homologous chromosomes at the cluster site. In support of this a bouquet stage has been observed during prophase I in haploid rye (Santos *et al.*, 1994).

14.10 CONCLUSIONS

The intriguing feature of the model of the pairing process presented is that it can easily deal with different types of chromosome distribution within the premeiotic nucleus and allows even for efficient pairing of rearranged chromosomes.

Karyotypic evolution reflects the reshuffling of chromosomal segments, leading to variations in size and number. It is therefore dependent on a conserved mechanism which ensures efficient homologue pairing, even in states when, during a step in karyotype evolution, heterozygosity between a rearranged (thus novel) chromosomes and its 'conservative' partner is present in a meiocyte. Despite certain variations on the theme, telomeres and centromeres are functionally highly conserved chromosome domains (Blackburn and Szostak, 1984; Biessman and Mason, 1994). While centromeres act to ensure separation of duplicated chromosomes during mitosis and of recombined ones during metaphase I of meiosis, telomeres are the best candidates for ensuring pairing interactions of replicated chromosomes by their clustering at earliest meiotic prophase. Processes acting during chromosome evolution must therefore spare the highly conserved chromosome domains at the very ends. This might explain the conservation of meiotic telomere behaviour, despite the high diversity of chromosome numbers and morphologies observed (see e.g. White, 1973).

ACKNOWLEDGMENTS

I wish to thank Prof. H. Zankl, Kaiserslautern, for continuous support, J. Loidl, Vienna, and A. Regelin, Kaiserslautern, for stimulating critical comments on the MS; T. Cremer and S. Dietzel for communication of data prior to their publication. I apologize to those investigators whose work has been only cited through reviews.

REFERENCES

Albini, S.M. and Jones, G.H. (1987) Synaptonemal complex spreading in *Allium cepa* and *A. fistulosum*. *Chromosoma*, **95**, 324—38.

Albini, S.M. and Schwarzacher T. (1992) In situ localization of two repetitive DNA sequences to surface spread pachytene chromosomes of rye. *Genome*, **35**, 551—9.

Armstrong, S.J., Kirkham, A.J. and Hultén, M.A.J. (1994) XY chromosome behaviour in the germ-line of the human male: a FISH analysis of spatial orientation,chromatin condensation and pairing. *Chromosome Research*, **2**, 445—52.

Ashley, T. (1979) Specific end to end attachment of chromosomes of *Ornithogallum virens*. *Journal of Cell Science*, **38**, 357—67.

Ashley, T. and Pocock, N. (1981) A proposed model of chromosomal organization in nuclei at fertilization. *Genetica*, **55**, 161—9.

Ashley, T., Cacheiro, N., Russell. L.B. and Ward D.C. (1993) Molecular characterization of a pericentric inversion in mouse chromosome-8 implicates telomeres as promoters for meiotic recombination. *Chromosoma*, **102**, 112—20.

Bähler, J., Hagens, G., Holzinger. G. *et al.* (1994) Possible meiotic checkpoint at pachytene revealed in *Saccharomyces cerevisiae* cells lacking the homologous pairing protein $p175^{SEP1}$. *Chromosoma*, **103**, 129—41.

Bélár, K. (1928) Chromosomenreduktion. Die cytologischen Grundlagen der Vererbung. In *Handbuch der Vererbungswissenschaft*, (eds. E. Baur and M. Hartmann) Geb. Borntraeger, Berlin. Vol. 1, pp.168—201.

Bellvé, A. R., Clarke, F.M., Bhatnagar, Y.M. and O_Brian, D.A. (1977a) Dissociation of the mouse testis and characterization of isolated spermatogenic cells. *Journal of Histochemistry and Cytochemistry*, **25**, 480—94.

Bellvé, A. R., Caviccia, J.C., Millette, C.F. *et al.* (1977b) Spermatogenic cells of the prepuberal mouse. *Journal of Cell Biology*, **74**, 68—85.

Bennett, M.D., Toledo L.A. and Stern H. (1979) The effect of colchicine on meiosis in *Lilium speciousum* cv. Rosmede. *Chromosoma*, **72,** 175—89.

Bennett, M.D. (1983) The spatial distribution of Chromosomes. Kew Chromosome Conference II, pp. 239—61.

Bennett, M.D. (1984) Premeiotic events and meiotic chromosome pairing. *Symposia of the Society of Experimental Biology*, **38**, 87—121.

Bernardi, G. (1989) The isochore organization of the human genome. *Annual Review of Genetics*, **23**, 637—61.

Bishop, D.K. (1994) RecA homlogs Dmc1 and Rad51 interact to form multiple nuclear complexes prior to meiotic chromosome synapsis. *Cell*, **79,** 1081—92.

Biessmann, H. and Mason, J.M. (1994) Telomere repeat sequences. *Chromosoma*, **103**, 154—61.

Blackburn, E.H. and Szostak, J.W. (1984) The molecular structure of centromeres and telomeres. *Annual Review of Biochemistry*, **53**, 163—94.

Blackburn, E.H. (1994) Telomeres: no end in sight. *Cell*, **77,** 621—3.

Boiko, M. (1983) Human meiosis VIII. Chromosome pairing and formation of the synaptonemal complex in Oocytes. *Carlsberg Research Communications*, **48**, 457—83.

Boiko, M. (1990) Synaptic adjustment of inversion loops in *Neurospora crassa*. *Genetics*, **124**, 593—8.

Boyle, A.L. and Ward, D.C. (1992) Isolation and initial characterization of a large repeat sequence element specific to mouse chromosome 8. *Genomics*, **12**, 517—25.

Brown, W. V. and Stack, S. M. (1968) Somatic pairing as a regular preliminary to meiosis. *Bulletin of the Torrey Botanical Club*, **95**, 369—78.

Bussey, H., Kaback, D.B., Zhong W.W. *et al.*, (1995) The nucleotide sequence of chromosome I from *Saccharomyces cerevisiae*. *Proceedings of the National Academy of Science USA*, **92**, 3809—13.

Carpenter, A. (1975) Electron microscopy of meiosis in *Drosophila melanogaster* females. *Chromosoma*, **51**, 157—82.

Carpenter, A. (1987) Gene conversion, recombination nodules, and initiation of meiotic synapsis. *BioEssays*, **6**, 232—6.

Chandley, A. C. (1986) A model for effective pairing and recombination at meiosis based on early replicating sites (R-bands) along chromosomes. *Human Genetics*, **72**, 50—7.

Chemes, H.E., Fawcett, D.W. and Dym, M. (1978) Unusual features of the nuclear envelope in human spermatogenic cells. *Anatomical Record*, **192**, 493—512.

Chen, Q., Rearlman R. and Moens, P.B. (1992) Isolation and characterization of a cDNA encoding synaptonemal complex proteins. *Biochemistry and Cell Biology*, **70**, 1030—8.

Cheng, E.Y. and Gartler, S.M. (1994) A fluorescent in situ hybridization analysis of X chromosome pairing in human female meiosis. *Human Genetics*, **94**, 389—94.

Cherry, L.M. and Hsu, T.C. (1984) Antitubulin immunofluorescence studies of spermatogenesis in the mouse. *Chromosoma*, **90**, 265—74.

Chevret, E., Rousseaux, S., Monteil *et al.* (1995) Meiotic segregation of of X and Y chromosomes and chromosome 1 analyzed by three-color FISH in human interphase spermatozoa. *Cytogenetics and Cell Genetics*, **71**, 126—30.

Chigashige, Y., Ding, D.Q., Funabiki, H. *et al.* (1994) Telomere-led premeiotic chromosome movement in fission yeast. *Science*, **264**, 270—3.

Church, K. (1976) Arrangement of chromosome ends and axial core formation during early meiotic prophase in the male grasshopper *Brachystola magna* by 3D, EM reconstruction. *Chromosoma*, **58**, 365—76.

Cole, K.D., Kandala, J.C. amd Kistler, W.S. (1986) Isolation of the gene for the testis-specific H1 histone variant H1t. *Journal of Biological Chemistry*, **261**, 7178—83.

Craig, J.M. and Bickmore, W.A. (1994) The distribution of CpG islands in mammalian chromosomes. *Nature Genetics*, **7**, 376—82.

Cremer, T., Kurz, A., Zirbel, R. et al. (1993) The role of chromosome territories in the functional compartmentalization of the cell nucleus. *Cold Spring Harbor Symposia on Quantitative Biology*, **58**, 777—92.

Darlington, C.W. (1931) Meiosis. *Biol. Reviews*, **6**, 221—64.

Dawe, K.R., Sedat, J.W., Agard, D.A. and Cande, Z.W. (1994) Meiotic chromosome pairing in Maize is associated with a novel chromatin organization. *Cell*, **76**, 901—12.

de Boer, P. and de Jong, J.H. (1989) Chromosome pairing and fertility in mice. In *Fertility and Chromosome Pairing: Recent Studies in Plants and Animals*, (ed. C.B. Gillies) CRC Press, Boca Raton, Florida, pp. 37—76.

de Boni, U. (1994) The interphase nucleus as a dynamic structure. *International Review of Cytology*, **150,** 149—71.

de Lange, T. (1992) Human telomeres are attached to the nuclear matrix. *EMBO Journal*, 11, 717—24.

de Rooij, D.G. (1988) Regulation of the proliferation of spermatogonial stem cells. *Journal of Cell Science*, Suppl. **10,** 181—94.

Dernburg, A.F., Sedat, J.W., Cande, W.Z. and Bass, H.W. (1995) The cytology of telomeres. In *Telomeres*, (eds E.H. Blackburn and C.W. Greider) Cold Spring Harbor Monograph Series: CSH Press (in press).

Dietrich, A.J.J. and de Boer, P. (1983) A sequential analysis of the development of the synaptonemal complex in spermatocytes of the mouse by electron microscopy using hydroxyurea and agar filtration. *Genetica*, **61,** 119—29.

Dietrich, A.J.J., Kok, E., Offenberg, H.H. *et al.* (1992) The sequential appearance of the synaptonemal complex during meiosis of the female rat. *Genome*, **35,** 492—7.

Dobson, M.J., Pearlman, R.E., Karaiskakis, A. *et al.* (1994) Synaptonemal complex proteins: occurence, epitope mapping and chromosome disjunction. *Journal of Cell Science*, **107,** 2749—60.

Dresser, M. and Giroux, C. (1988) Meiotic chromosome behaviour in spread preparations of yeast. *Journal of Cell Biology*, **106**, 567—73.

Driscoll, C.J. and Darvey, N.L. (1970) Chromosome pairing: effect of colchicine on an isochromosome. *Science*, **169,** 290—1.

Dujon, B., Alexandraki, D., André, B., *et al.*, (1994) Complete DNA sequence of yeast chromosome XI. *Nature*, **369**, 371—8.

Eisen, G. (1900) The spermatogenesis of Batrachoseps. *Journal of Morphology*, **17**, 1—117.

Fawcett, D.W. and Chemes, H.E. (1979) Changes in distribution of nuclear pores during differentiation of the male germ cells. *Tissue & Cell*, **11**, 147—62.

Feldman, H., Aigle, M., Alijnovic G., *et al.*, (1994) Complete DNA sequence of yeast chromosome II. *EMBO Journal*, **13**, 5795—809.

Ferguson, M.L.F. and Ward, D.C. (1992) Cell cycle dependent chromosomal movement in pre-mitotic human T-lymphocyte nuclei. *Chromosoma*, **101,** 557—65.

Franke, W.W. (1971) Relationship of nuclear membranes with filaments and microtubules. *Protoplasma*, **73,** 263—92.

Fuchikawa, K., Inagaki, H., Naruge, T. *et al.* (1994) cDNA cloning and functional chracterization of a meiosis specific protein (MNS1) with apparent nuclear association. *Chromosome Research*, **2**, 99—113.

Fussell, C.P. (1987) The rabl orientation: A prelude to synapsis. In *Meiosis*, (ed P.B. Moens) Academic Press, Orlando, pp. 275—99

Gelei, J. (1905) Weitere Studien über die Oogenese des *Dendrocoelum lacteum.* III. Die Konjugation der chromosomen in der Literatur und meine Befunde. *Archiv f. Zellforsch.*, **16**, 300—65.

Gibson, J.B. and Moses, M.J. (1986) Effects of vinblastine sulfate on the synaptonemal complex in *Mus musculus*. *Genetics*, **113,** 66.

Gillies, C.B. (1974) The nature and extent of synaptonemal complex formation in haploid barley. *Chromosoma*, **48,** 441—53.

Gillies, C.B. (1975) Synaptonemal Complex and chromosome pairing. *Annual Review of Genetics*, 9, 91—109.

Gilson, E.T., Laroche, T. and Gasser, S.M. (1993) Telomeres and the functional architecture of the nucleus. *Trends in Cell Biology*, **3**, 128—34.

Giroux, C. (1988) Chromosome synapsis and meioti recombination. In *Genetic Recombination*, (eds R. Kucherlapati and G.R. Smith) American Society of Microbiology, Washington D.C., pp. 497—527.

Glamann, J. (1986) Crossing over in the male mouse as analysed by recombination nodules and bars. *Carlsberg Research Communications*, **51,** 143—162.

Goetz, P., Chandley, A.C. and Speed, R.M. (1984) Morphological and temporal sequence of meiotic prophase develpoment at puberty in the male mouse. *Journal of Cell Science*, **65,** 249—63.

Goldman, A.S.H. and Hultén, M.A. (1992) Chromosome in situ suppression hybridisation in human male meiosis. *Journal of Medical Genetics*, **29,** 98—102.

Goldman, A.S.H. and Hultén, M.A. (1993) Meiotic analysis by FISH of a human male 46,XY,t(15;20)(q11.2;q11.2) translocation heterozygote; quadrivalent configuration, orientation at first meiotic segregation. *Chromosoma*, **102,** 102—11.

Guitart, M., Coll, M.D., Ponsa, M. and Egozcue J. (1985) Sequential study of synaptonemal complexes in mouse spermatocytes by light and electron microscopy. *Genetica*, **67**, 21—30.

Haaf, T. and Schmid, M. (1991) Chromosome topology in mammalian interphase nuclei. *Experimental Cell Research*, **192**, 325—32.

Hasekampf, C.A. (1984) Synaptonemal complex formation in pollen mother cells of *Tradescantia*. *Chromosoma*, **90,** 275—84.

Hawley, R.S. (1988) Exchange and chromosomal segregation in eukaryotes. In *Genetic Recombination* (eds R. Kucherlapati and G.R. Smith) American Society of Microbiology, Washington D.C., pp 497—527.

Hawley, R.S. and T. Arbel (1993) Yeast Genetics and the fall of the classical view of meiosis. *Cell*, **72**, 301—3.

Heng, H.H.Q., J. Chamberlain, X.-M. Shi *et al.* (1995) Effect of chromosome position on meiotic chromatin loop size. *Chromosome Research*, **3**, 116.

Hiraoka T. (1952) Observational and experimental studies of meiosis with special reference to the bouquet stage. XIV. Some considderations on a probable mechanism of the bouquet formation. *Cytologia*, **17**, 292—9.

Hiraoka T. and Fuchikawa Y. (1993) An electron microscopic and morphometric analysis on the cell polarity in the synaptic stage of meiosis. *Cytologia*, **58,** 77—84.

Hiraoka, Y., Dernburg, A.F., Parmerlee, S.J. *et al.* (1993) The onset of homologous chromosome pairing during *Drosophila melanogaster* embryogenesis. *Journal of Cell Biology*, **120**, 591—600.

Holmquist, G. P. (1992) Chromosome bands, their chromatin flavors, and their functional features. *American Journal of Human Genetics*, **51**, 17—37.

Hughes-Schrader, S. (1943) Polarization, kinetochore movements and bivalent structure in the meiosis of male mantids. *Biological Bulletin*, **85**, 265—300.

Jinks-Robertson, S. and Petes, T.D. (1985) High-frequency meiotic gene conversion between repeated genes on non-homologous chromosomes. *Proceedings of the National Academy of Sciences USA*, **82**, 3350—4.

Johannisson, R. and Winking, H. (1994) Synaptonemal complexes of chains and rings in mice heterozygous for multiple Robertsonian translocations. *Chromosome Research*, **2**, 137—45.

Johnston, M., Andrews, S., Brinkman, R., *et al.*, (1994) Complete DNA sequence of yeast chromosome VIII. *Science*, **265**, 2077—82.

John, B. (1969) Myths and mechanisms in meiosis. *Chromosoma*, **54**, 295—325.

John, B. (1990) *Meiosis*. Cambridge University Press.

Jones, G.H., and Croft, J.A. (1986) Surface spreading of synaptonemal complexes in locust. II. Zygotene pairing behaviour. *Chromosoma*, **93**, 489—95.

Kleckner, N., Padmore, R. and Bishop, D.K. (1991) Meiotic chromosome metabolism: one view. *Cold Spring Harbor Symposia on Quantitative Biology*, **56**, 729—43.

Kleckner, N. and Weiner, B. (1993) Potential advantages of unstable interactions for pairing in meiotic, somatic, and premeiotic cells. *Cold Spring Harbor Symposia on Quantitative Biology*, **58**, 553—65.

Klein, F., Laroche T., Cardenas, M.E. *et al.* (1992) Localization of Rap1 and topoisomerase II in nuclei and chromosomes of yeast. *Journal of Cell Biology*, **117**, 935—48.

Klein, H. (1995) Choose your partner. Chromosome pairing in yeast. *BioEssays*, **16**, 869—71.

Kluin, P.M. and de Rooij, D.G. (1981) A comparison between the morphology and cell kinetics of gonocytes and adult type undifferentiated spermatogonia in the mouse. *International Journal of Andrology*, **4**, 475—93.

Kohli, J. (1994) Telomeres lead chromosome movement. *Current Biology*, **4**, 724—7.

Kubai D.F. and Ris H. (1969) Division in the dinoflagellate *Gyrodinium cohnii* (Schiller.) *Journal of Cell Biology*, **40**, 508—28.

Lammers, J.H.M. Offenberg, H.H., van Aalderen, M. *et al.* (1994) The gene encoding a major component of the lateral elements of synaptonemal complexes of the rat is related to X-linked lymphocyte-regulated genes. *Molecular and Cellular Biology*, **14**, 1137—46.

Lichten, M., Borts, R.H. and Haber, J.E. (1987) *Genetics*, **115**, 223—46.

Lichter, P., Cremer, T., Borden, J. *et al.* (1988) Delineation of individual human chromosomes in metaphase and interphase cells by in situ suppression hybridization using recombinant DNA-libraries. *Human Genetics*, **80**, 224—34.

Lichter, P., Boyle, A.L., Cremer, T., and Ward D.C. (1991) Analysis of genes and chromosomes by non-isotopic in situ hybridization. *Genet. Anal. Techn. Appl.*, **8**, 24—35.

Lima-de-Faria A. (1983) *Molecular evolution and organization of the chromosome*. Elsevier, Amsterdam, New York, Oxford.

Loidl, J. (1989) Colchicine action at meiotic prophase revealed by SC-spreading. *Genetica*, **78**, 195—203.

Loidl, J. (1990) The initiation of meiotic chromosome pairing: the cytological view. *Genome*, **33**, 759—78.

Loidl, J. (1991) Meiotic chromosome synapsis in haploid yeast. *Chromosoma*, **100**, 221—8.

Loidl, J., Klein, F. and Scherthan, H. (1994) Homologous pairing is reduced but not abolished in asynaptic mutants of yeast. *Journal of Cell Biology*, **125**, 1191—200.

Loidl, J. (1994) Cytological aspects of meiotic recombination. *Experientia*, **50**, 285—94.

Maguire, M.P. (1984) The mechanism of meiotic chromosome pairing. *Journal of Theoretical Biology*, **106**, 605—15.

Maguire, M.P. (1988) Interactive meiotic systems. In *Chromosome structure and function* (ed J.P. Gustavson) Plenum Press, New York, pp. 605—15.

Maguire, M.P. and Riess, R.W. (1994) The relationship of homologous synapsis and crossing over in a Maize inversion. *Genetics*, **137**, 281—8.

Meistrich, M.L., Bucci, L.R., Trostle-Weige, P.K. and Brock W.A. (1985) Histone variants in rat spermatogonia and and primary spermatocytes. *Dev. Biol.*, **112**, 230—40.

Metz, C.W. (1916) Chromosome studies on the Diptera. II. The paired association of chromosomes in the Diptera, and its significance. *Journal of Experimental Zoology*, **21**, 213—79.

Meuwissen, R.L.J., Offenberg, H.H., Dietrich, A.J.J. *et al.* (1992) A coiled-coil related protein specific for synapsed regions of meiotic prophase chromosomes. *EMBO Journal*, **11**, 5091—100.

Moens, P.B. (1969) The fine structure of meiotic chromosome polarization and pairing in *Locusta migratoria* spermatocytes. *Chromosoma*, **28**, 1-25.

Moens, P.B. (1974) Quantitative electron microscopy of chromosome organization at meiotic prophase. *Cold Spring Harbor Symposia on Quantitative Biology*, **38**, 99—107.

Moens, P.B. (1994) Molecular perspectives of chromosome pairing at meiosis. *BioEssays*, **16**, 101—6.

Moens, P.B. Bernelot-Moens, C., Spyropulos, B.(1989) Chromosome core attachment to the nuclear envelope regulates synapsis in *Choelatis* (Orthoptera). *Genome*, **32**, 601—10.

Moens, P.B. and Pearlman, R.E. (1988) Chromatin organization at meiosis. *BioEssays*, **9**, 151—3.

Moens, P.B. and Pearlman, R.E. (1989) Satellite DNA I in chromatin loops of rat pachytene chromosomes and in spermatids. *Chromosoma*, **98**, 287—94.

Moens, P.B. and Pearlman, R.E. (1990) Telomere and centromere DNA are associated with the cores of meiotic prophase chromosomes. *Chromosoma*, **100**, 8—14.

Moses, M.J. (1968) Synaptinemal complex. *Annual Review of Genetics*, **2**, 363—412.

Moses, M.J., Dresser, M.E. and Poorman, P.A. (1984) Composition and role of the synaptonemal complex. *Symposia of the Society of Experimental Biology*, **38**, 245—70.

Moss, S.B., Burnham, B.L. and Bellvé, A.R. (1993) The differential expression of lamin epitopes during mouse spermatogenesis. *Molecular Reproduction and Development*, **34**, 164—74.

Moyzis, R.K., Buckingham, J.M., Cram, C.S. *et al.* (1988) A highly concerved DNA sequence, (TTAGGG)n, present at the telomeres of human chromosomes. *Proceedings of the National Academy of Sciences USA*, **83**, 6622—8.

Nag, D.K., Scherthan, H., Rockmill, B. *et al.* (1995) Heteroduplex formation and homolog pairing in yeast meiotic mutants. *Genetics*, **141**, 75—86.

O'Keeffe, C., M. Hultén, S. Armstrong, and C. Tease (1995) Analysis of chromosome pairing in mouse female meiosis. *Chromosome Research*, **3** Suppl.1, 119.

Oliver, S.G., Van der Aart Q.J.M., Agostini-Carbone M.L. *et al.*, (1992) Complete DNA sequence of yeast chromosome III. *Nature*, **357**, 38—46.

Oud, J.L. and Reutlinger, A.H.H. (1981) Chromosome behaviour during early meiotic prophase of mouse spermatocytes. *Chromosoma*, **83**, 395—407.

Padmore, R., Cao, L. and Kleckner, N. (1991) Temporal comparison of recombination and synaptonemal complex formation during meiosis in *S. cerevisiae*. *Cell*, **66**, 1239—56.

Palladino, F., Laroche, T., Gilson, E. *et al.* (1993) SIR3 and SIR4 Proteins are required for the positioning and integrity of yeast telomeres. *Cell*, **75**, 543—55.

Pardue, M.L, and Gall, J.G. (1970) Chromosomal localization of mouse satellite DNA. *Science*, **168**, 1356—8.

Parvinen, M. and Söderström, K.-O. (1976) Chromosome rotation and formation of synapsis. *Nature*, **260**, 534—5.

Pearlman, R.E., Tsao, N. and Moens, P.B. (1992) Synaptonemal complexes from DNase treated rat spermatocytes contain (GT)n and Line sequences but no MARs/SARs. *Genetics*, **130**, 865—72.

Penman, S. (1995) Rethinking cell structure. *Proceedings of the National Academy of Science USA*, **92**, 5251—7.

Pinkel, D., Landegent, J., Collins, C. *et al.* (1988) Fluorescence in situ hybridization with human chromosome specific libraries: detection of trisomy 21 and translocations of chromosome 4. *Proceedings of the national Academy of Science USA*, **85**, 9138—42.

Poccia, D.L., Simpson, M.V. and Green, G.R. (1987) Transitions in histone variants during sea urchin spermatogenesis. *Developmental Biology*, **121**, 445—53.

Puertas, M.J., de la Peña, A., Estades, B. and Merino. F. (1984) Early sensitivity to colchicine in developing anthers of rye. *Chromosoma*, **89**, 121—6.

Radman, M. and Wagner, R. (1993) Mismatch recognition in chromosomal interactions and speciation. *Chromosoma*, **102**, 369—73.

Ragg, S., Härle, M. and Scherthan, H. (1995) Analysis of sex chromosome distribution in the gondal tissue of a 42,X/47,XYY mosaic by fluorescence in situ hybridization. *Modern Pathology*, **8**, 295—8.

Rasmussen, S.W. and Holm, P.B. (1978) Human Meiosis II. Chromosome pairing and recombination nodules in human spermatocytes. *Carlsberg Research Communications*, **43**, 275—327.

Rasmussen S.W. and Holm, P.B. (1980) Mechanics of Meiosis. *Hereditas*, **93**, 187—216.

Rhoades, M.M. (1961) Meiosis. In *The Cell: Biochemistry, Physiology, and Morphology, Vol. 3: Meiosis and Mitosis* (eds J. Brachet and A.E. Mirsky) Academic Press, New York, pp. 1—75.

Rickards, G.K. (1975) Prophase chromosome movements in living house cricket spermatocytes and their relation to prometaphase, anaphase and granule movements. *Chromosoma*, **49**, 407—55.

Riggs, C.D. (1994) Molecular cloning of cDNAs encoding variants of meiotin-1. *Chromosoma*, **103**, 251—61.

Roeder, G.S. (1990) Chromosome synapsis and genetic recombination: their roles in meiotic chromosome segregation. *Trends in Genetics*, **6**, 385—9.

Roeder, G.S. (1995) Sex and the single cell: meiosis in yeast. *Proceedings of the National Academy of Science USA*, (in press).

Rockmill. B., Sym. M., Scherthan, H. and Roeder G.S. (1995) Two RecA homologs facilitate meiotic chromosome synapsis by establishing connections between homologous chromosomes. *Genes and Development*, (in press).

Ross-McDonald, P. and Roeder, G.S. (1994) Mutation of a meiosis-specific MutS homolog decreases crossing over but not mismatch correction. *Cell*, **79**, 1069—80.

Salonen, K., Paranko, J. and Parvinen, M. (1982) A colcemid sensitive mechanism involved in regulation of chromosome movements during meiotic pairing. *Chromosoma*, **85**, 611—8.

Santos, J.L., A.L. del Cerro, and M. Diez. (1993) Spreading synaptonemal complexes from the grasshopper *Chorthippus jacobsi*: pachytene and zygotene observations. *Hereditas*, **118**, 235—41.

Santos, J.L., Jiménez, M.M. and Díez, M. (1994) Meiosis in haploid rye:extensive synapsis and low chiasma frequency. *Heredity*, **73**, 580—8.

Scherthan, H., Loidl, J., Schuster, T. and Schweizer, D. (1992) Meiotic chromosome condensation and pairing in *Saccharomyces cerevisiae* studied by chromosome painting. *Chromosoma*, **101**, 590—5.

Scherthan, H., Schweizer, D. and Loidl, J. (1993) Delineation of individual chromosomes of *Saccharomyces cerevisiae* by two colour in situ hybridisation. *Trends in Genetics*, **9**, 41.

Scherthan, H., Bähler, J. and Kohli, J. (1994) Dynamics of chromosome organization and pairing during meiotic prophase of fission yest. *Journal of Cell Biology*, **127**, 273—85.

Scherthan, H. and Cremer, T. (1994) Methodology of Non Isotopic In Situ-Hybridization in Embedded Tissue Sections. In *Methods in Molecular Genetics*, (ed K.W. Adolph) Academic Press, San Diego, **5**, 223—38.

Scherthan, H. (1995) Chromosome evolution in muntjac revealed by centromere, telomere and whole chromosome paint probes. *Kew Chromosome Conference IV* (ed P.E. Brandham), 267—80.

Scherthan, H., Weich, S., Schwegler, H. *et al.* (1995) Centromere and telomere movements during early meiotic prophase of mouse and man are associated with the onset of chromosome pairing. submitted.

Schreiner, A. and Schreiner, K.E. (1905) Über die Entwickelung der männlichen Geschlechtszellen von *Myxine glutinosa* L. II. Die Centriolen und ihre Vermehrungsweise. *Archives de Biologie*, **21**, 315—55.

Selig, S., Okumura, K., Ward, D.C. and Cedar, H. (1992) Delineation of replication time zones by fluorescence in situ hybridization. *EMBO Journal*, **11**, 1217—25.

Smithies, O. and Powers, P.A. (1986) Gene conversions and their relation to to homologous chromosome pairing. *Phil. Trans. R. Soc. Lond. B*, **312**, 291—302.

Speed, R.M. (1982) Meiosis in the foetal mouse ovary. *Chromosoma*, **85**, 427—37.

Speed, R.M., Vogt, P., Köhler, M.R. *et al.* (1993) Chromatin condensation behaviour of the Y chromosome in human testis: I. Evidence for decondensation of distal Yq in germ cells prior tp puberty with a switch to sertoli cells in adults. *Chromosoma*, **102**, 421—7.

Stack, S.M. (1984) Heterochromatin the synaptonemal complex and crossing over. *Journal of Cell Science*, **71**, 159—76.

Stack, S.M. and Brown, W.V. (1969) Somatic pairing as a regular preliminary to meiosis. *Nature*, **222**, 1275—6.

Sukegawa, J. and Blobel, G. (1993) A nuclear pore complex protein that contains zinc finger motifs and faces the nucleoplasm. *Cell*, **72**, 29—38.

Sym, M., Engebrecht, J. and Roeder, G.S. (1993) ZIP1 is a synaptonemal complex protein reqired for meiotic chromosome synapsis. *Cell*, **72**, 365—78.

Sym, M. and Roeder, G.S. (1994) Crossover interference is abolished in the absence of a synaptonemal complex protein. *Cell*, **79**, 283—92.

Terasawa. M, Shinohara, A., Hotta, Y. *et al.* (1995) Localization of Rec-A like recombination proteins on chromosomes of the lily at various stages of meiosis. *Genes and Development*, **9**, 925—34.

Therman, E. and Sarto, G.E. (1977) Premeiotic and early meiotic stages in the pollen mother cells of eremus and in human embrionic oocytes. *Human Genetics*, **35**, 137—51.

Venster, B., Smith, A., Krohne, G. and Benavente, R. (1993) Presence of a nuclear lamina in pachytene spermatocytes of the rat. *Journal of Cell Science*, **104**, 557—63.

Vergowen, R.P.F.A., Jacops, S.G.P.M., Huiskamp, R. *et al.* (1991) Proliferative activity of gonocytes, Sertoli cells and interstitial cells during testicular development in mice. *Journal of Reproduction and Fertility*, **93**, 233—43.

Vergowen, R.P.F.A., Huiskamp, R., Bas, R.J. *et al.* (1993) Postnatal development of testicular cell populations in mice. *Journal of Reproduction and Fertility*, **99**, 233—43.

Virkki, N. (1974) The Bouquet. *Journal of Agriculture of the University of Puerto Rico*, **58**, 479—85.

von Wettstein, D., Rasmussen, S.W. and Holm, P.B. (1984) The synaptonemal complex in genetic segregation. *Annual Review of Genetics*, **18**, 331—413.

Vourc'h, C., Taruscio, D., Boyle, A.L. and Ward, D.C. (1993) Cell cycle-dependent distribution of telomeres, centromeres, and chromosome-specific subsatellite domains in the interphase nucleus of mouse lymphocytes. *Experimental Cell Research*, **205**, 142—51.

Wagenaar, E.B. (1969) End-to-end attachments in mitotic interpahse and their possible significance to meiotic chromosome pairing. *Chromosoma*, **26**, 410—26.

Walters, M.S. (1970) Evidence on the time of chromosome pairing in *Lilium longiflorum* 'Croft'. *Chromosoma*, **29**, 375—418.

Wandall, A. and Svendsen, A. (1985) Transition from somatic to meiotic pairing and progressional changes of the synaptonemal complex. *Chromosoma*, **92**, 254—64.

Weimer, R., Haaf, T., Poot, M. and Schmid, M. (1992) Characterization of centromere arrangements and test for random distribution in G_0, G_1, S, G_2, G_1, and early S' phase in human lymphocytes. *Human Genetics*, **88**, 673—82.

Weiner, B. and Kleckner, N. (1994) Chromosome pairing via multiple interstitial interactions before and during meiosis in yeast. *Cell*, **77**, 977—91.

Wilken, N., Kossner, U., Senécal, J.-L. *et al.* (1993) Nup180, a novel nuclear pore complex protein localizing to the cytoplasmic ring and associated fibrils. *Journal of Cell Biology*, **123**, 1345—54.

White, M.J.D. (1973) *Animal Cytology and Evolution*. Cambridge University Press, London.

Wilson, E.B. (1925) *The Cell in Heredity and Development*, 3rd edition. MacMillan, New York.

Zickler, D. and L.W. Olson. (1975) The synaptonemal complex and the spindle plaque during meiosis in yeast. *Chromosoma*, **50**, 1—23.

Zickler, D. (1977) Development of the synaptonemal complex and 'recombination nodules' during the meiotic prophase in the seven bivalents of the fungus *Sordaria macrospora* Auersw. *Chromosoma*, **61**, 289—316.

15

Univalent orientation in living meiocytes

E. Rebollo and P. Arana

15.1 INTRODUCTION

The spindle is adapted to bipartite units. When the two chromatids of a mitotic chromosome or the two chromosomes of a bivalent in meiosis attach to a different pole at prometaphase (bipolarity), tension forces of opposite directions develop, the orientation is stabilized and persists until anaphase. Then partner chromatids or chromosomes are correctly distributed to daughter nuclei.

Chromosomes lacking an attached partner in the first meiotic division (univalents) have certain difficulties in acquiring stability. Sister kinetochores behave as a unit in the first meiotic division (syntely) so none of the unipolar syntelic orientations of a univalent can be stabilized by opposite tension forces. This instability leads to a series of random orientational and reorientational movements that may end in reductional segregation of the whole chromosome to one pole or to equational separation of chromatids at anaphase I (Östergren, 1951; Dietz, 1956; Nicklas, 1961). Equational segregation is a consequence of amphitelic (mitotic-like) orientation that may occur at the end of prometaphase. Amphitelic orientations are bipolar and stable and imply disruption of the side-to-sideness of sister kinetochores that is the rule at the first meiotic division (Wagenaar and Bray, 1973; Goldstein, 1981). Amphioriented chromosomes lie perpendicular to the spindle axis, cease interpolar movements and lag at anaphase I (Bauer *et al.*, 1961). Lagging normally has deleterious consequences for the cells: if the chromatids do not separate, laggards may not be included in the daughter nuclei or may provoke spindle collapse yielding unreduced secondary meiocytes. Even when the

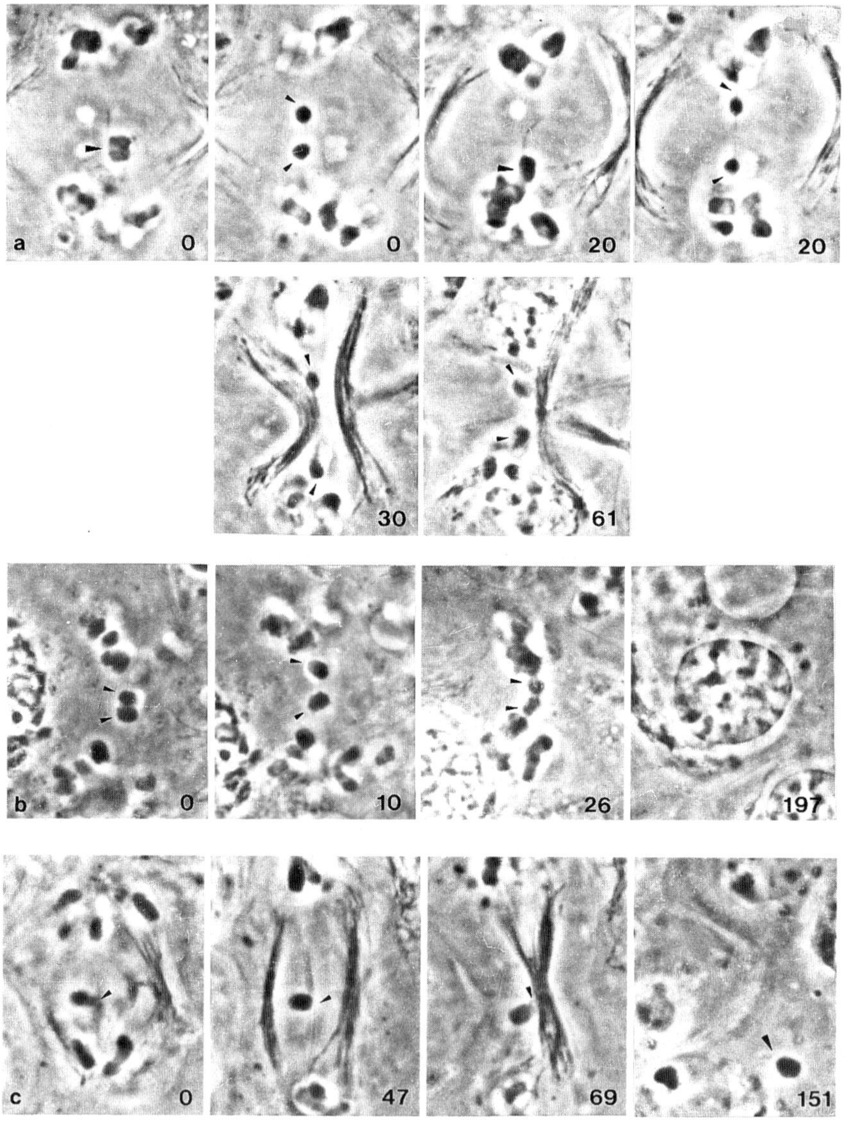

Fig. 15.1 Segregation abnormalities of univalents. **1a** Two anaphase I laggards (0 min prints) showing reductional and equational divisions respectively (20 min prints). Separated chromatids are not included in the daughter nuclei (30 and 61 min prints). **1b** Spindle collapse in a spermatocyte with one anaphase I laggard. An unreduced nucleus is formed (197 min print). **1c** Migration failure of a lagging chromatid at anaphase II.

chromatids succeed in separation, they often lag and are lost at anaphase II (Fig. 15.1).

The complements of many species, however, possess chromosomes that, despite their univalent condition, have evolved specific mechanisms to achieve regular transmision. Some of these mechanisms allow the correct segregation of two or more sex univalents so as the sex determination system is self perpetuated (see Wolf, 1994 for review). Other chromosomes lacking any possible partner have also developed mechanisms for regular transmission: in grasshoppers with sex determinism XX—XO, males possess a single sexual univalent that flawlessly segregates to one pole at anaphase I and separates chromatids at anaphase II. As irregular segregation of any member of the normal complement produces genetic imbalance with deletereous effects on individual fitness, these chromosomal mechanisms have been probably acquired by natural selection.

Some of these XX—XO grasshopper species possess B chromosomes. B chromosomes are dispensable genetic elements that do not crossover with the members of the normal complement and form widespread polymorphisms in many plant and animal species (Camacho *et al.*, p.301—318). Most Bs studied are considered 'selfish' elements since they are maintained in the populations by preferential transmission mechanisms despite their harmful effect on the fitness of the bearers (see Shaw and Hewitt, 1990, for review). Thus the success of a B chromosome in being established in a natural population depends directly on the effectiveness of its own transmission mechanisms (Henriques-Gil and Arana, 1990; Lopez-León *et al.*, 1992). As Bs in meiosis appear mostly as univalents, their behaviour should achieve a certain degree of regularity that is acquired by 'non phenotypic selection', instead of natural selection.

A third, radically different situation is found in univalents deriving from autosomes of the normal complement. In normal conditions, the genetic control of crossing over ensures the formation of at least one chiasma per bivalent preventing the generation of univalents. Autosomal univalency is deleterious not only for the risk of segregation abnormalities but also because random syntelic segregation of two homologous univalents at meiosis I would yield 50% aneuploid gametes. Thus, it is unlikely that the behaviour of univalents of autosomes has been moulded by any kind of selection, either natural or non-phenotypic.

We have analyzed the orientation behaviour at prometaphase I of different univalents in grasshopper species with an XX—XO determinism of sex and B chromosomes: *Eyprepocnemis plorans* (Rebollo and Arana, 1995), *Locusta migratoria*, *Heteracris littoralis* and *Omocestus burri*. In two of them: *E. plorans* and *L. migratoria*, autosomal univalents were induced by heat treatment at early meiotic stages when pairing and crossing over occur (Henderson, 1963). As two or more univalents of different nature may be present within the same spindle their orientation behaviour can be directly compared.

The choice between reductional or equational segregation at either meiotic division is determined at the chromosomal level. When Nicklas (1977), in grasshopper living spermatocytes, transferred prometaphase I bivalents to second division spindles and *vice versa*, chromosomes segregated according to their spindles of origin. In yeast single division mutants, different chromosomes have different preferences for reductional or equational segregation. When centromeric regions are transferred to heterologous chromosomes, these acquire the segregation tendency of the new centromere (see Simchen and Hugerat, 1993, for review).

Although chromosomal segregations can be readily addressed in fixed material, living cultured cells are necessary to disclose the dynamic strategies that a chromosome displays to achieve transmission regularity, as the same final segregational outcome may be the result of very different orientational behaviours (Ault, 1986).

So, our system allows us to analyze to what extent orientation and segregation features are controlled by chromosomal, rather than cellular characteristics.

15.2 VELOCITY ESTIMATES

In our experiments, living cultured spermatocytes at prometaphase I were tracked until anaphase. A series of photographs at different focal levels to show centrioles and centromeric regions of all univalents was taken at approximately 15 min intervals and also whenever a reorientation movement occurred. The cells were reconstructed from images projected in a photographic enlarger. To estimate the distance covered by each chromosome in each interval, the midpoint of the cellular axes in two consecutive series were superimposed and the straight distance between the centromere positions of each chromosome was measured; velocities were then calculated for each interval.

As moving chromosomes may show velocity changes, pauses and switches between polar and antipolar movements (Skibbens *et al.*, 1993) our velocity estimates, which are are an average for a 15 min interval, do not provide accurate measurements of the real velocity chromosomes can reach. Thus, the possibility that slow chromosomes can move at a high speed for short periods of time cannot be ruled out. However velocities so calculated are adequate to estimate the net displacement of chromosomes in the spindle — the most outstanding feature of chromosome movements we are addressing to — and to perform comparisons between chromosomes.

15.3 'STATIC' *VERSUS* 'DYNAMIC' CHROMOSOMES

All the univalents we analyzed move between the poles and may also show lateral displacements, pauses and even antipolar movements without reorientation. Their position may be either central or peripheral in the spindle and change during prometaphase. Reorientations may occur at any time and position in the spindle. However when the behaviour of univalents is compared, certain differences become evident (Fig. 15.2); some chromosomes show very short net displacements and rarely reorient, hence they do not display a great variation in their positions in the spindle during prometaphase (see, for instance, X and B_1 of *E. plorans* in Fig. 15.2). These chromosomes may employ the whole prometaphase in travelling between the equator and the pole. Other chromosomes undergo long, sustained interpolar trips, hence they occupy a greater diversity of positions in the spindle (B_5 and autosomal univalents of *E. plorans* in Fig. 15.2). This behaviour is generally linked to a high frequency of reorientations. We consider the first type of chromosomes as 'static', as opposed to the second group of 'dynamic' chromosomes.

Table 15.1 shows a synthesis of the main characteristics of the dynamic behaviour of the different chromosomes analyzed. Maximum, minimum and mean velocities were calculated for each chromosome in each cell, taking into account all the velocity estimates, then, for illustrative purposes, they were averaged for each chromosome type. The number of reorientations refers to the number of chromosomes analyzed. The frequency of anaphase I amphioriented laggards and

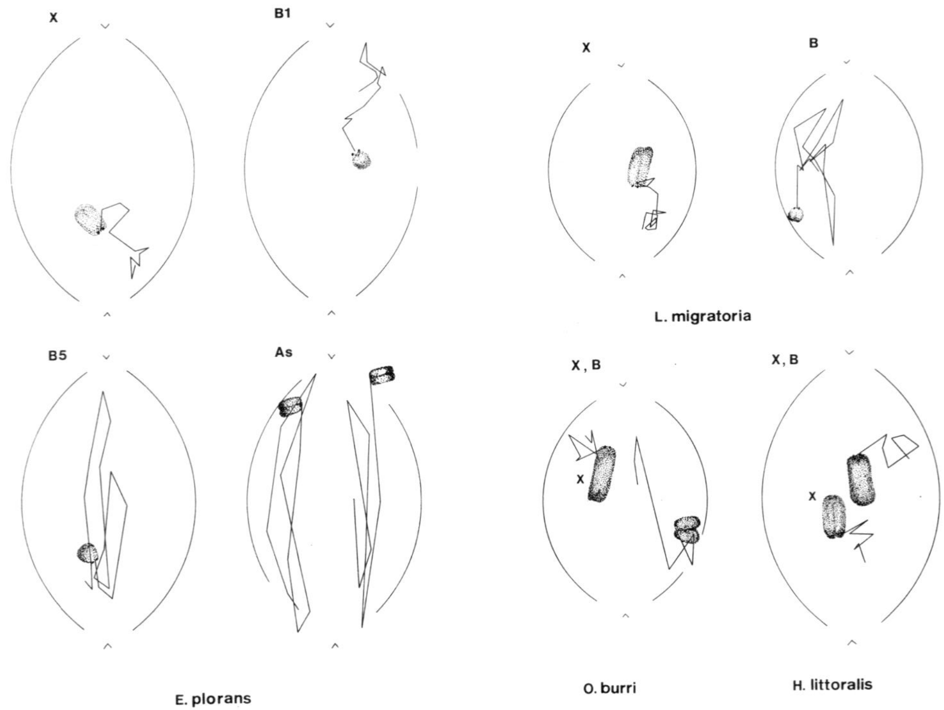

Fig. 15.2 Schematic drawings of the typical trajectories described by different univalent types during prometaphase. Each straight segment represents one interval of approximately 15 minutes. Junctions between two consecutive segments are the successive positions of the chromosomes when they were measured. Note the differences between 'static' and 'dynamic' chromosomes. (Partially reproduced from Rebollo and Arana, 1995.)

equational segregations are also shown, and appear to be more common in dynamic chromosomes. All univalents present similar minimum velocities but different mean and maximum velocities. This reflects the fact that dynamic chromosomes may stay still for certain periods of time, but when they move, they can move fast. Maximum velocities are our best estimation of the dynamic capability of a chromosome but do not reflect some other aspects of chromosome behaviour like the relative importance of static and dynamic periods, continuity or discontinuity of interpolar motions and duration of reorientations. Mean velocities are not useful since the distributions of velocity are not normal so, in the comparisons between chromosomes, the 15 min velocity estimates were used in the non-parametric rank tests Kruskall—Wallis and Wilcoxon—Mann—Whitney.

In general, chromosomes equivalent in nature from different species share certain characteristics. Thus, sex univalents are fairly static whereas autosomal univalents are very dynamic, and the greatest variation in both velocity and frequency of reorientations is found amongst the B chromosomes, which probably constitute the most heterogeneous group.

15.4 X CHROMOSOMES

The X chromosomes of the four species are quite different in size, but all of them appear faint and somewhat decondensed under phase contrast. They are slow and rarely reorient, and they are always reductional at anaphase I (Table 15.1). The X chromosomes of *E. plorans*, *L. migratoria* and *O. burri* are capable of reorientation at any time during prometaphase so the existence of specific anchoring mechanisms through microtubules impinging in the X chromatin, like that described by Ault (1984) in *Melanoplus sanguinipes*, can be ruled out. Reorientations have not been scored for the X of *H. littoralis*. However, unlike the X of *M. sanguinipes*, it does not stay at the pole, but rather moves slowly during the whole prometaphase (Fig. 15.2). In figures 15.3, 15.4 and 15.5, prometaphase movements of the X chromosomes of *E. plorans*, *L. migratoria* and *O. burri* are shown.

X chromosomes of *E. plorans* were classified according to the population of origin and the chromosomal constitution of the individuals. Thus they are labelled as X(B1), X(B2), X(B5) and X(Bn) from the type of B chromosomes that the cells carry. X chromosomes from cells without Bs are labelled X(0) and X from cells with

Table 15.1 Orientational and segregational features of the different univalents analysed

Species	*Chromo-some*	*Velocity (a)*			*Number of chromosomes*	*Number of reorientations*	*Anaphase laggards*	*Equational divisions*
		Maximum	*Minimum*	*Mean*				
E. plorans	X(0)	0.38 (0.23—0.56)	0.05 (0.00—0.10)	0.16 (0.12—0.20)	5	3	0	0
	X(B1)	0.30 (0.14—0.60)	0.03 (0.00—0.06)	0.13 (0.09—0.21)	9	4	0	0
	X(B2)	0.31 (0.11—0.60)	0.04 (0.00—0.07)	0.14 (0.08—0.28)	8	1	0	0
	X(B5)	0.27 (0.19—0.45)	0.02 (0.00—0.05)	0.11 (0.09—0.17)	9	3	0	0
	X(A)	0.27 (0.12—0.43)	0.05 (0.01—0.09)	0.13 (0.07—0.19)	5	1	0	0
	X(Bn)	0.36 (0.22—0.59)	0.08 (0.00—0.18)	0.18 (0.12—0.23)	5	2	0	0
	B1	0.40 (0.13—1.03)	0.04 (0.00—0.11)	0.16 (0.09—0.22)	9	5	0	0
	B2	0.31 (0.08—0.53)	0.06 (0.00—0.14)	0.15 (0.09—0.21)	8	4	0	0
	B5	0.70 (0.31—1.25)	0.05 (0.00—0.10)	0.26 (0.13—0.43)	9	25	5	0
	A	1.05 (0.37—1.43)	0.13 (0.04—0.38)	0.48 (0.17—0.74)	10	36	4	1
	Bn	0.71 (0.23—1.10)	0.06 (0.03—0.10)	0.29 (0.14—0.48)	5	17	3	0

L. migratoria	X	0.35 (0.17—0.58)	0.03 (0.00—0.07)	0.14 (0.08—0.26)	13	3	0	0
	B	0.74 (0.48—1.31)	0.07 (0.00—0.12)	0.31 (0.21—0.42)	12	48	9	7
	A	0.93 (0.53—1.46)	0.04 (0.01—0.09)	0.32 (0.21—0.53)	10	40	7	3 (b)
O. burri	X	0.25 (0.07—0.55)	0.04 (0.02—0.12)	0.12 (0.04—0.24)	11	4	0	0
	B	0.56 (0.13—1.04)	0.07 (0.00—0.14)	0.25 (0.11—0.44)	14	33	6	2
H. littoralis	X	0.30 (0.16—0.54)	0.05 (0.00—0.16)	0.17 (0.08—0.31)	8	0	0	0
	B	0.22 (0.15—0.32)	0.07 (0.02—0.14)	0.14 (0.07—0.21)	8	0	0	0

(a) averaged values, ranges of variation in brackets, (b) two cells collapsed autosomal univalents X(A). A Kruskall—Wallis test including all *E.plorans* X chromosomes did not reveal significant differences between velocities.

15.5 B CHROMOSOMES

B chromosomes constitute a very heterogeneous group as their origin, behaviour, endophenotypic or exophenotypic effects on the bearers, and modes of transmission are highly variable (see Jones and Rees, 1982).

When we consider meiotic behaviour, this variability becomes evident even between B chromosomes of the same species. In *E. plorans*, more than 40 B chromosome types have been found; B_1, B_2 and B_5 are the most widespread and determine three distinct areas in the geographical distribution of the species in the Iberian Peninsula. The size of B_1 and B_5 is approximately half of the X while B_2 and B_n are smaller; they also appear decondensed in phase contrast. B_1, the most widespread, is supposed to be the original B that has been substituted by its derivatives B_2 and B_5 in the regions where these appeared by mutation (Henriques-Gil *et al.*, 1984a ; Henriques-Gil and Arana, 1990). These B chromosomes retain a certain degree of homology since they can form chiasmata when two different Bs are present in the same individual (Henriques-Gil *et al.*, 1984b). Rare new B variants, like B_n, are occasionally found, indicating that the polymorphism is engaged in an active process of change. The frequency of equational segregation at anaphase I for the three main B types (B_1, B_2 and B_5), as estimated from fixed secondary spermatocytes, is quite low (about 4%) and their transmission rate deduced from experimental crosses is 0.5. No data about positive or negative effects on fitness nor accumulation mechanisms have been found to date for these Bs. (For a model to explain the maintenance of the *E. plorans* polymorphism see Camacho *et al.,* p.301—318).

When velocities of the Bs are compared to those of their respective X chromosomes by Wilcoxon's tests, B_1 and B_2 show no significant differences, neither in their maximum velocities ($Z=1.14$ and $Z=0.19$ respectively) nor in the overall comparisons using all velocity estimates ($Z=1.46$ and $Z=1.12$). B_5 (Fig. 15.3) and B_n are clearly faster than their respective Xs ($Z=6.81$ and $Z=2.69$ respectively in the overall comparisons), but maximum velocities are only significatively higher for B_5 ($Z=3.36$). Similar results are obtained when these chromosomes are

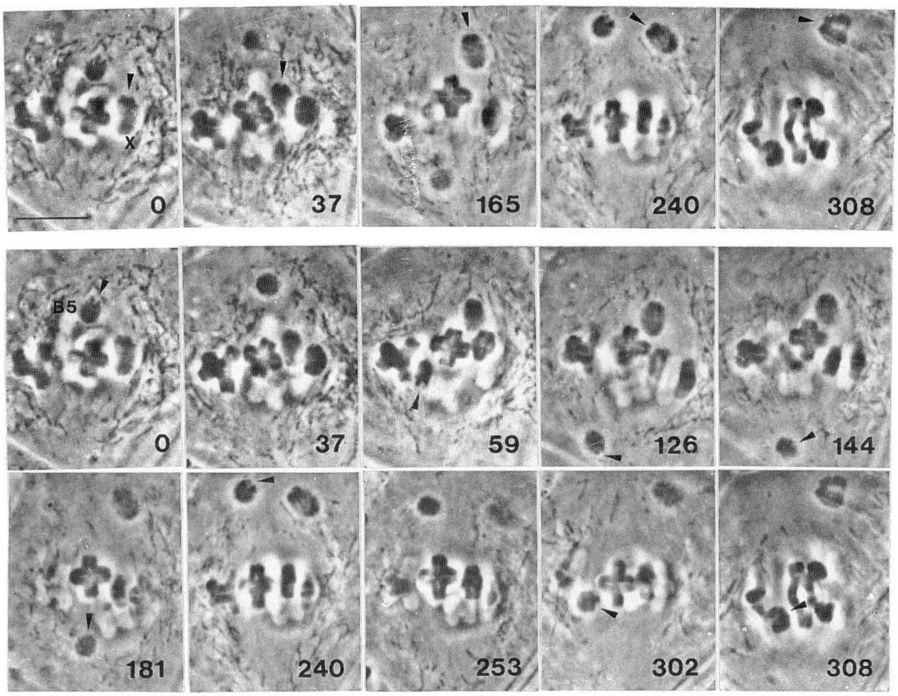

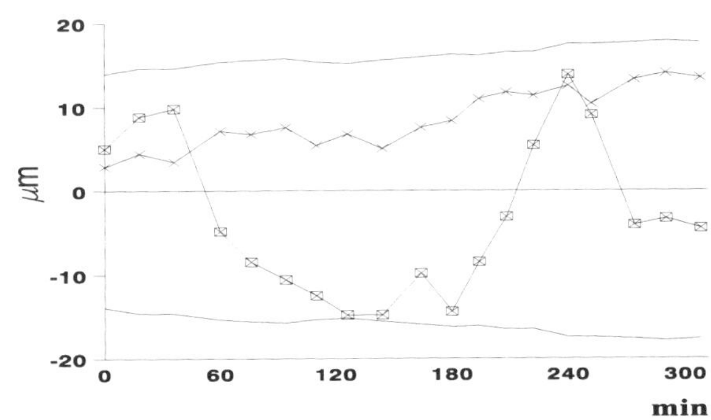

Fig 15.3 Prometaphase movements of the X (upper row) and B_5 (lower two rows) of *E.plorans*. Arrowheads mark the centromeric ends of chromosomes. The graph shows the position of chromosomes *versus* time, the upper and lower continuous lines represent the position of spindle poles. The X (×) does not reorient and B_5 (⊠) reorients three times (37, 144 and 253 min prints). The final orientation is amphitelic (308 min print), but segregation is always reductional (not shown). (Reproduction from Rebollo and Arana, 1995, granted by permission from *Chromosoma*).

compared to B_1 and B_2 (table 15.1).

B_1 and B_2 resemble the X in the frequency of reorientations and in the absence of amphioriented laggards at anaphase I, whereas B_5 and B_n reorient more often than the X and may lag at anaphase I (see Fig. 15.2 for B_1 and B_5). Nevertheless, these laggards do not divide equationally nor fail in migration but always end up undivided in one of the poles. In figure 15.3 the prometaphase movements of B_5 and the X of the same cell are shown.

The small, telocentric B chromosome of *L.migratoria* is very frequent in Spanish populations. It posseses an efficient mechanism for preferential transmission through mitotic nondisjunction and accumulation in the male germ line and meiotic drive in females (Pardo *et al.*, 1994). At prometaphase, it is a very dynamic chromosome (see Figs. 15.2 and 15.4). Velocity comparisons with the X are significant for maximum velocities ($Z=19.32$) and when all estimates are used ($Z=10.03$). This accessory also reorients more often than the X, may lag at anaphase, and these laggards divide equationally with a considerable frequency (table 15.1).

The B chromosome of *O.burri* is frequent in the populations where is found and shows an accumulation mechanism through meiotic drive in the female (Santos *et al.*, 1993). It is an isochromosome which generally forms a ring during prometaphase, releasing its arms at the anaphase onset (Figs. 15.2 and 15.5). It often moves in a position perpendicular to the spindle axis. This B is more dynamic than the X, (see table 15.1), maximum velocities are higher (Wilcoxon's test: $Z=2.73$) and the overall velocity comparisons are also significant ($Z=4.46$). It shows frequent reorientations, sometimes lags at anaphase I and these laggards may divide equationally.

The B chromosome of *H.littoralis* is restricted to a small area within the Iberian distribution of this species but it is quite frequent in these populations. It also shows meiotic drive in the female (Cano and Santos, 1989). It is very similar to the X in size, appearance and behaviour: it moves slowly (comparisons with the X are not significant), rarely reorients, does not lag and is always reductional at anaphase (table 15.1).

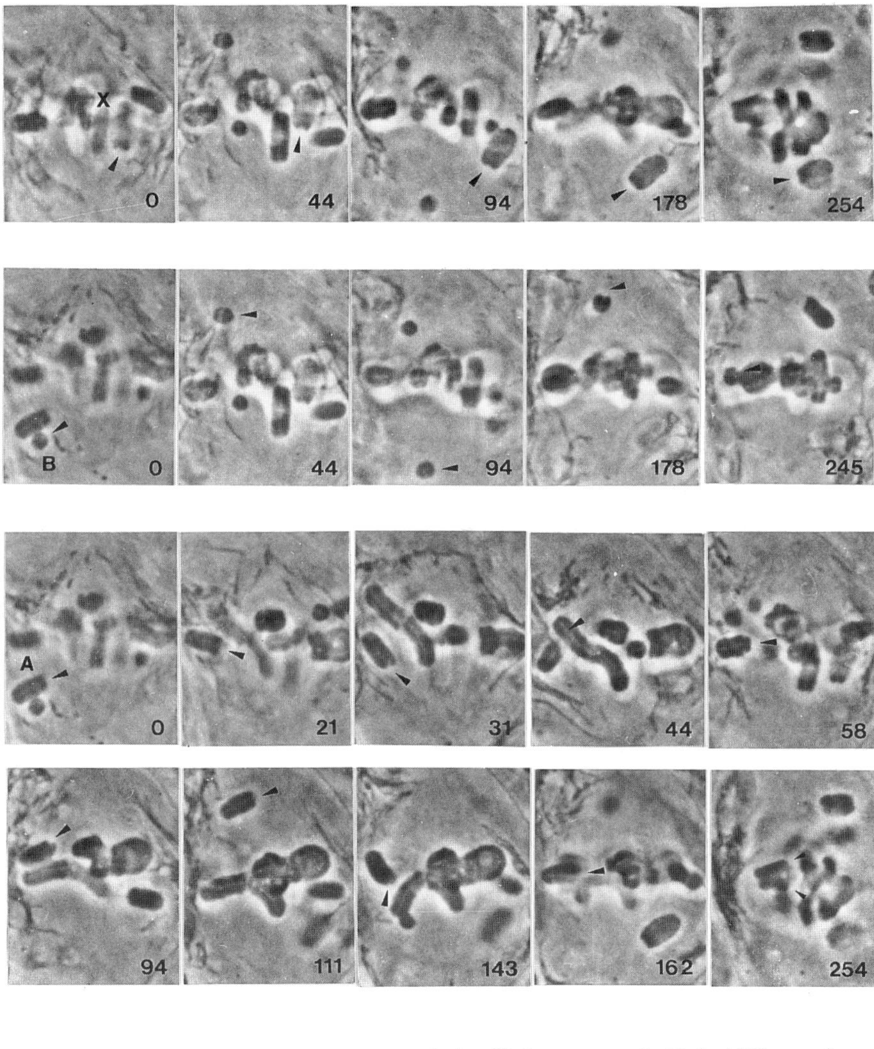

Fig. 15.4 Prometaphase movements of the X (upper row), B (middle row) and one autosomal univalent (two lower rows) of *L.migratoria*. Arrowheads mark the centromeric ends of chromosomes. The X does not reorient, the B reorients three times (44, 94 and 178 prints) and the autosomal univalent firstly reorients twice (21 and 44 min prints), pauses at the equator (58 min print) and then reorients once more (111 min print) before amphitelic stabilisation at the equator (162 and 254 min prints).

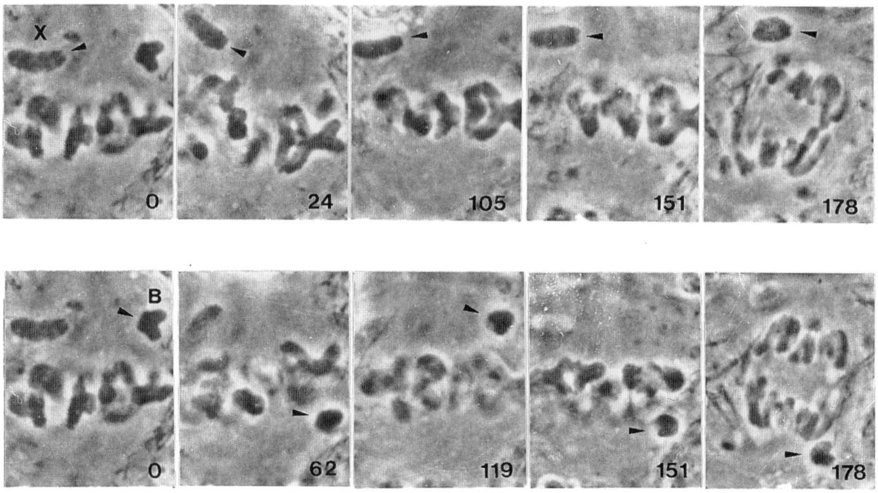

Fig. 15.5 Prometaphase movements of the X (upper row) and the B chromosome (lower row) of *O. burri*. Arrowheads mark the cenromeric ends of chromosomes. The X does not reorient though it shows some oscillations (24 min print), the B chromosome reorients three times (0, 62 and 119 min prints).

15.6 AUTOSOMAL UNIVALENTS

In *E. plorans*, all the cells tracked had one pair of medium-sized autosomal univalents and in one they were spontaneous. Univalents of *L. migratoria* were all heat-induced and the cells analyzed *in vivo* carried one or two pairs of univalents, although only one pair of medium size was tracked per cell. Autosomal univalents of *E. plorans* and *L. migratoria* are darker than the X and B chromosomes in phase contrast. The two homologues within the spindle behave independently of each other.

In both species, autosomal univalents are very dynamic and differences from their respective X chromosomes are clear (see table 15.1). In *E. plorans*, they are even faster and reorient more often than B_5, which is also a dynamic chromosome (Wilcoxon's tests: $Z=2.37$ for maximum velocities and $Z=5.88$ for the overall comparisons). Amphiorientations and lagging at anaphase are frequent but, in the cells tracked throughout prometaphase, only one out of four laggards divided equationally (a spontaneous univalents; table 15.1).

Autosomal univalents of *L. migratoria* (Fig. 15.4) do not show differences in terms of velocity and reorientations from the B chromosome. Like those of *E. plorans*, they may lag at anaphase I and divide equationally (table 15.1).

15.7 THE EFFECTS OF HEAT ON CHROMOSOME BEHAVIOUR

Heat shock during the pairing phase may produce both asynapsis and desynapsis (Peacock, 1968; Loidl, 1989). This in turn, may alter the subsequent orientation and segregation behaviour through an effect on sister chromatid cohesion. Concerning orientation, we found no differences in the pattern of prometaphase movements between the two spontaneous univalents of *E. plorans* and the heat-induced univalents, so, instability is not due to chromosome damage produced by the treatment. This is also supported by the normal behaviour of the X chromosome in these cells. Spontaneous autosomal univalents may also amphiorient and divide equationally. However, the sample is too small to draw definitive conclusions.

A more extensive study of segregation could be performed since a substantial number of spontaneous autosomal univalents were found in several heterozygotes for an A—B translocation in which the non translocated autosome and one additional pair of medium-sized autosomes often appeared as univalents.

The choice of induced univalents for this comparison was difficult since the presence of anaphase laggards may provoke spindle collapse (Fig. 15.1); cytokinesis is inhibited and an unreduced restitution nucleus is formed. These cells may enter a diploid but otherwise normal second division. The probability of spindle collapse increases with the number of laggards; in cells with more than four laggards spindle collapse is the rule. In heat treatments, the frequency of univalents reaches a peak approximately at the tenth day after recovery and then decreases for two or three days. Thus, to analyse the segregation behaviour of induced autosomal univalents we chose individuals that were already in the decreasing phase where spermatocytes had generally only one or two pairs of autosomal univalents. Spindle collapses occasionally ocurred in these cells, but they were not considered in the comparisons.

Table 15.2 Segregation of spontaneous and induced autosomal univalents

	Species		
	E. plorans	*E. plorans*	*L. migratoria*
Type of univalents	Spontaneous	Induced	Induced
Syntelic orientation	13	23	12
Amphitelic laggards	29 (69%)	59 (71%)	84 (88%)
Reductional division	26	30	20
Equational division	3	20	61
Migration failure	13	33	16
Equational laggards	10%	34%	71%
Total nº of univalents	42	82	96

Spontaneous univalents adopt amphitelic orientation as often as heat-induced univalents but equational divisions are much less frequent. Instead, spontaneous autosomal univalents either move undivided to one of the poles or stay at the equator after cytokinesis and are not included in the daughter nuclei (see table 15.2 and Fig. 15.1).

The results obtained in heat-treated spermatocytes of *L. migratoria* are similar to those of *E. plorans*; cells with a high number of laggards collapse, while in cells with two or four autosomal univalents, these amphiorient with a high frequency and most divide equationally (table 15.2).

From these data, it can be concluded that although prometaphase movements seem not to be affected by heat treatment, there is an effect on segregation. Lagging at anaphase implies that amphioriented chromosomes are resistant to chromatid separation. This resistence can be so strong as to prevent chromosome migration and even cellular division. Laggards may or may not be actively engaged in the collapse of the spindle; for instance kinetochore motors may keep depolymerizing the chromosome fibres anchored to the poles, preventing or overcoming spindle elongation, but collapse may simply result from laggards remaining at the equator and disturbing cytokinesis after cellular signals have determined the reorganization of the cytoskeleton to enter interphase.

Although we do not know the origin of spontaneous univalents, such a major disturbance of the pairing and crossing-over phases as is heating, seems to reduce the degree of resistence to chromatid separation of laggards, thus supporting a relationship between pairing and/or crossing over and sister chromatid cohesion (Maguire, 1978).

15.8 CONCLUSIONS

The main conclusion emerging from this study is that the behaviour of chromosomes in the spindle is controlled at the chromosomal rather than at the cellular level. The use of living cells allowed us to demonstrate that specific chromosome characteristics are decisive not only for the modes of segregation at anaphase, as demonstrated in yeast (Sharon and Simchen, 1990), but also for prometaphase dynamics.

There is a certain parallelism between the meiotic behaviour of the chromosomes and their degree of adaptation to univalence. Sex chromosomes are all very regular in transmision and, in the species analyzed, they coincide in being slow and reorienting rarely. This seems to be a common strategy for sex univalents but it is not the only way to achieve regular transmission since the X of *Melanoplus differentialis* (Nicklas, 1961; Ault, 1986) is very fast and reorients often yet, it is flawlessly reductional at the first anaphase.

Bs, as expected, compose the most heterogeneous group in terms of orientation behaviour. Examples of accessories as static as the X chromosomes and as dynamic as autosomal univalents may be found, and some others lie in between. In general, faster chromosomes also reorient more often and lag at anaphase I, suggesting that these chromosome traits are interrelated. However, there are relevant differences in the modes of segregation that may have an impact in the effectiveness of transmission. For instance, the B_5 of *E. plorans* lags at anaphase I as does the B chromosome of *L. migratoria* but, unlike it, B_5 does not divide equationally. Some capacity to hinder chromatid separation in anaphase laggards may lie in the centromeric region of

B_5. This chromosome shows that dynamic chromosomal characteristics can be fast changing as it is partially homologous to B_1 and B_2 and yet shows very different meiotic behaviour. The comparison between these three B variants also illustrates how several strategies may be used to reach the same level of efectiveness of transmission (Rebollo and Arana, 1995).

We expected that accessories with accumulation mechanisms would not have been subject to strong selective pressure for meiotic segregation regularity in the males. In agreeing with this, the most common and widespread B variants of *E. plorans* do not possess accumulation mechanisms (Camacho *et al.*, p. 301—318) and are very effective in transmision. By contrast the B chromosomes of *L. migratoria* and *O. burri* can afford a certain degree of meiotic instability (and possible loss) since they are compensated by accumulation mechanisms. However, the B chromosome of *H. littoralis* is regular in meiosis and also has drive in the female. Additional data about the evolutionary stage of the polymorphisms would be needed to fully understand these variations. A newly arisen accessory chromosome may possess a higher or lower degree of meiotic regularity but its fate in the population will depend on many other factors including the acquisition of drive mechanisms.

The meiotic irregularity of autosomes agrees with the idea that their behaviour as univalents has not been selected for regular transmission. Not only are they poorly transmitted but also may provoke cellular abnormalities like unreduced gametes. Again, a relationship between high velocity, frequent reorientations and amphiorientation appears.

The causes for all these differences in orientation and segregation behaviour of chromosomes probably lies at the kinetochore/ centromere level since kinetochores mediate the chromosome interactions with the spindle and are also the site for the motors of movement (see McIntosh and Pfarr, 1991 and Rieder, 1991, for review) while centromeric chromatin plays an important role in determining syntelic segregation in the first meiotic division (reviewed by Miyazaki and Orr-Weaver, 1994). Kinetochore characteristics possibly having an impact on chromosome behaviour are: kinetochore degree of exposure, velocity of kinetochore maturation, degree of activity of kinetochore motors or regulation of the switches between different phases of kinetochore activity (Skibbens *et al.*, 1993). The pattern of movements of chromosomes such as the X, B_1 and B_2 of *E.plorans*, with pauses, oscillations and antipolar displacements without reorientation, agrees with a model of a kinetochore with frequent phase switches. By contrast, the kinetochores of B_5 and autosomal univalents might remain in the poleward phase for longer periods. Nevertheless, a close examination of kinetochore structure and evolution during prometaphase and a more detailed analysis of *in vivo* movements would be needed to address the possible causes of the variability of chromosome dynamic behaviour.

Finally, although centromeres of different univalents behave differently in the same spindle, this does not imply that cellular characteristics do not have an impact on chromosome behaviour. The B_1 accessory of *E. plorans* appears amphitelically-oriented in all metaphase I-arrested oocytes of females although segregation is reductional (Henriques-Gil *et al.*, 1989). It remains to be seen how sex and autosomal univalents behave in these circumstances.

ACKNOWLEDGEMENTS

This work was supported by grants PB 92-0079-C03-02 and PB 92-0812 from the DGICYT and PR 189/92-4057 from the Universidad Complutense.

REFERENCES

Ault, J.G. (1984) Unipolar orientation stability of the sex univalent in the grasshopper *Melanoplus sanguinipes*. *Chromosoma,* **89**, 201—5.

Ault, J.G. (1986) Stable versus unstable orientations of sex chromosomes in two grasshopper species. *Chromosoma,* **93**, 298—304.

Bauer, H., Dietz, R. and Röbbelen, C. (1961) Die Spermatocytenteilungen der Tipuliden.III. Das Bewegungsverhalten der chromosomen in Translokationheterozygoten von *Tipula oleracea*. *Chromosoma,* **12**, 116—89.

Cano, M.I. and Santos, J.L. (1989) Cytological basis of the B chromosome accumulation mechanism in the grasshopper *Heteracris littoralis*. *Heredity,* **62**, 91—5.

Dietz, R. (1956) Die Spermatocytenteilungen der Tipuliden. II. Graphische Analyse der Chromosomenbewegung Während der Prometaphase I im Leben. *Chromosoma,* **8**, 183—211.

Goldstein, L.S.B. (1981) Kinetochore structure and its role in chromosome orientation during the first meiotic division in male *D. melanogaster*. *Cell,* **25**, 591—602.

Henderson, S.A. (1963) Temperature and chiasma formation in *Schistocerca gregaria*. I. An analysis of the response at a constant 40°C. *Heredity,* **18**, 77—94.

Henriques-Gil, N. and Arana, P. (1990) Origin and substitution of B-chromosomes in the grasshopper *Eyprepocnemis plorans*. *Evolution,* **44**, 747—53

Henriques-Gil, N., Santos, J.L. and Arana, P. (1984a) Evolution of a complex B-chromosome polymorphism in the grasshopper *Eyprepocnemis plorans*. *Chromosoma,* **89**, 290—93.

Henriques-Gil, N., Arana, P. and Santos, J.L. (1984b) Factors affecting the meiotic pairing behaviour of B chromosomes in the grasshopper *Eyprepocnemis plorans* (Acrididae, Orthoptera). *Canadian Journal of Genetics and Cytology*, **26**, 664—8.

Henriques-Gil, N., Cano, M.I. and Santos, J.L. (1989) Sex-dependent meiotic behaviour of B chromosomes in the grasshopper *Eyprepocnemis plorans*. *Heredity,* **62**, 217—21.

Jones, R.N. and Rees H (1982) *B chromosomes*. Academic Press, London.

Loidl, J. (1989) Effects of elevated temperature on meiotic chromosome synapsis in *Allium ursinum*. *Chromosoma,* **97**, 449—58.

Lopez-León, M.D., Cabrero, J., Camacho, J.P.M., Cano, M.I. and Santos, J.L. (1992) A widespread chromosome polymorphism maintained without apparent drive. *Evolution,* **46**, 529—39.

Maguire, M.P. (1978) A possible role for the synaptonemal complex in chiasma maintenance. *Experimental Cell Research*, **112**, 297—308.

McIntosh, J.R. and Pfarr, C.M. (1991) Mitotic motors. *Journal of Cell Biology,* **115**, 577—85.

Miyazaki, W.Y. and Orr-Weaver, T.L. (1994) Sister-chromatid cohesion in mitosis and meiosis. *Annual Review of Gene*tics, **28**, 167—87.

Nicklas, R.B. (1961) Recurrent pole-to-pole movements of the sex chromosome during prometaphase I in *Melanoplus differentialis* spermatocytes. *Chromosoma,* **12**, 97—115.

Nicklas, R.B. (1977) Chromosome distribution: experiments on cell hybrids and in vitro. *Philosophical Transactions of the Royal Society, Series B*., **227**, 267—76

Östergren, G. (1951) The mechanism of co-orientation in bivalents and multivalents. *Hereditas,* **37**, 85—156.

Pardo M.C., Lopez-León, M.D., Cabrero, J. and Camacho, J.P.M. (1994) Transmission analysis of mitotically unstable chromosomes in *Locusta migratoria*. *Genome,* **37**, 1027—34.

Peacock, W.J. (1968) Replication, recombination, and chiasma in *Goniaea australasiae* (Orthoptera: Acrididae). *Genetics,* **65**, 593—617.

Rebollo, E. and Arana, P. (1995) A comparative study of orientation at behavior of univalent in living grasshopper spermatocytes. *Chromosoma*, **104**, 56—67.

Rieder, C.L. (1991) Mitosis: towards a molecular understanding of chromosome behaviour. *Current Opinion in Cell Biology,* **3**, 1—8.

Santos, J.L., del Cerro, A.L., Fernández, A. and Díez, M. (1993) Meiotic behaviour of B chromosomes in the grasshopper *Omocestus burri*: A case of drive in females. *Hereditas,* **118**, 139—43.

Sharon, G. and Simchen, G. (1990) Centromeric regions control autonomous segregation tendencies in single-division meiosis of *Saccharomyces cerevisiae*. *Genetics,* **125**, 487—94.

Shaw, M.W. and Hewitt, G.M. (1990) B chromosomes, selfish DNA and theoretical models: what next? in *Oxford Surveys in Evolutionary Biology*, **7**. (eds. D. Futuyuma and J. Antonovics), pp. 197—223.

Simchen, G. and Hugerat, Y. (1993) What determines whether chromosomes segregate reductionally or equationally in meiosis? *Bioessays,* **15**, 1—8.

Skibbens, R.V., Skeen, V.P. and Salmon, E.D. (1993) Directional instability of kinetochore motility during chromosome congression and segregation in mitotic newt lung cells: A push-pull mechanism. *Journal of Cell Biology*, **122**, 859—75.

Wagenaar, E.B. and Bray, D.F. (1973) The ultrastructure of kinetochores of unpaired chromosomes in a wheat hybrid. *Canadian Journal of Genetics and Cytology*, **15**, 801—6.

Wolf, K.W. (1994) How meiotic cells deal with non-exchange chromosomes. *Bioessays,* **16**, 107—14.

16

Proteins controlling sister-chromatid cohesion

T.L. Orr-Weaver, D.P. Moore, A.W. Kerrebrock, S.E. Bickel and D.W. Wyman

16.1 INTRODUCTION

Prior to the metaphase/anaphase transition of cell division the replicated copies of each chromosome, the sister chromatids, are attached to each other. We use the term 'sister-chromatid cohesion' to describe this physical association between the sister chromatids. Sister-chromatid cohesion is essential in mitosis to ensure that sister chromatids attach to microtubules from opposite spindle poles. Thus when segregation occurs, each daughter cell receives one copy of each chromosome.

Meiosis is a specialized cell division in which two rounds of chromosome segregation follow a single doubling of the DNA (Fig. 16.1). In the first meiotic division the homologues pair and segregate. Meiosis II is like mitosis in that the sister chromatids travel to opposite poles. The sister chromatids must remain attached throughout meiosis I in order to delay their segregation until the second meiotic division. In addition, it is likely that sister-chromatid cohesion acts indirectly to hold the homologues together until anaphase I by stabilizing the chiasmata that attach homologues.

Premature separation of sister chromatids appears to be the primary defect in several human diseases. In cells from patients with the inherited disease Roberts Syndrome, cohesion at the centromere regions is aberrant, and the sister chromatids prematurely separate in mitosis (German, 1979; Van Den Berg and Francke, 1993). Missegregation of chromosomes during human meiosis results in

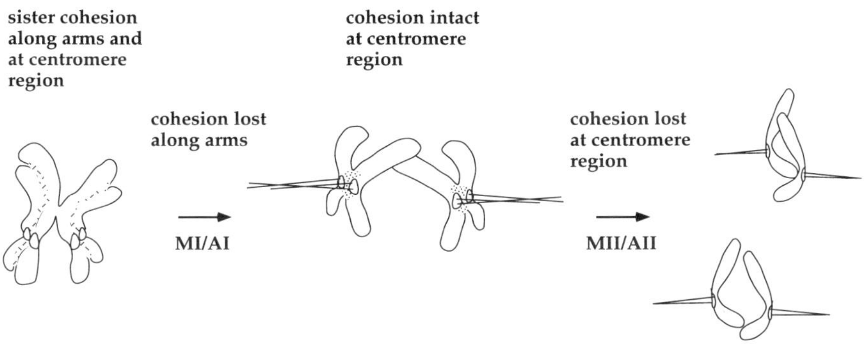

Fig. 16.1 Diagram of the cytology of sister-chromatid cohesion in meiosis. Early in meiosis I, the sister chromatids are associated along their entire length (stippled). Arm cohesion appears to be important for stabilizing the chiasma that hold the homologues together (crossed structure). The kinetochores are shown as open circles, and the black lines indicate microtubules. At the metaphase I/ anaphase I transition (MI/AI), as the homologues segregate arm cohesion is dissolved so that by anaphase I the sister chromatids are held together only at their centromere regions (dots). At the metaphase II/ anaphase II transition (MII/AII), centromere cohesion is lost as the sister chromatids segregate.

spontaneous abortions or severe birth defects. Premature separation of sister chromatids has been observed in oocytes and thus may be one of the ways trisomy arises (Angell, 1995; Angell *et al.*, 1994).

Cytological examination of the association between sister chromatids provides insights into potential mechanisms for cohesion. In mitosis, the sister chromatids are associated along their length, both on the arms and at the centromere region. However, the cohesion in the vicinity of the centromere appears to be tighter than on the arms. In analyzing mitotic chromosomes by scanning electron microscopy, by late prophase or early metaphase Sumner observed single chromatid arms that were distinct but closely apposed (Sumner, 1991). In contrast, in the region of the centromeres the two sister chromatids could not be distinguished. In higher eukaryotes the chromatin differs between the arms and the centromere region. The arms are largely composed of euchromatic DNA, while the centromeres are embedded in heterochromatin.

Unlike the sudden loss of both arm and centromere cohesion that occurs in a single step at the metaphase/ anaphase transition of mitosis, loss of cohesion in meiosis occurs in two steps (Fig. 16.1). In meiosis I, the sister chromatids are held together along their length as in mitosis. At the metaphase I/ anaphase I transition cohesion along the chromatid arms is lost as the homologues segregate (Fig. 16.1). The sister chromatids remain associated at their centromere regions until the metaphase II/ anaphase II transition (Fig. 16.1). This sequential loss of cohesion in meiosis suggests that there may be different mechanisms for cohesion on the arms and at the centromere in meiosis. A meiosis-specific function may be needed to hold the sister chromatids at their centromere from anaphase I until metaphase II. Alternatively, the same mechanism of cohesion could be occurring at both sites, but its release could be regulated so that cohesion persists at the centromere until anaphase II.

Several lines of evidence establish that sister-chromatid cohesion is needed for stable attachment of chromosomes to opposite poles of the spindle (for review see Bickel and Orr-Weaver, 1996). Both in meiosis and in mitosis when the chromosomes are attached to opposite spindle poles there is tension. This tension results from forces directed toward the opposite spindle poles being counteracted by attachment between the two homologues in meiosis I and between the two sister chromatids in meiosis II and mitosis. This tension is needed for stable bipolar attachment of the chromosomes. Clearly, sister-chromatid cohesion is essential to establish tension and bipolar attachment in mitosis and meiosis II. Moreover, it is likely to be necessary also in meiosis I. In most organisms the homologues are held together by chiasmata (for review see Ross *et al.*, 1996). The chiasmata in turn need to be held in place, and sister-chromatid cohesion appears to contribute to chiasma stability (for review see Maguire, 1993).

It is likely that proteins act as a glue to hold sister chromatids together. There are also predicted to be regulatory proteins that time the release of cohesion at the metaphase/anaphase transition. These cohesion proteins can be identified by several criteria. Cytologically, proteins that attach sister chromatids together would be expected to localize to points of association between the sister chromatid and to dissociate when the sister chromatids separate in anaphase. The Inner Centromere Proteins (INCENPs) and the Centromere Linking Proteins

(CLiPs) are vertebrate proteins that show this pattern of localization on mitotic chromosomes (Cooke *et al.,* 1987; Earnshaw and Bernat, 1991; Earnshaw and Cooke, 1991; MacKay *et al.*, 1993; Rattner *et al.*, 1988). In addition, even though the synaptonemal complex dissociates in prophase I of meiosis, some components of it persist (Esponda and Krimer, 1979; Moens and Church, 1979). The mammalian Cor1 protein is a component of the axial elements that remains at the kinetochore until anaphase II (Dobson *et al.*, 1994). Thus based on their localization, these proteins are candidates for promoting cohesion.

An alternative approach to identify cohesion proteins is to isolate their genes by a mutant phenotype. Mutations in a gene encoding a protein needed for sister-chromatid cohesion would result in premature separation of the sister chromatids in mitosis, meiosis, or both. In addition to the human cases described above, mutations resulting in premature separation of sister chromatids have been identified in *Drosophila*, maize, tomato, *Sordaria*, and yeast (Clayberg, 1959; Davis, 1971; Goldstein, 1980; Kerrebrock *et al.*, 1992; Maguire, 1978; Maguire *et al.*, 1991; Maguire *et al.*, 1993; Mason, 1976; Molnar *et al.*, 1995; Moreau *et al.*, 1985; Rockmill and Roeder, 1988; Rockmill and Roeder, 1990). The *Drosophila* genes *mei-S332* and *ord* have been extensively characterized, and recently their protein products have been identified (Bickel *et al.*, 1996; Kerrebrock *et al.*, 1995).

Identifying the proteins involved in sister-chromatid cohesion is critical to an understanding of the molecular basis of cohesion. Once essential proteins have been isolated, it will be possible to determine whether they act as a glue holding the sister chromatids together or whether they regulate the release of cohesion. Moreover, analyzing alterations in these proteins as cohesion is dissolved at anaphase will provide further insight into the control of the metaphase/anaphase transition.

16.2 CONSEQUENCES OF MUTATIONS IN THE *DROSOPHILA MEI-S332* AND *ORD* GENES

Mutations in the *mei-S332* and *ord* genes result in premature separation of the sister chromatids in female and male meiosis. In *ord* mutant females, the chromosomes missegregate in meiosis in ratios predicted to result from premature separation of the sister chromatids early in meiosis followed by random segregation through both meiotic divisions (Mason, 1976; Miyazaki and Orr-Weaver, 1992). This conclusion was confirmed cytologically in male meiosis by the demonstration that the sister chromatids separated as early as prometaphase I (Goldstein, 1980; Miyazaki and Orr-Weaver, 1992). In addition to causing loss of sister-chromatid cohesion, mutations in *ord* reduce recombination levels (Mason, 1976; Miyazaki and Orr-Weaver, 1992). Although nondisjunction occurs in the mitotic divisions in the germline of *ord* mutant males, there are no effects on somatic mitosis (Lin and Church, 1982; J. Wu, W. Miyazaki, and T. Orr-Weaver, unpublished results). Thus *ord* appears to function specifically in the germline. The phenotypes of *ord* mutations indicate that the gene is necessary for cohesion both at the arms and the centromere. The fact that *ord* mutations affect several processes suggests that *ord* may play a regulatory role rather than directly acting as a cohesion glue (Bickel and Orr-Weaver, 1996).

Mutations in the *mei-S332* gene also cause premature sister-chromatid separation, but these mutations differ from *ord* in that cohesion is not lost until anaphase I (Davis, 1971; Goldstein, 1980; Kerrebrock *et al.*, 1992). This corresponds to the time in wild-type cells at which the sister chromatids are held together only at their centromere regions (Fig. 16.1). Mutations in *mei-S332* have no effect on mitotic chromosome segregation (Kerrebrock *et al.*, 1995). Therefore it was postulated that the MEI-S332 protein is required only in meiosis and specifically at the centromere region to hold sister chromatids together.

16.3 THE MEI-S332 AND ORD PROTEINS

We isolated the *mei-S332* and *ord* genes by positional cloning (Bickel *et al.*, 1996; Kerrebrock *et al.*, 1995). Deficiencies were generated that deleted the genes, and these were used to delineate the genomic regions containing the genes. The genomic intervals were cloned by genomic walking, and we used transformation rescue to identify the transcription unit corresponding to each gene. Both *ord* and *mei-S332* encode novel proteins that are not homologous to any proteins in the existing databases or to each other.

Although ORD is a pioneer protein, analysis of the nature and position of mutations in the protein sequence provided insights into the function of the protein (Bickel *et al.*, 1996). All of the mutations characterized so far reside in the C-terminal half of the protein. Two

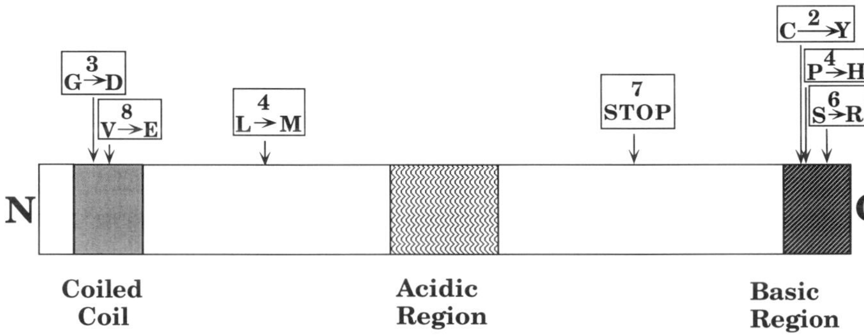

Fig. 16.2 The MEI-S332 protein and mutations. The box represents the 401 amino acids of the MEI-S332 protein. The relative positions of the coiled coil, acidic, and basic domains are indicated as well as the changes caused by the *mei-S332* mutations. The number of each *mei-S332* mutation is shown together with the resulting amino acid change.

strong mutations cause stop codons that truncate the protein. The other four mutations result in missense changes and single amino acid substitutions. Since these mutations affect both cohesion and recombination, this domain of the protein must be required for both functions. In addition, some of the *ord* mutations exhibit unusual genetic properties that are most simply explained by the C-terminal half of the protein binding other proteins to promote cohesion (Bickel *et al.*, 1996).

The mutations in the MEI-S332 protein also have pointed out important functional domains (Fig. 16.2) (Kerrebrock *et al.*, 1995). The MEI-S332 protein has a coiled coil of the leucine zipper class at its N-terminus (Berger *et al.*, 1995; Lupas *et al.*, 1991), an acidic region in the middle of the protein, and a basic region at the C-terminus. The protein contains PEST sequences that may be important for its degradation (Rechsteiner 1988; Rogers *et al.*, 1986). The most severe mutations delete the C-terminal third of the protein.

The strongest *mei-S332* mutations affect chromosome segregation in both female and male meiosis, but weaker alleles affect predominantly

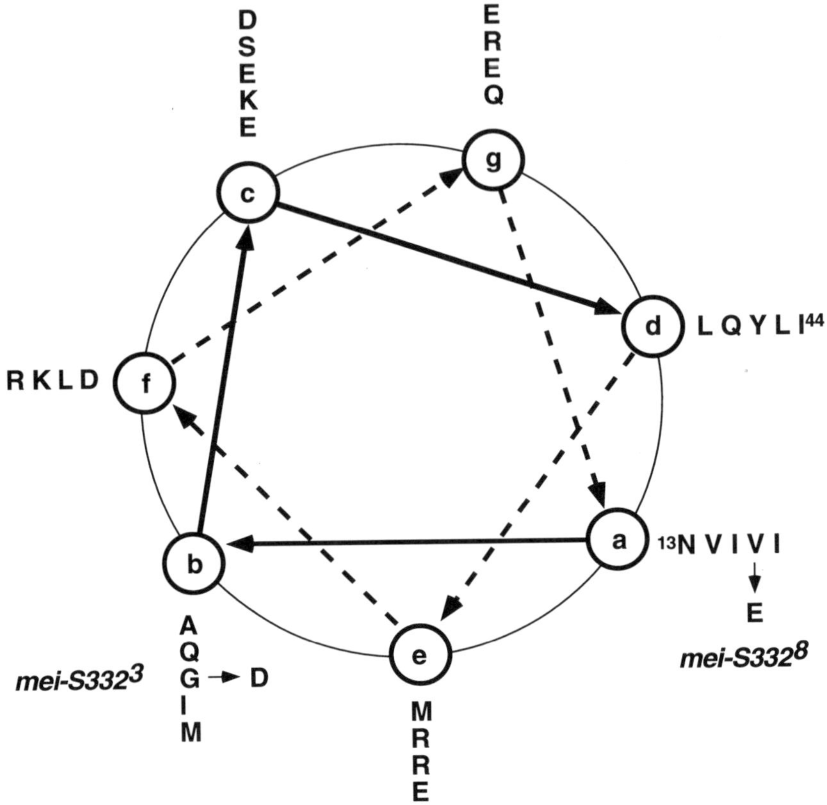

Fig. 16.3 The coiled coil domain of MEI-S332. The coiled coil domain is shown in a helical wheel representation. The amino acids at the a and d positions form the dimer interface and are hydrophobic. The amino acid changes resulting from the *mei-S332*[3] and *mei-S332*[8] mutations are shown.

one sex or the other (Kerrebrock *et al.*, 1992). This is of interest because meiosis I is under separate genetic control in *Drosophila* females and males (for review see Orr-Weaver, 1995). Although *mei-S332* is required for sister-chromatid cohesion in both sexes, many of the proteins with which it interacts could be different in females and males. Thus the sex-specific mutations could identify regions of the MEI-S332 protein that interact with sex-specific proteins.

The two male-specific mutations, *mei-S332*[3] and *mei-S332*[8], cause amino acid substitutions in the coiled-coil domain at the N-terminus of the protein (Fig. 16.2). The strongest of these, *mei-S332*[8], changes a valine to a glutamic acid at a position in the coiled coil that must be hydrophobic in order for dimers to form (Fig. 16.3) (O'Shea *et al.*, 1991). This mutation would be predicted to disrupt dimerization and thus be severe. The coiled coil may be particularly important for cohesion in male meiosis, possibly by forming a heterodimer with a male meiotic protein. *mei-S332*^{2} and *mei-S332*^{6}, the two mutations that affect chromosome segregation predominantly in female meiosis, result in amino acid changes in the basic region at the C-terminus of the protein (Fig 16.2). The basic domain cannot be solely required for female meiosis, since the *mei-S332*^{4} mutation affects both females and males. However, this domain may play a more critical function in female meiosis than in male meiosis.

16.4 LOCALIZATION OF *MEI-S332* ON MEIOTIC CHROMOSOMES

The phenotype of *mei-S332* mutations suggested that the protein is needed at the centromere regions of meiotic chromosomes to promote cohesion. It was important therefore to determine whether the MEI-S332 protein bound to meiotic chromosomes and, if so, when and where.

We localized the MEI-S332 protein by fusing it to the green fluorescent protein (GFP) from jellyfish (Kerrebrock *et al.*, 1995). The GFP protein is intrinsically fluorescent and can be used to localize a protein to which it is fused without relying on secondary detection methods such as antibodies (Chalfie *et al.*, 1994; Wang and Hazelrigg, 1994). We fused the GFP coding sequences to the N-terminus of MEI-S332. The fusion gene was transformed into Drosophila on a *mei-S332* genomic fragment. Thus the fusion protein was expressed under the control of the normal *mei-S332* regulatory elements. The MEI-S332-GFP is fully functional, because the fusion gene complemented the mutant phenotype in transformed flies. Since it is active, the localization of the fusion protein is likely to be the same as that of the MEI-S332 protein.

The MEI-S332-GFP protein is localized specifically to the centromere region of meiotic chromosomes (Fig 16.4) (Kerrebrock *et al.*, 1995). We examined the distribution of MEI-S332-GFP during male meiosis because it is easy to identify the meiotic stages in spermatocytes, and individual chromosome arms and centromere regions can be distinguished. Prior to chromosome condensation in

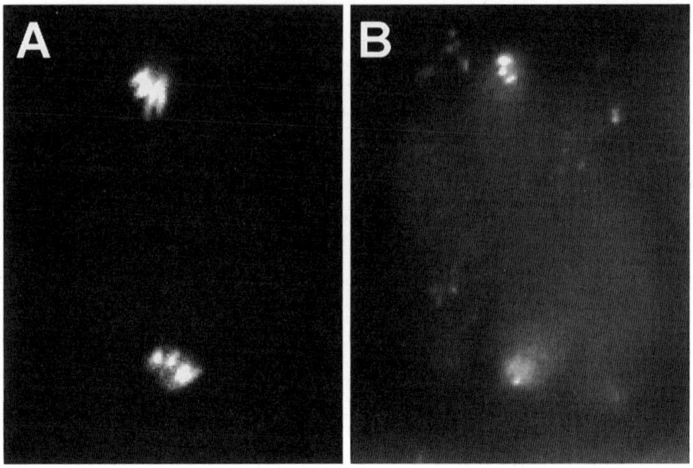

Fig. 16.4 Centromere localization of the MEI-S332-GFP protein in an anaphase I cell. (A) DNA stain of a primary spermatocyte from a fly expressing MEI-S332-GFP. (B.) GFP fluorescence from the same cell. The anaphase I configuration of the chromosomes can be seen, with the centromere regions closest to the poles. In the bottom spindle pole, some of the centromeres are out of the plane of focus. The MEI-S332-GFP protein is localized on the centromere region of each of the four chromosomes.

prophase I, the MEI-S332-GFP protein is cytoplasmic. As the chromosomes condense the cytoplasmic staining diminishes greatly, and MEI-S332-GFP becomes localized to a specific site on each chromosome. In anaphase I, the centromere regions can be identified because they move towards the poles ahead of the arms. From analyzing anaphase I cells, it is clear that MEI-S332-GFP is localized at the centromere region. Fig. 16.4 shows a late anaphase I cell in which the chromosomes have reached the poles. The centromere regions are closest to the poles, and MEI-S332-GFP is located there. MEI-S332-GFP remains at the centromere region until metaphase II. Strikingly, it is no longer detectable at anaphase II when cohesion must be lost to allow the sister-chromatids to segregate. Thus the localization of MEI-S332-GFP suggests that the MEI-S332 protein binds to the centromere regions to hold sister chromatids together and that it either delocalizes or is degraded to permit sister-chromatid separation.

16.5 CONCLUSIONS

The identification of the ORD and MEI-S332 proteins is an important step towards understanding the molecular basis of sister-chromatid cohesion. The localization of MEI-S332-GFP to the centromere regions and its disappearance as the sister chromatids separate implies that MEI-S332 directly holds the sister chromatids together. It will be interesting to determine whether the protein is degraded or modified at the metaphase II/ anaphase II transition. Analysis of the localization of ORD will provide insights into its role in sister-chromatid cohesion and recombination. In addition, identifying proteins that interact with MEI-S332 and ORD will provide a means to isolate other components of sister-chromatid cohesion. This will be facilitated by the fact that mutations in MEI-S332 and ORD suggest parts of these proteins that may bind other proteins.

ACKNOWLEDGMENTS

We thank H. Protzman for help with the preparation of figures. A.K. was supported by a postdoctoral fellowship from The Medical Foundation, and S.B. was a postdoctoral fellow of the Damon Runyon-Walter Winchell Cancer Research Fund. This research was supported by grants from the National Science Foundation and the American Cancer Society.

REFERENCES

Angell, R.R. (1995) Meiosis I in human oocytes. *Cytogenetics and Cell Genetics,* **69**, 266–72.

Angell, R.R., Xian, J,. Keith, J. *et al.* (1994) First meiotic division abnormalities in human oocytes: mechanism of trisomy formation. *Cytogenetics and Cell Genetics,* **65**, 194–202

Berger, B., Wilson, D.B., Wolf, E. *et al.* (1995) Predicting coiled coils using pairwise–residue correlations. *Proceedings of the National Academy of Science USA,* **92**, 8259—63.

Bickel, S.E. and Orr-Weaver. T.L. (1996) Holding chromatids together to ensure they go their separate ways. *BioEssay*, in press.

Bickel, S.E., Wyman, D.W., Miyazaki, W.Y. *et al.* (1995) Identification of ORD, a *Drosophila* protein required for sister-chromatid cohesion. *EMBO Journal*, in press.

Chalfie, M., Tu, Y., Euskirchen, G. *et al.* (1994) Green fluorescent protein as a marker for gene expression. *Science,* **263**, 802–5.

Clayberg, C. (1959) Cytogenetic studies of precocious meiotic centromere division in Lycopersicon esculentum Mill. *Genetics,* **44**, 1335–46.

Cooke, C., Heck, M. and Earnshaw, W. (1987) The inner centromere protein (INCENP) antigens: Movement from inner centromere to midbody during mitosis. *Journal of Cell Biology,* **105**, 2053–67.

Davis, B. (1971) Genetic analysis of a meiotic mutant resulting in precocious sister-centromere separation in *Drosophila melanogaster*. *Molecular and General Genetics,* **113**, 251–72.

Dobson, M.J., Pearlman, R.E., Karaiskakis, A. *et al.*, (1994) Synaptonemal complex proteins: occurrence, epitope mapping, and chromosome disjunction. *Journal of Cell Science,* **107**, 2749–60.

Earnshaw, W. and Bernat, R. (1991) Chromosomal passengers: Toward an integrated view of mitosis. *Chromosoma,* **100**, 139–46.

Earnshaw, W.C. and Cooke, C.A. (1991) Analysis of the distribution of the INCENPs throughout mitosis reveals the existence of a pathway of structural changes in the chromosomes during metaphase and early events in cleavage furrow formation. *Journal of Cell Science,* **98**, 443–61.

Esponda, P. and Krimer, D.B. (1979) Development of the synaptonemal complex and polycomplex formation in three species of grasshoppers. *Chromosoma,* **73**, 237–45.

German, J. (1979) Robert's syndrome. I. Cytological evidence for a disturbance in chromatid pairing. *Clinical Genetics,* **16**, 441–7.

Goldstein, L.S.B. (1980) Mechanisms of chromosome orientation revealed by two meiotic mutants in *Drosophila melanogaster*. *Chromosoma,* **78**, 79–111.

Kerrebrock, A.W., Miyazaki, W.Y., Birnby, D. and Orr-Weaver, T.L. (1992) The *Drosophila mei-S332* gene promotes sister-chromatid cohesion in meiosis following kinetochore differentiation. *Genetics,* **130**, 827–41.

Kerrebrock, A.W., Moore, D.P., Wu, J.S. and Orr-Weaver, T.L. (1995) MEI-S332, a Drosophila protein required for sister-chromatid cohesion, can localize to meiotic centromere regions. *Cell,* **83**, 247—56.

Lin, H.P. and Church, K. (1982) Meiosis in *Drosophila melanogaster* III. The effect of orientation disruptor (*ord*) on gonial mitotic and the meiotic divisions in males. *Genetics,* **102**, 751–70.

Lupas, A., Van Dyke, M., Stock, J. (1991) Predicting coiled coils from protein sequences. *Science,* **252**, 1162–4.

MacKay, A.M., Eckley, D.M., Chue, C. and Earnshaw, W.C. (1993) Molecular analysis of INCENPS (inner centromere proteins): Separate domains are required for association of microtubules during interphase and with the central spindle during anaphase. *Journal of Cell Biology,* **123**, 373–85.

Maguire, M.P. (1978) Evidence for separate genetic control of crossing over and chiasma maintenance in maize. *Chromosoma,* **65**, 173–83.

Maguire, M.P. (1993) Sister chromatid association at meiosis. *Maydica,* **38**, 93–106.

Maguire, M.P., Paredes, A.M. and Riess, R.W. (1991) The desynaptic mutant of maize as a combined defect of synaptonemal complex and chiasma maintenance. *Genome,* **34**, 879–87.

Maguire, M.P., Riess, R.W. and Paredes, A.M. (1993) Evidence from a maize desynaptic mutant points to a probable role of synaptonemal complex central region components in provision for subsequent chiasma maintenance. *Genome,* **36**, 797–807.

Mason, J.M. (1976) Orientation disruptor (*ord*): A recombination-defective and disjunction-defective meiotic mutant in *Drosophila melanogaster*. *Genetics,* **84**, 545–72.

Miyazaki, W.Y. and Orr-Weaver, T.L. (1992) Sister-chromatid misbehavior in Drosophila *ord* mutants. *Genetics,* **132**, 1047–61.

Moens, P.B. and Church, K. (1979) The distribution of synaptonemal complex material in metaphase I bivalents of *Locusta* and *Chloealtis* (Orthoptera: Acrididae). *Chromosoma,* **73**, 247–54.

Molnar, M., Bahler, J., Sipiczki, M. and Kohli, J. (1995) The *rec8* gene of *Schizosaccharomyces pombe* is involved in linear element formation, chromosome pairing and sister-chromatid cohesion during meiosis. *Genetics,* **141**, 61–73.

Moreau, P.J.F., Zickler, D. and Leblon, G. (1985) One class of mutants with disturbed centromere cleavage and chromosome pairing in *Sordaria macrospora*. *Molecular and General Genetics,* **198**, 189–97.

O'Shea, E.K., Klemm, J.D., Kim, P.S. and Alber, T. (1991) X-ray structure of the GCN4 leucine zipper, a two-stranded, parallel coiled coil. *Science,* **254**, 539–44.

Orr-Weaver, T.L. (1995) Meiosis in Drosophila: Seeing is believing. *Proceedings of the National Academy of Science USA*, **92**, 10443–9.

Rattner, J.B., Kingwell, B.G. and Fritzler, M.J. (1988) Detection of distinct structural domains within the primary constriction using autoantibodies. *Chromosoma,* **96**, 360–7.

Rechsteiner, M. (1988) Regulation of enzyme levels by proteolysis: therole of PEST regions. *Advances in Enzyme Regulation,* **27**, 135–51.

Rockmill, B. and Roeder, G.S. (1988) *RED1*: A yeast gene required for the segregation of chromosomes during the reductional division of meiosis. *Proceedings of the National Academy of Science USA,* **85**, 6057–61.

Rockmill, B and Roeder, G.S. (1990) Meiosis in asynaptic yeast. *Genetics,* **126**, 563–74.

Rogers, S., Wells, R. and Rechsteiner, M. (1986) Amino acid sequences common to rapidly degraded proteins: The PEST hypothesis. *Science,* **234**, 364–8.

Ross, L.O., Bascom-Slack, C.A. and Dawson, D.S. (1996) Chiasmata, crossovers, and meiotic chromosome segregation. *Advances in Genetics*, in press.

Sumner, A. (1991) Scanning electron microscopy of mammalian chromosomes from prophase to telophase. *Chromosoma,* **100**, 410–8.

Van Den Berg, D.J. and Francke, U. (1993) Roberts syndrome: A review of 100 cases and a new rating system for severity. *American Journal of Medical Genetics,* **47**, 1104–23.

Wang. S. and Hazelrigg, T. (1994) Implications for bcd mRNA localization from spatial distribution of exu protein in *Drosophila* oogenesis. *Nature,* **369**, 400–3.

17

Expanding genetic maps: reevaluation of the relationship between chiasmata and crossovers

B.S. Gill, K.S. Gill, B. Friebe and T.R. Endo

17.1 INTRODUCTION

The paradigm of a 1:1 relationship between cytological chiasmata and genetic crossovers is widely accepted (see Jones, 1987, for review). Extensive literature on meiosis will not be reviewed here, but John (1990) provides a good review of its various aspects. It was Darlington (1932, 1934), who compared chiasma frequencies with genetic map lengths in maize, both for individual chromosomes and the whole genome, and found a good correlation between the two estimators of recombination. With the recent explosion in genetic map construction in diverse organisms using DNA restriction fragment length polymorphisms (RFLPs), genetic maps based on RFLPs far exceed the length expected based on chiasma counts. Nilsson *et al.,* (1993) summarized data from different crop plants documenting the discrepancy between the estimated number of crossovers from chiasma counts or genetic maps (Table 17.1). In every case, chiasma counts underestimated crossover frequencies calculated from recombination data.

Furthermore, an historical survey of the development of maize genetic linkage maps shows that in general, maps increased in length as more markers were added and that expansion was particularly large in RFLP maps (Table 17.2, Nilsson *et al.*, 1993). The length of the most dense classical linkage maps were within the limits of estimates based on chiasma counts. However, the length of most high-density,

Table 17.1 Estimated number of crossovers in various plant species (from Nilsson *et al.*, 1993).

	Estimated no. crossovers per meiosis from:	
Species	*Chiasma counts*	*Genetic maps*
Brassica campestris	10.0—18.5	37.0
Brassica oleracea	12.8—14.8	22.2
Hordeum vulgare	13.5—15.6	22.7
Lactuca sativa	14.6—20.7	28.1
Lycopersicon esculentum	16.2—17.0	25.5
Oryza sativa	18.9—27.6	36.7
Pisum sativum	10.3—18.1	29.3
Solanum tuberosum	13.2—14.1	20.7
Zea mays	17.4—25.0	35.8

RFLP-based maps exceeded the estimates based on chiasma counts. They noted three explanations, which alone or in combination, may account for the above data. First, error in chiasma counts, especially closely spaced ones. Second, chiasmata and crossovers may not show a 1:1 relationship because of chiasma terminalization and nonchiasmate crossing over. Third, recombination frequencies may be biased upward, because of errors in the data or assumptions used to construct the maps.

The debate has been rejoined by the cytologists. Tease and Jones (1995) have ruled out the possibility of closely spaced chiasmata or their terminalization from an analysis of BrdU-labeled bivalents. Whereas there is little doubt that all chiasmata result from crossovers, we must resolve the apparent contradiction between recombination and chiasma frequency. It is worthwhile to review the different approaches used to analyze the relationship between chiasmata and crossovers and their potential limitations.

Table 17.2 Historical development of Zea mays linkage maps (From Nilsson *et al.*, 1993).

	Genetic length (cM) estimated from:					
		Linkage data				
Chromosome	*Chiasma counts*	*1934**	*1950**	*1976**	*1990**	*1990†*
1	187	102	156	161	176	238
2	163	58	128	155	155	239
3	150	92	121	128	167	194
4	148	80	111	143	137	174
5	148	44	72	87	107	235
6	110	52	64	68	78	169
7	123	50	96	112	112	131
8	123	20	28	28	42	173
9	110	52	71	138	140	132
10	98	68	57	99	95	115

*Based on the analysis of gene markers. †Based on the analysis of RFLP markers.

Mainly, three approaches have been used: cytological, genetic and the cytogenetic approach combining both cytological and genetic observations.

Whereas the cytological approach has provided much useful data, its limitations must be considered. The resolution of the microscope is a real problem. The closely spaced chiasmata, if they do occur, will not be resolved under the microscope. If double crossovers occur in a submicroscopic region, they will not be resolved. There are also technical limitations. For example, chromatid labeling is the most direct method of localizing chiasmata and crossovers (Tease and Jones,

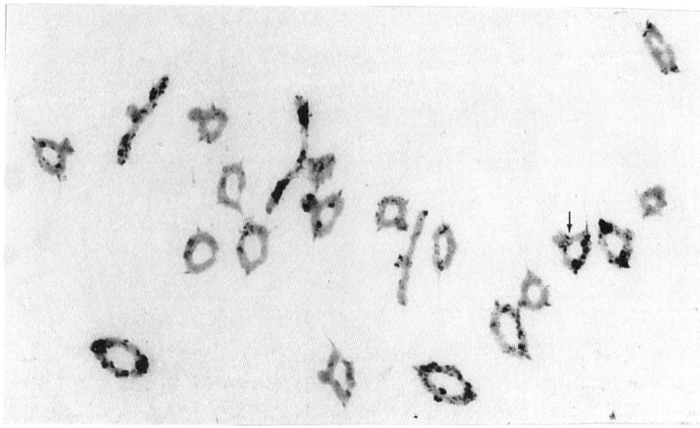

Fig. 17.1 A typical C-banded cell at diakinesis to early metaphase I stage of meiosis of *Triticum aestivum* showing 18 ring and 3 rod bivalents. Note the subterminal to terminal localization of chiasmata in all bivalents. In chromosome 5B (arrow), the chiasma is terminal in the short arm, but is localized proximal to a C-band at FL 0.55 (see Fig. 17.5) in the long arm. Is there another chiasma in the 5BL arm distal to the C-banded material? It is difficult to tell.

1995). However, half the bivalents are noninformative, presumably involving exchanges among similarly-labeled chromatids. Observations are limited to monochiasmate bivalents. Sister chromatid exchanges and anomalous types of configurations that are difficult to interpret are also observed. Organisms such as grasshoppers, with favorable cytology where unequivocal chiasma counts can be made, unfortunately lack linkage data.

The genetic map length is commonly used to calculate cross-overs. However, assumptions underlying map construction may be faulty and errors in phenotypic classification inflate the length of the maps (see Nilsson *et al.*, 1993). Ideally, one would like to apply a unified cytological and genetic, i.e. cytogenetic approach, to this problem. Unfortunately, only a limited number of cytological and genetical markers were used in most studies (Curtis and Lukaszewski, 1991). This introduces error in the data because genetic recombination is not fully assayed as many recombination events are not detected.

We believe wheat (*Triticum aestivum* L.) offers a model cytogenetic system to analyze the relationship between cytological chiasmata and genetic crossovers. The wheat chromosomes are large in size (average

Chinese Spring (CS) × 5B disomic substitution from *Triticum dicoccoides* in 'CS'

↓

monotelo 5BL × F_1

↓

- 41 chromosome plants were selected and allowed to self
- 42 chromosome progeny possessing recombinant chromosome in disomic condition were selected for each plant and used for mapping

Fig. 17.2. The crossing scheme used to generate a population of recombinant substitution lines. The two parents differed only for chromosome 5B, and only one cycle of meiosis was allowed before the recombinant products were isolated individually as separate lines.

size ≅ 10μm), and contain diagnostic C-bands for individual chromosome identification in mitotic and meiotic stages (Gill *et al.*, 1991). Marker chromosomes (telocentrics) and C-banding can be used to estimate chiasma frequency in individual chromosome arms (Fig. 17.1). Intercultivar and interspecific single chromosome substitutions can be produced experimentally and are used to produce single chromosome recombinant lines by appropriate crosses (Law, 1966). The recombinant lines represent a single-chromosome, single-meiotic cycle population and provide an accurate estimate of the precise distribution of crossovers in the target chromosome (Fig. 17.2). Most important, numerous single-break, terminal deletion lines are available and can be used to estimate precise distribution of genes, and crossovers in specific chromosome segments (Endo and Gill, 1996). Numerous chromosomally engineered stocks also are available for assaying frequencies of chiasmata and recombination along the chromosome length (Friebe *et al.*, 1995; Lukaszewski, 1995a, b).

We have used the wheat cytogenetic system to analyze chiasmata and recombination distribution along the chromosome length, and the results will be reviewed here.

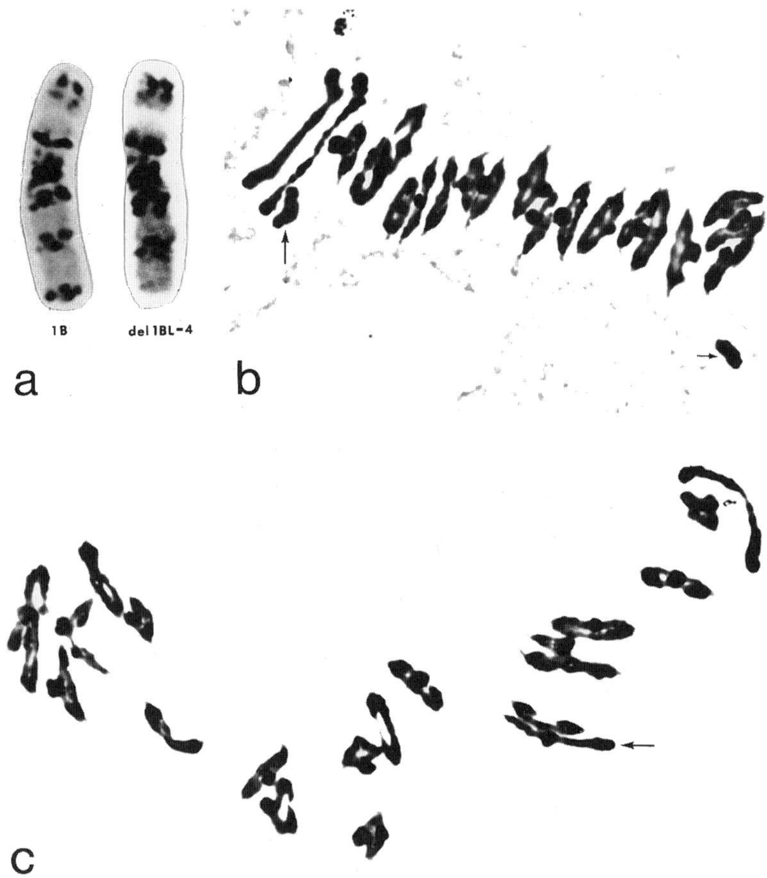

Fig. 17.3 Metaphase I chiasmate association is drastically reduced in the arm of a bivalent heteromorphic for even a tiny terminal deletion. **a.** A normal chromosome 1B and a deletion 1B (1BL-4) with breakpoint at fraction length 0.93 in the long arm (distal 0.07% , including the two prominent terminal C-bands are deleted). Meiotic MI chiasmate association of deletion chromosome 1BL-4 with telocentric 1BL was assayed. **b.** In 90% of the cells, univalents of 1BL-4 and telosomic 1BL (arrows) were observed. **c.** Only 10% of the cells showed heteromorphic bivalents (arrow). The chiasmate association is restored in the deletion arm in del1BL-4 homozygous plants (Tyagi and Gill, unpublished results).

17.2 TELOMERIC REGION HOMOLOGY IS ESSENTIAL FOR PAIRING AND CHIASMA FORMATION

Several types of data are available that indicate that telomeric homology is essential for initiation of pairing and chiasmate association at metaphase 1 stage of meiosis. In wheat, chiasmata are terminally localized (Fig. 17.1), which in itself indicates the important role of telomeric regions as sites of chiasma formation and maintenance. In wheat, in bivalents heteromorphic for terminal deletions that otherwise share up to 90% homology, MI chiasmate association is drastically reduced (Curtis *et al.*, 1991; Endo *et al.*, 1991; see also Fig. 17.3). It is not known if chromosome pairing and recombination also are reduced drastically. In deletion homozygotes, almost complete chiasmate association is restored.

In chromosome combinations where terminal telomeric regions in one arm share 5—10% homology, but the remaining chromosome regions in a bivalent are nonhomologous, almost 100% chiasmate association is observed (Fig. 17.4; also see Maguire, 1989). These results show that homology in terminal telomeric regions is critical for the initiation of chromosome pairing, chiasma formation and maintenance. Under a microscope, only one chiasma can be resolved in such bivalents. However, in most wheat maps, the genetic length of such small regions often exceeds 50 recombination units (Delaney *et al.*, 1995a, b; Mickelson-Young *et al.*, 1995; Gill *et al.*, 1993, 1996; Werner *et al.*, 1992, Hohmann *et al.*, 1994). This indicates that double crossovers often do occur exceeding expected estimates based on chiasma counts.

17.3 DISTRIBUTION OF RECOMBINATION IS NONRANDOM AND RESTRICTED TO GENE-RICH REGIONS FREQUENTLY FOUND IN TELOMERIC REGIONS

For the past few years, we have been isolating a vast array of deletion stocks, at present numbering more than 400 and involving all 21 chromosomes of wheat (Endo and Gill, 1966). We have constructed physical maps of the 21 chromosomes of wheat by mapping RFLP probes onto the deletion stocks (reviewed in Gill and Gill, 1994; see also references cited in the previous paragraph). These data show that RFLP markers, which are mostly cDNAs and, hence, expressed genes are nonrandomly distributed and more numerous in distal regions of

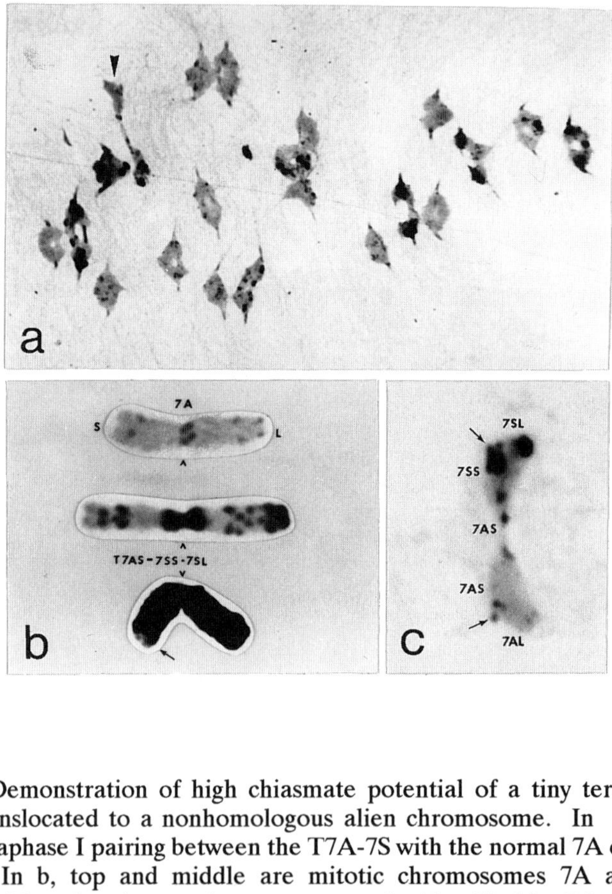

Fig. 17.4 Demonstration of high chiasmate potential of a tiny terminal wheat segment translocated to a nonhomologous alien chromosome. In a and c, the meiotic metaphase I pairing between the T7A-7S with the normal 7A chromosome is shown. In b, top and middle are mitotic chromosomes 7A and T7A-7S, respectively. Note that the two chromosome can be identified by C-banding during meiosis (a and c). In b, the bottom chromosome shows the genomic *in situ* pattern of the T7A-7S (middle chromosome) using total *Triticum speltoides* DNA as probes. The arrow indicated the breakpoint of the translocation. The tiny unlabeled segment is from 7A (top chromosome in b.) and shows chiasmate association with 7A in 97% of the cells in MI of meiosis (c). The labeled segment is from an alien chromosome and is nonhomologous.

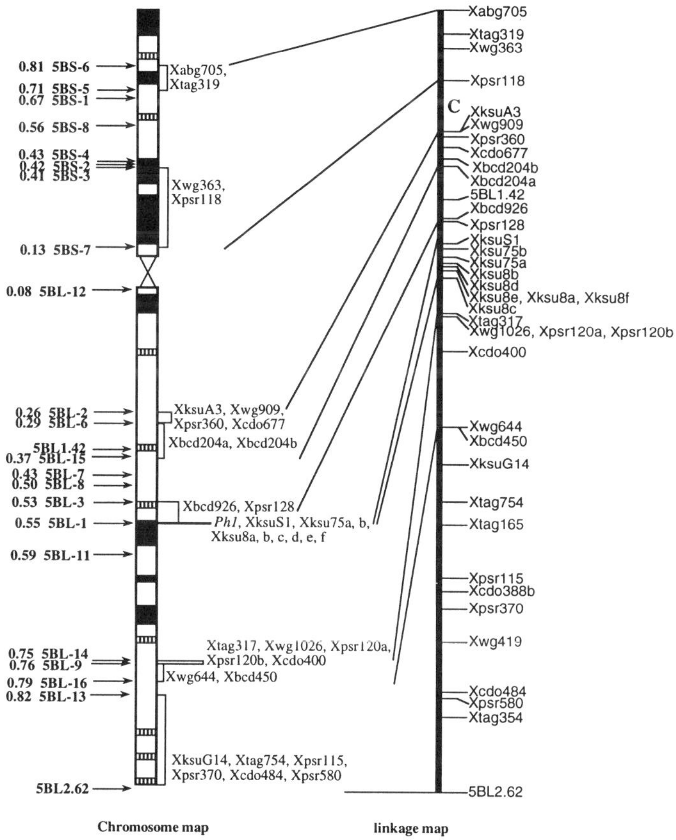

Fig. 17.5 A cytogenetic ladder map constructed by joining a common set of markers between a chromosome map and a genetic linkage map of 5B of wheat. The chromosome map consists of a C-banded idiogram with deletion breakpoints and fraction length distances from the centromere for each arm indicated to the left. Note nonrandom distribution of genes and restriction of recombination to gene island regions. Double crossovers are common in gene island regions. If each crossover matures into a chiasma, then such closely juxtaposed chiasmata will not be resolved under the microscope (reproduced from Gill *et al.* 1995).

chromosomes. When we compare physical maps with genetic maps by joining commonly mapped probes to construct the so-called cytogenetic ladder maps (CLMs), we find that recombination is restricted to gene-rich regions. An example of a CLM for chromosome 5B of wheat is shown (Fig. 17.5). There are no markers and very little recombination in the proximal centromeric region spanning 26% of 5BL arm. Similarly, there is another marker-poor region between 59% and 75% of 5BL arm and there is little recombination in that region. It is clear that most recombination is restricted to the gene-rich regions, especially the distal 25% of 5BL arm. The data reveal the existence of structural and functional domains along the chromosome axis corresponding to genetically inactive and active regions, respectively. Recombination is an inherent function of the functionally active gene-rich regions.

17.4 CHIASMA COUNTS UNDERESTIMATE RECOMBINATION

We analyzed chiasma formation in 129 MI pollen mother cells in an F_1 hybrid of the cultivar Chinese Spring (CS) and a substitution line of CS containing chromosome 5B from an emmer wheat (*Triticum dicoccoides*). We observed an average of 1.1 chiasmata for chromosome 5B. Based on chiasma data, the calculated genetic length of 5B chromosome is 56 recombination units (Gill *et al.* 1995). We calculated the observed genetic length of the 5B by mapping 38 RFLP markers and two polymorphic C-bands in a mapping population of 136 recombinant lines of 5B (Table 17.3). The observed genetic length of 5B is 118 recombination units, two times greater than expected based on chiasma counts. The data provide conclusive evidence that in this particular hybrid, chiasma counts did underestimate recombination by a factor of two.

Table 17.3 Number of crossovers for chromosome 5B estimated from recombination frequency per gamete.

Recombination events	*Number of gametes*	*Estimated number of crossovers*
none	44	0
one	46	46
two	28	56
three	15	45
four	2	8
six	1	6
Total	136	161

Note: In addition, there were 29 novel recombination events where the switch in the data type was for only one marker (Reproduced from Gill, B.S., *et al.*, 1995).

17.5 DEMONSTRATION OF DOUBLE CROSSOVERS IN SUBMICROSCOPIC REGIONS

The CLM's allow us to look at the frequency of crossovers in specific chromosome regions. The selected array of gametes with double crossovers in the FL 0.75—0.82 region of 5BL arm are shown in Fig. 17.6. We estimate the physical size of this region at about 0.6 microns. If each crossover formed a chiasma, then such closely spaced double chiasmata will not be resolved under the microscope. Then there are novel recombination events involving exchange of single markers, which appear to be gene conversion type of events. One gamete showed six crossovers (Fig. 17.6), whereas a maximum of two chiasmata (expected two crossovers per chromatid) were observed. The evidence strongly suggests that either observed chiasmata sometimes may depict more than one recombination event or not all chiasmata mature into crossovers.

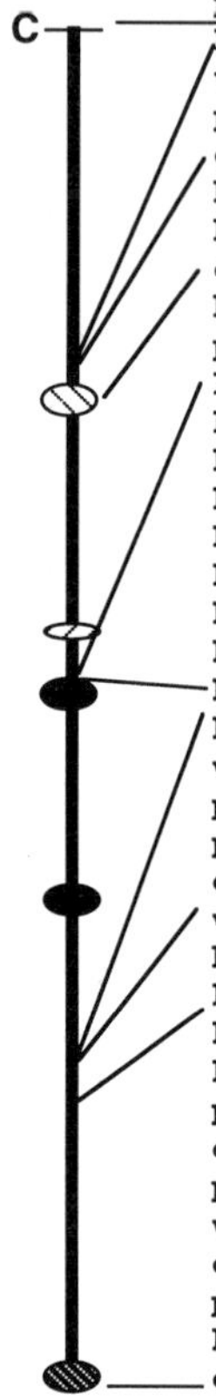

abg705	A	A	**A**	**A**	B	**B**	A	A	**A**	A
ku319	A	A	**A**	B	B	A	A	**B**	B	A
wg363	A	A	-	B	B	A	A	**B**	B	A
psr118	A	A	B	B	B	A	A	-	B	A
ksuA3	A	A	B	B	B	A	A	A	B	A
wg909	A	A	B	B	B	A	A	A	B	A
psr360	A	A	B	B	B	A	A	A	B	A
cdo677	A	A	B	B	B	A	A	**B**	B	A
bcd204b	A	**B**	B	B	B	A	A	**B**	B	A
bcd204a	A	A	B	-	-	A	A	**B**	B	A
cband1	A	A	B	-	-	-	A	A	-	A
bcd926	A	-	B	B	B	A	A	A	B	A
psr128	A	A	B	B	B	A	A	**B**	B	A
ksuS1	A	A	B	B	B	A	A	A	B	A
ksu75b	A	A	B	B	B	A	A	A	B	A
ksu75a	A	A	B	B	B	A	A	A	B	A
ksu8b	A	A	B	B	B	A	A	A	B	A
ksu8d	A	A	B	B	B	A	A	A	B	A
ksu8e	A	A	B	B	B	A	A	A	B	A
ksu8f	A	A	B	B	B	A	A	A	B	A
ksu8a	A	-	B	B	B	A	A	A	B	A
ksu8c	A	-	B	B	B	A	A	A	B	A
ku317	A	**B**	B	**A**	**A**	**B**	A	A	B	A
wg1026	A	**B**	B	**A**	**A**	**B**	**B**	A	B	A
psr120a	A	**B**	B	**A**	**A**	**B**	**B**	A	B	A
psr120b	A	**B**	B	**A**	**A**	**B**	**B**	A	B	A
cdo400	A	A	**A**	B	**A**	**B**	**B**	A	**A**	**B**
wg644	A	A	**A**	B	B	A	**B**	A	**A**	A
bcd450	A	A	**A**	B	B	A	**B**	A	**A**	A
ksuG14	**B**	A	**A**	B	B	A	-	-	-	-
ku754	A	A	B	B	B	A	A	**B**	B	A
ku165	A	A	B	B	B	A	A	**B**	B	A
psr115	A	**B**	B	B	B	A	A	**B**	B	A
cdo388b	A	**B**	B	B	B	A	A	**B**	B	A
psr370	A	**B**	**A**	B	B	A	A	-	B	A
wg419	A	**B**	**A**	B	B	A	A	A	B	A
cdo484	A	**B**	**A**	B	**A**	A	A	A	B	A
psr580	A	**B**	**A**	B	**A**	A	A	A	B	A
ku354	A	**B**	**A**	B	**A**	A	A	A	B	A
cband2	A	**B**	**A**	-	-	-	A	A	-	A

Fig. 17.6 The actual genetic mapping data for the DNA and C-band markers, along with their physical location on 5B chromosome. The DNA markers are chromosome 5-specific probes analyzed by the gel blot DNA hybridization procedure. The 'As' represent Chinese Spring and the 'Bs' represent *Triticum dicoccoides* alleles. The mapping population was developed in such a way that the two parents of the population differed for only one chromosome and only one cycle of recombination was allowed. The mapping data of 10 single-chromosome recombinant lines are given. Each plant represents a gametic product of a single recombination cycle. Each column represents marker data for one plant and the crossover regions are marked bold. The gene conversion type events are boxed. Note: double crossovers and gene conversion-type events in a tiny region at FL 0.75—0.82 (see Fig. 17.5). Plant 8 from the left has six crossovers and a gene conversion-type recombination event.

17.6 CONCLUDING REMARKS

The cytogenetic approach and DNA-based RFLP markers provide strong tools to analyze some basic aspects of meiosis. The RFLPs not only provide an unlimited source of markers for genetic linkage mapping, but, in addition, reveal the molecular mechanisms of recombination because DNA sequences are also assayed. Deletion stocks and other cytological markers, as well as chromosome painting (Jiang and Gill, 1994), provide excellent physical markers for analyzing various aspects of meiosis as demonstrated here. These new data are challenging long-held, basic concepts of the meiotic process. It is in this context that the relationship between chiasmata and crossovers needs to be reexamined. Our data provide evidence that more recombination occurs than can be estimated from chiasma counts. This confirms the trend revealed by RFLP linkage maps (Nilsson *et al.*, 1993). However, one key difference should be pointed out. The classical recombination studies utilized highly inbred parents, whereas most RFLP mapping involves highly divergent parents, even interspecific crosses. There is evidence that different patterns of recombination may be observed in wide crosses (Cox, 1995). In our studies, we utilized a chromosome derived from a wild progenitor of wheat. We need to obtain similar data on more inbred parents, which is much more difficult because of the low level of polymorphism.

ACKNOWLEDGEMENTS

Research supported in part by USDA-NRI competitive grants. The authors thank John Raupp for editing the manuscript. The senior author thanks the organizing committee for the invitation to attend the conference and financial support. Kansas Agricultural Experiment Station contribution no. 96-363-B.

REFERENCES

Cox, E.C. (1995) Recombination, mutation and the origin of species. *BioEssays*, **17** (9), 747—9.

Curtis, C.A. and Lukaszewski, A.J. (1991) Genetic linkage of C-bands and storage protein genes in chromosome 1B of tetraploid wheat. *Theoretical and Applied Genetics*, **81**, 245—52.

Curtis, C.A., Lukaszewski, A.J. and Chrzastek, M. (1991) Metaphase I pairing of deficient chromosome and genetic mapping of deficiency breakpoints in common wheat. *Genome*, **34**, 553—60.

Darlington, C.D. (1932) *Recent Advances in Cytology*, Churchill, London.

Darlington, C.D. (1934) The origin and behaviour of chiasmata VII. *Zea mays. Zietschrift fur Induktiv und Abstrammung Vererbungslehre*, **67**, 96—114.

Delaney, D.E., Nasuda, S., Endo, T.R. *et al.* (1995a) Cytologically based physical maps of the group-2 chromosomes of wheat. *Theoretical and Applied Genetics*, **91**, 568—73.

Delaney, D.E., Nasuda, S., Endo, T.R. *et al.* (1995b) Cytologically based physical maps of the group-3 chromosomes of wheat. *Theoretical and Applied Genetics*, **91**, 780—2.

Endo, T.R. and Gill B.S. (1996) The deletion stocks of common wheat, *Journal of Heredity*, in press.

Endo, T.R., Mukai, Y., Yamamoto, M. and Gill, B.S. (1991) Physical mapping of a male-fertility gene of common wheat. *Japanese Journal of Genetics*, **66**, 291—5.

Friebe, B., Jiang, J., Raupp, W.J. and Gill, B.S. (1995) Molecular cytogenetic analysis of radiation-induced alien genetic transfers in wheat. In *Proceedings of the 8th International Wheat Genetics Symposium* (eds Z.S. Li and Z.Y. Xin), Beijing, China, China Agriculture Scientech Press, Beijing, pp. 519—29.

Gill, B.S., Friebe, B. and Endo, T.R. (1991) Standard karyotype and nomenclature system for description of chromosome bands and structural aberrations in wheat (*Triticum aestivum*). *Genome*, **34** (5), 830—9.

Gill, B.S., Gill, K.S. and Friebe, B. (1995) Cytogenetic ladder maps and cereal chromosome structure, function, and manipulation, in *Classical and Molecular Cytogenetic Analysis. Proceedings of a U.S.-Japan Symposium* (eds. W.J. Raupp and B.S. Gill), Kansas Agriculture Experiment Station Report 95-352-D, Manhattan, pp. 129—35.

Gill, K.S. and Gill, B.S. (1994) Mapping in the realm of polyploidy: the wheat model. *BioEssays*, **16** (11), 841—6.

Gill, K.S., Gill, B.S. and Endo, T.R. (1993) A chromosome region-specific mapping strategy reveals gene-rich telomeric ends in wheat. *Chromosoma*, **102**, 374—81.

Gill, K.S., Gill, B.S., Endo, T.R. and Boyko, E. (1996) Identification and high-density mapping of gene-rich regions of chromosome group 5 of wheat. *Genetics*, in press.

Hohmann, U., Endo, T.R., Gill, K.S. and Gill, B.S. (1994) Comparison of genetic and physical maps of group 7 chromosomes from *Triticum aestivum* L. *Molecular and General Genetics*, **245**, 694—53.

Jiang, J, and Gill B.S. (1994) Nonisotopic *in situ* hybridization and plant genome mapping: the first 10 years. *Genome*, **37**, 717—25.

John, B. (1990) *Meiosis*, Cambridge University Press, England.

Jones, G.H. (1987) Chiasmata, in *Meiosis* (ed. P. Moens), Academic Press, London. pp. 213—44.

Law, C.N. (1966) The location of genetic factors affecting a quantative character in wheat. *Genetics*, **53**, 478—98.

Lukaszewski, A.J. (1995a) Chromatid and chromosome type breakage-fusion-bridge cycles in wheat (*Triticum aestivum* L.). *Genetics*, **140**, 1069—85.

Lukaszewski, A.J. (1995b) Manipulation of the genome by chromosome breakage, in *Classical and Molecular Cytogenetic Analysis. Proceedings of a U.S.-Japan Symposium* (eds. W.J. Raupp and B.S. Gill), Kansas Agriculture Experiment Station Report 95-352-D, Manhattan, pp. 136—9.

Maguire, M.P. (1989) Near-normal chiasma formation and maintenance in a short distal translocated segment in maize. *Genetics*, **79**, 129—32.

Mickelson-Young, L.A., Endo, T.R. and Gill, B.S. (1995) Cytogenetic ladder maps of the group 4 chromosomes of wheat. *Theoretical and Applied Genetics*, **90**, 1007—11.

Nilsson, N-O., Sall, T. and Bengtsson, B.O. (1993) Chiasma and recombination data in plants: are they compatible? *Trends In Genetics*, **9** (10), 344—7.

Tease, C. and Jones, G.H. (1995) Do chiasmata disappear? An examination of whether closely spaced chiasmata are liable to reduction or loss. *Chromosome Research*, **3**, 162—8.

Werner, J.E., Endo, T.R. and Gill, B.S. (1992) Toward a cytogenetically based physical map of the wheat genome. *Proceedings of the National Academy of Science USA*, **89**, 11307—11.

PART FIVE

EVOLUTIONARY DYNAMICS

Juan Pedro Camacho

Juan Vega

Ilia Leitch

Milton Gallardo

18

Evolution of a near-neutral B chromosome

J.P.M. Camacho, J. Cabrero, M.D. López-León, and M.W. Shaw

18.1 INTRODUCTION

Among the constellation of selfish genetic elements, including transposons, segregation distorters and some cytoplasmic factors, which make up eukaryote genomes, B chromosomes are in fact the earliest example to be widely studied (Jones, 1985; Shaw and Hewitt, 1990). The term B chromosome was first used by Randolph (1928) in maize and more than 1300 plant and nearly 500 animal species have since been reported to carry them (Jones and Puertas, 1993).

An updated definition of B chromosomes was provided by J.P.M. Camacho and J.S. Parker during the First B chromosome Conference held in Miraflores de la Sierra (Madrid, Spain) in September 1993: they are 'dispensable supernumerary chromosomes that do not recombine with the A chromosomes and follow their own evolutionary pathway' (Beukeboom, 1994). This definition emphasizes the two only properties that, for the moment, are applicable to all Bs hitherto known: dispensability and absence of chiasma formation with the As. There is, however, another universal B chromosome characteristic, i.e. they do not constitute a Mendelian system, in strict terms, because they are not always in pairs in diploid organisms. The most Mendelian a B can be is showing a mean transmission ratio close to 0.5, and thus lacking drive. Although there are some Bs showing Mendelian transmission ratio, e.g. those in the plant *Allium schoenoprasum* (Bougourd and Parker, 1979) and the grasshopper *Eyprepocnemis plorans* (López-León *et al.*, 1992a), most Bs hitherto studied show drive by means of several different mechanisms (Jones, 1991).

The two most important properties determining the maintenance of a B in natural populations are its inheritance mode and its effects on the fitness of the individuals carrying it. Two models have traditionally been proposed: i) the 'parasitic' model states that B chromosomes maintain themselves at the expense of drive independently of the harmful effects they might have on carrier fitness (Östergren, 1945), and ii) the 'heterotic' model by which Bs lack drive but maintain themselves through the beneficial effects they have on carrier fitness, at least in low numbers (White, 1973). We propose a third model, the 'near-neutral' one, by which a B might lack drive and not be appreciably beneficial or harmful for individual fitness, at least in low numbers. As we will see, a near-neutral B can neither increase when very rare nor be maintained indefinitely in natural populations, so that near-neutrality is a transient stage of B chromosome evolution.

Most Bs hitherto well studied seem to fit the parasitic model (see Shaw and Hewitt, 1990, for review), although there is at least one case fitting the heterotic model, i.e. in *Allium schoenoprasum* (Holmes and Bougourd, 1989, 1991; Plowman and Bougourd, 1994), and another fitting the near-neutral model, i.e. in *Eyprepocnemis plorans* which constitutes the subject of the present paper.

18.2 THE B CHROMOSOME SYSTEM IN *E.PLORANS*: MOLECULAR NATURE AND ORIGIN

The grasshopper *E. plorans* shows a very complex polymorphism for more than 40 different B chromosome variants distributed along the Mediterranean coasts in the Iberian Peninsula (Camacho *et al.*, 1980, Henriques-Gil *et al.*, 1982a, 1984; Henriques-Gil and Arana, 1990), North Africa (Henriques-Gil, 1984) and Italy (López-Fernández *et al.*, 1992). The three most frequent variants are named B_1, B_2 and B_5. The most widespread variant, B_1, is the most probable original B (Henriques-Gil *et al.*, 1984), but has been supplanted by other variants in some Spanish regions (Henriques-Gil and Arana, 1990).

Cytogenetic characterization of B_2 by silver impregnation showed that it possesses a latent nucleolar organiser region (NOR) on the distal third that was exceptionally active in a male carrying the B fused to the longest autosome (Cabrero *et al.*, 1987). G, C, N and fluorescence banding showed that this region in the B_2 is composed of GC-rich DNA whereas the two proximal thirds are composed of AT-rich DNA (Camacho *et al.*, 1991). *In situ* digestion with MspI and HpaII showed that the distal third of B_2 was methylated in spermatocytes, where the NOR is inactive, but was not methylated in the fused B which showed NOR activity (López-León *et al.*, 1991). However DNA-methylation is not the only cause of the inactivity of the B NOR, because it is neither methylated nor active in embryos (López-León *et al.*, 1995a).

Double fluorescent *in situ* hybridization (FISH) with two DNA probes, i.e. a 180 bp tandemly repeated DNA and ribosomal DNA, showed that the distal third of B_2 is composed of ribosomal DNA, thus confirming the presence of the latent NOR, and the two proximal thirds are made up of the 180 bp repetitive DNA (López-León *et al.*, 1994). But the most important finding of this investigation was the observation that both DNA probes were present in both A and B chromosomes and that the order of the two probes in respect to the centromere was only coincident in the B and the X, which indicated that the B most likely was derived from the proximal region of the X with later amplification of the two DNA sequences involved.

The 180 bp repetitive DNA seems to be very homogeneous because all the five clones sequenced in both directions showed exactly the same sequence. The basic repeat has little similarity with any sequence in the data banks. It is very unlikely to encode a protein (López-León *et al.*, 1995b).

18.3 VARIABLE EFFECTS ON ENDOPHENOTYPIC TRAITS

Endophenotypic effects of B chromosomes in *E.plorans* have been investigated at two levels: chiasma frequency and NOR activity. The effect of the B on chiasma frequency has varied between experiments, perhaps because this effect depends on the background genotype and hence population (Camacho *et al.*, 1980; Henriques-Gil *et al.*, 1982a,b; Cano *et al.*, 1987). Changes in chiasma frequency are likely to have only very small effects on fitness (Shaw and Hewitt, 1990). The effect of the B on the activity of the NORs located on the standard chromosomes, i.e. X, S_9, S_{10} and S_{11} has also been studied, but again the results differed between populations: whereas in spermatocytes of 1B males from Salobreña there were significantly more NORs which were active than in 0B males (Cabrero *et al.*, 1987), in Jete there were similar numbers (López-León *et al.*, 1995c). Although the amount of available rRNA per cell might affect fitness, rRNA cistrons are present

Table 18.1 Viability from embryo to adult

		Frequency of individuals with			
Population	*Stage*	*0B*	*1B*	*2B+*	*N*
Jete	embryo	0.40	0.41	0.19	2091
	adult	0.42	0.43	0.15	193
Salobreña	embryo	0.49	0.37	0.14	1765
	adult	0.43	0.39	0.18	138

on almost all chromosomes in *E.plorans* (López-León *et al.*, 1994) so a minimum level of rRNA per cell is probably guaranteed in spite of slight variations in the number of active NORs.

18.4 NO EFFECTS ON MORPHOLOGICAL TRAITS

Camacho *et al.* (1980) performed a comparative analysis of five external morphological traits between non-B and B-carrying individuals and failed to find any significant difference. We have recently made a more extensive analysis in about 500 individuals from the Salobreña population (A. Martín-Alganza, M.D. López-León, J. Cabrero and J.P.M. Camacho, unpublished). The traits analysed were somatic weight, gonadal weight, number of units in the gonad (testis follicles or ovarioles), thorax length, thorax width, tegmina length, femur length, tibia length, number of segments in the antennae, somatic condition (defined as the standardized residuals of the regression of somatic weight on body size represented by thorax length), gonadal condition (defined similarly), and fluctuating asymmetry (as a measure of developmental instability) in bilateral traits: tegmina, femur, tibia and antenna. In females, several other traits were measured, i.e. number of developed eggs, egg length, egg diameter, mean volume per egg, total egg volume and proportion of ovarioles producing an egg. No differences related to the presence of B chromosomes were observed for any of these traits.

Table 18.2 Clutch size (number of eggs per pod) and egg fertility (proportion of eggs containing an embryo) in 55 females from Jete and 51 from Salobreña collected in 1990. ANOVA showed non significant differences between karyotypes in all cases (N: number of females; S.E.: standard error)

Population	*Item*	*Karyotype*	*N*	*Mean*	*S.E.*
Jete	Clutch size	0B	24	43.58	2.24
		1B	21	41.05	1.59
		$2B^+$	10	37.40	2.65
	Fertility	0B	24	0.97	0.02
		1B	21	1.00	0.00
		$2B^+$	10	0.97	0.03
Salobreña	Clutch size	0B	23	42.55	2.46
		1B	24	43.44	2.65
		$2B^+$	4	39.25	5.82
	Fertility	0B	23	0.92	0.03
		1B	24	0.92	0.02
		$2B^+$	4	0.91	0.06

18.5 NO FITNESS EFFECTS

Viability from embryo to adult was assessed in two natural populations, i.e. Jete and Salobreña, by comparing genotypic frequencies in embryos and adults of the same generation. As tables 18.1 and 18.2 show, viability, clutch size and female fertility did not differ between 0B, 1B and $2B^+$ individuals. Similarly, no differential mating success was detected in relation with the presence of Bs in Jete and Salobreña (López-León *et al.*, 1992b), and a new and larger sample in Salobreña corroborates this result (table 18.3). Thus no significant B effects on fitness have been detected, within the accuracy possible so far.

Table 18.3 Analysis of mating success in relation to the number of B chromosomes in the Salobreña population sampled in 1992

		Number of individuals with					
Sex	*Status*	*0B*	*1B*	*2B+*	*Total*	$\chi^2_{(2\ df)}$	*p*
Female	non-mated	50	54	25	129		
	mated	45	62	22	129	1.01	0.60
	Total	95	116	47	258		
Male	non-mated	35	47	28	110		
	mated	43	46	38	127	1.13	0.57
	Total	78	93	66	237		

18.6 A B CHROMOSOME LACKING DRIVE

The average transmission ratio per B chromosome (k_B) is very close to the Mendelian expectation of 0.5, for B_1, B_2 and B_5. Transmission was estimated by three methods: by analysing male and female meiosis; by progeny analyses in controlled crosses; and by analysis of broods laid by gravid females mated in the wild (López-León *et al.*, 1992a). All methods suggested the same conclusion, with only slight variation between the averages for different individuals. The system remains non-Mendelian since in individuals with 2Bs, the Bs were paired in only 0.36–0.57 of spermatocytes and 0.15–0.44 of ovocytes. Despite this, the frequency of meiotic nondisjunction of Bs was lower than 3% in all individuals studied.

18.7 COULD A NEAR-NEUTRAL B INVADE A POPULATION?

A computer simulation was made of a population in which there appears a single individual carrying one B which does not affect the carrier fitness and is transmitted with k_B=0.5. This showed that such

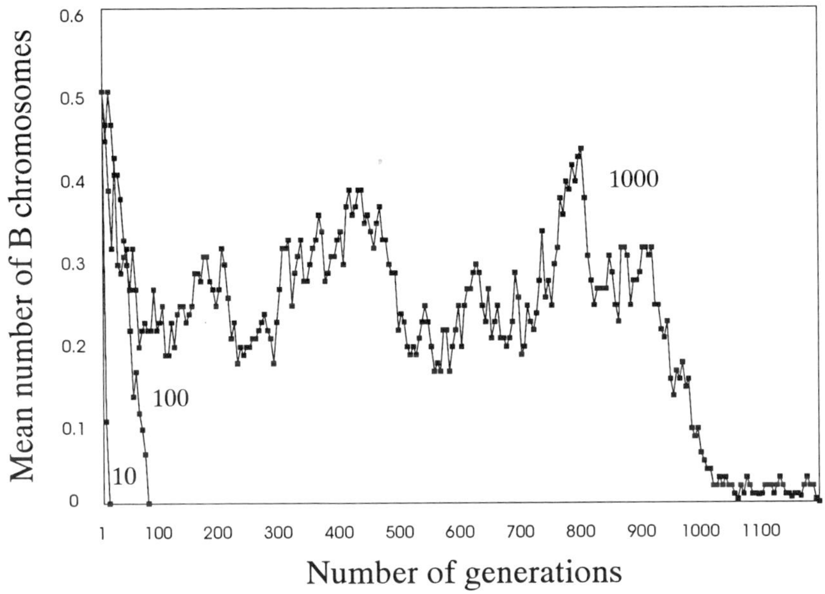

Fig. 18.1 Random walk of a near-neutral B to extinction in three populations composed of 10, 100 and 1000 individuals with initial frequency of Bs equal to 0.5.

a B is eliminated in less than 18 generations, on average, independently of population size. Thus, unless a B possesses initial drive it cannot establish a polymorphism. This is in agreement with the hypothesis advanced by López-León *et al.* (1992a) that *E.plorans* Bs initially had drive, which is supported by the results of the interpopulational crosses performed by Herrera *et al.* (1996). When 1B females were crossed with 0B males from the same population, the B was transmitted in a Mendelian ratio (k_B=0.48), but when these same females were crossed to a 0B male from a non-B population, then the B was significantly accumulated (k_B=0.63). This result suggests that non-B populations may lack drive suppressor genes located on the As which are common in populations in which Bs have been present for a long time.

What about a B which has become near-neutral after reaching a moderate frequency (e.g. 0.5) in the population? Simulation analyses have shown that this B is condemned to extinction (Fig. 18.1) simply because it cannot fix and drift will eliminate it after a more or less long random walk (Shaw, 1984). A correlation analysis showed that the time to extinction depends strongly on the size of freely interbreeding populations ($r=0.97$, $N=11$, $t=12.95$, $p<0.001$). It does not, however, depend strongly on the number of populations or the migration rate between them (below $N_em=10$).

These, and other simulations exploring the evolution of drive suppressor genes, suggest that the 'life cycle' of a near-neutral B chromosome would consist of three quite distinct phases: i) invasion of the B, for which strong drive is necessary, so that this phase would be accomplished in a few or few 10s of generations; ii) drive suppression, taking 10s to 100s of generations, and iii) loss of the B by drift, taking 100s to 1000s of generations.

18.8 AN INTRAGENOMIC CONFLICT: HOW COULD THE A GENOME NEUTRALIZE THE B?

Since B chromosomes tend to accumulate in the genome regardless of its effect on the phenotype, they are considered selfish genetic elements (see Shaw and Hewitt, 1990, and Werren and Beukeboom, 1993, for review). The presence of a B favours two kinds of mutations in the A genome. Any allele on the As suppressing B accumulation will be favoured by natural selection if the B has harmful effects on the individuals carrying it, because the B imposes a genetic load on the population. Also, mutations will be favoured which ameliorate the disadvantages incurred by individuals with Bs. On the other hand, variant B chromosomes with increased drive or decreased deleterious phenotypic effects will be favoured.

Evidence of this coevolutionary arms race between the A and B chromosomes is now available. Thus, drive suppressor genes have been shown in at least two species: the grasshopper *Myrmeleotettix maculatus* (Shaw and Hewitt, 1985; Shaw *et al.*, 1985) and the mealy bug *Pseudococcus affinis* (Nur and Brett, 1985; 1987; 1988), and they are postulated to exist in *E.plorans* (Herrera *et al.*, 1996). In addition, some kind of genetic control of B transmission has been shown in maize (Carlson, 1969), rye (Romera *et al.*, 1991; Jiménez *et al.*, 1995) and *Aegilops speltoides* (Cebria *et al.*, 1994). At least in *P.affinis* it has been shown that drive suppressor genes usually have damaging pleiotropic effects, for which reason these genes are not very frequent in natural populations of this species. But in a species where suppressor genes had slight effects on fitness, these genes could be widespread and the B drive could be suppressed completely, as seems to be the case for B_1, B_2 and B_5 in *E.plorans*. No evidence has so far been provided to show changes in the A genome which ameliorate harmful effects of Bs, but this is actually a difficult task because such alleles would tend to fix rapidly. Thus, in addition to the difficulties inherent to all studies to quantify fitness (Endler, 1986), it would be hard to distinguish an altered A genome from a mutated B which was less harmful to its carriers.

Several cases have been reported of B chromosomes spontaneously fused to A chromosomes, all of them in *E.plorans*. The first two cases were reported by Henriques-Gil *et al.* (1983), but at least three other have been found by Cabrero *et al.* (1987), López-León *et al.* (1995a) and P. Arana and E. Rebollo (personal communication). It potentially represents a way for Bs to obviate the conflict with the As. The possible evolutionary pathway would be to become a polymorphic supernumerary segment which eventually might reach fixation. No evidence exists for this hypothesis, and the absence of polymorphisms for fused Bs in natural populations and the orientation and transmission problems shown by the fused-B heteromorphic bivalents during male meiosis (P. Arana, personal communication) indicate that integration of Bs in the A genome through translocation is very unlikely.

The existence of genetic variation in the Bs has been shown in several species, mainly at the cytogenetic level. This is the case, for instance, in the grasshoppers *Chortoicetes terminifera* (Webb and Nehaus, 1979), *Myrmeleotettix maculatus* (Shaw, 1983) and *E.plorans* (López-León *et al.*, 1993b), and the plants *Rumex acetosa* (Parker and Wilby, 1989), rye (see Jones and Puertas, 1993, for review) and *Allium schoenoprasum* (Plowman and Bougourd, 1994). The most extreme variation is shown by the B system in *E.plorans*, where not only about 40 different B variants have been reported but new variants are continuously being generated since López-León *et al.* (1993) demonstrated the *de novo* origin of new B derivatives in controlled crosses where none of the parents carried them. These authors showed that the mutation rate for several B types was actually high (e.g. 0.002 for the iso-B_2), but their transmission ratio was never higher than 0.5 and some of them showed a very high tendency to elimination.

As Shaw (1984) and Shaw and Hewitt (1990) predicted, if genetic variation is present in a B, variants with increased drive or decreased

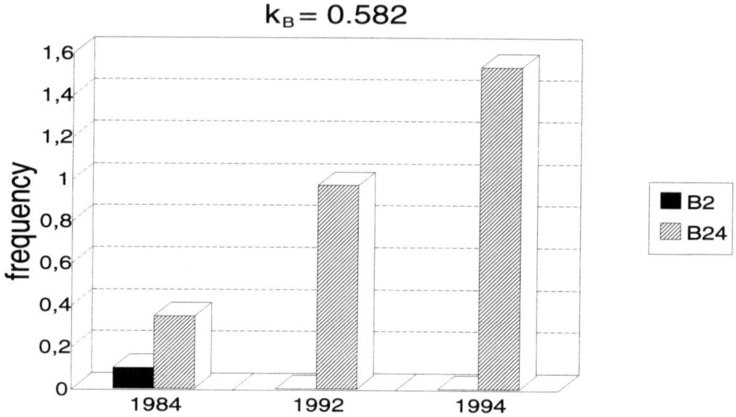

Fig. 18.2 Substitution of B_2 by B_{24} in the Torrox population. k_B= Mean transmission ratio of B_{24} in 22 crosses performed with individuals collected in 1992.

deleterious phenotypic effects will obviously increase. If a new B variant emerge with one of these two properties, it would replace the old version of the B. As postulated by Henriques-Gil *et al.* (1984) and Henriques-Gil and Arana (1990), this could explain the prevalence in *E.plorans* of at least three widespread B variants, including B_2 in Granada and B_5 in Fuengirola (Málaga). The first evidence for the mechanism by which such B substitutions could have taken place was reported by López-León *et al.* (1993b) who observed that a female from Jete carrying $1B_1+1B_2$ accumulated the B_2 but eliminated the B_1. This suggested that the substitution of B_1 for B_2 in the Granada populations involved differences in B transmission rate. A second piece of evidence that B substitutions in this species occur through superior transmission ability for the new B variant is that Henriques-Gil and Arana (1990) reported a B type, B_{24}, that was the most frequent in Torrox (Málaga), where it coexisted with a rare variant, B_2. As B_2 was both common and the main variant in all the surrounding populations, Henriques-Gil and Arana hypothesized that B_{24} was substituting B_2 in this locality. We have recently analysed transmission of B_{24} in samples of adult individuals in 1992 and 1994 (Camacho *et al.*, unpublished). Our results indicate that B_{24} shows a significant tendency to accumulate through females and that its frequency increased by 50% in only two years (Fig. 18.2). These results reinforce the idea that B substitutions in *E.plorans* take place mainly through superior performance in transmission of the new B derivative. In our fieldwork, we have not found the B_2, which suggests that the substitution has nearly been completed. This supports the idea proposed by Shaw and Hewitt (1990) that the replacement of a B by a new variant of higher drive would be fast.

18.9 CONCLUSIONS AND FUTURE PROSPECTS

The B chromosomes in *E.plorans* do not have large effects on fitness, and the three most widespread variants, i.e. B_1, B_2 and B_5, lack drive at present, so that these Bs fit the near-neutral model. The existence of B substitutions indicates that the near-neutral model has to be understood as a dynamic one in which selfish new variants can appear from time to time replacing the former B version because of their superior transmission performance. After a time, the A genome neutralizes this new B variant by evolving new drive suppressor genes and the cycle begins again and again.

Fig. 18.3 shows a scheme of the evolution of a near-neutral B in relation to drive. The birth of a non-beneficial B is only possible when it drives; only that way will it increase in frequency. But the evolution of drive suppressor genes in the A chromosomes neutralizes the B drive and it begins a random walk towards extinction. This may be the death of the B. However, in the course of the random walk, a new selfish B variant may emerge replacing the former B and constituting a true reincarnation of the B. The evolution of new drive suppressor genes neutralizes the new B variant and the cycle may take place over and over again. Thus, although a near-neutral B, as such, is condemned to extinction, its life as a polymorphism may be very long when the successive reincarnations of the B are considered.

It is even possible that the scheme shown in Fig. 18.3 reflects the potential evolutionary pathway of selfish Bs in many species, in most cases still being in the initial parasitic phase. In *E.plorans*, however, B chromosome evolution seems to have gone further reaching the near-neutral phase and several reincarnations and thus providing us with the most complete picture of the coevolution between these selfish genetic elements and the standard genome.

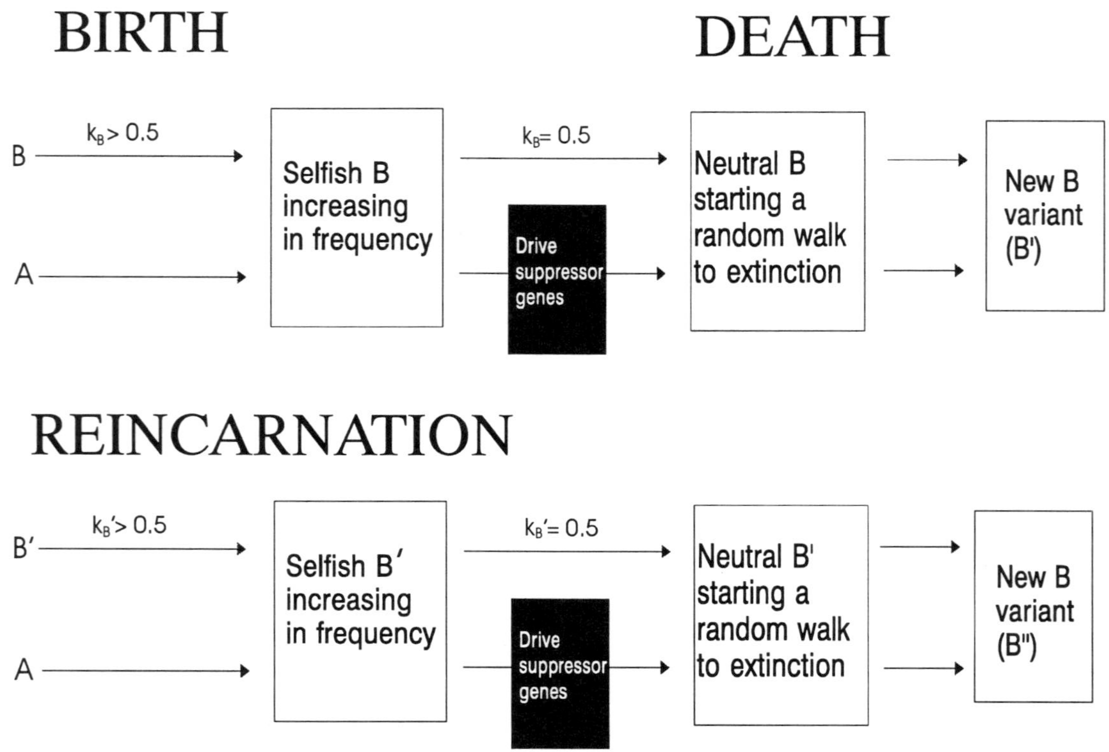

Fig. 18.3 Evolutionary pathway of a near-neutral B chromosome.

The cline for B chromosome frequency observed in the Segura river basin (Cabrero *et al.*, unpublished) might represent a transient stage of B colonization upstream in each of the four rivers analysed, or else a cline produced by environmental conditions. The absence of Bs in the most upstream populations analysed in the four rivers indicates that the B still has not reached these latitudes, presumably due to geographical barriers since the cline border coincides with a narrow pass in all rivers analysed. Anyway, the main support for this hypothesis would be the existence of B drive in the populations close to the border of B distribution since it would imply that these populations lack drive suppressors simply because the B was never there. This is a task awaiting future research. Another test for the hypothesis that the B is absent in some of these populations not because of environmental conditions but simply because the B has never been there is to introduce the B into a non-B population and to analyse the evolution of B frequency in successive generations. We are currently doing this experiment in two populations. If the B has never been in these populations, then it is unlikely that the B will find drive suppressors common, so it will increase in frequency. The results of interpopulation crosses by Herrera *et al.* (1996) suggest that the non-B populations in this zone lack suppressors, and the increase of a 50% in B frequency in Torrox in only two years, mentioned above, indicates that the increase in B frequency in a population showing drive could be very fast. We consider that to ascertain the mechanism by which the B drive is induced by the non-B population males is one of the most interesting and challenging tasks for future work.

To finish, we would like to mention some future prospects on the molecular aspects of B chromosome study. We know at present that B_2 in *E.plorans* is mainly made up of two types of DNA: ribosomal DNA (which is almost always inactive) and a 180 bp repeat (which is most likely functionless). This does not rule out, however, the B harbouring other DNA sequences, some of them even being specific. In the last few years the molecular dissection of B chromosomes has commanded a lot of effort from researchers and has shown that some Bs contain DNA sequences shared with the As, e.g. in *Glossina* spp. (Amos and Dover, 1981), *Crepis capillaris* (Jamilena *et al.*, 1994), *E.plorans* (López-León *et al.*, 1994) and *Drosophila subsilvestris* (Gutknecht *et al.*, 1995). Other Bs, however, have been shown to possess specific DNA sequences, as is the case in rye (Sandery *et al.*, 1990; Blunden *et al.*, 1993) and *Brachycome dichromosomatica* (John *et al.*, 1991), and there is even a B, in *Nasonia vitripennis*, possessing both shared and specific

DNA sequences (Eickbush *et al.*, 1992). In *E.plorans*, several tasks remain at the molecular level: to isolate other DNA sequences which presumably might be contained in the B, to analyse DNA methylation of the B-NOR at the molecular level, and to investigate the enormous tendency of Bs to mutate. Given the enthusiasm of the several research groups presently studying *E.plorans* Bs in Spain, in a few years we could get a very complete picture of this unique B system.

ACKNOWLEDGEMENTS

We are indebted to P. Arana, M.J. Puertas and J.L. Santos for the critical reading of the manuscript. This study was partially supported by grants from the Dirección General de Investigación Científica y Técnica (no. PB93—1108) and the Plan Andaluz de Investigación, Grupo no. 3122 (Spain).

REFERENCES

Amos, A. and Dover, G. (1981) The distribution of repetitive DNAs between regular and supernumerary chromosomes in species of *Glossina* (Tsetse): A two-step process in the origin of supernumeraries. *Chromosoma*, **81**, 673—90.

Beukeboom, L.W. (1994) Bewildering Bs: an impression of the 1st B-Chromosome Conference. *Heredity*, **73**, 328—36.

Blunden, R., Wilkes, T.J., Forster, J.W. *et al.* (1993) Identification of the E3900 family, a second family of rye B chromosome specific repeated sequences. *Genome*, **36**, 706—11.

Bougourd, S.M. and Parker, J.S. (1979) The B chromosome system of *Allium schoenoprasum*. II. Stability, inheritance and phenotypic effects. *Chromosoma*, **75**, 369—83.

Cabrero, J., Alché, J.D. and Camacho, J.P.M. (1987) Effects of B chromosomes of the grasshopper *Eyprepocnemis plorans* on nucleolar organiser regions activity. Activation of a latent NOR on a B chromosome fused to an autosome. *Genome*, **29**, 116—21.

Camacho, J.P.M., Carballo, A.R. and Cabrero, J. (1980) The B chromosome system of the grasshopper *Eyprepocnemis plorans* subsp. *plorans* (Charpentier). *Chromosoma*, **80**, 163—6.

Camacho, J.P.M., Cabrero, J., Viseras, E. *et al.* (1991) G-banding in two species of grasshoppers and its relationship to C, N and fluorescence banding techniques. *Genome*, **34**, 638–43.

Cano, M.I., Jones, G.H. and Santos, J.L. (1987) Sex differences in chiasma frequency and distribution in natural populations of *Eyprepocnemis plorans* containing B-chromosomes. *Heredity*, **59**, 237–43.

Carlson, W. (1969) Factors affecting preferential fertilization in maize. *Genetics*, **62**, 543–54.

Cebria, A., Navarro, M.L. and Puertas, M.J. (1994) Genetic control of B chromosome transmission in *Aegilops speltoides* (Poaceae). *American Journal of Botany*, **81**, 1502–7.

Eickbush, D.G., Eickbush, T.H. and Werren, J.H. (1992) Molecular characterization of repetitive DNA sequences from a B chromosome. *Chromosoma*, **101**, 575–83.

Endler, J.A. (1986) *Natural Selection in the Wild*. Monographs in Population Biology no. 21, Princeton University Press, Princeton, New Jersey, USA.

Gutknecht, J., Sperlich, D. and Bachmann, L. (1995) A species specific satellite DNA family of *Drosophila subsilvestris* appearing predominantly in B chromosomes. *Chromosoma*, **103**, 539–44.

Henriques-Gil, N. (1984) *El sistema de cromosomas accesorios de Eyprepocnemis plorans (Acrididae: Orthoptera)*. PhD thesis, Universidad Complutense de Madrid.

Henriques-Gil, N. and Arana, P. (1990) Origin and substitution of B chromosomes in the grasshopper *Eyprepocnemis plorans*. *Evolution*, **44**, 747–53.

Henriques-Gil, N., Santos, J.L. and Giráldez, R. (1982a) B chromosome polymorphism and interchromosomal chiasma interference in *Eyprepocnemis plorans* (Acrididae; Orthoptera). *Chromosoma*, **85**, 349–59.

Henriques-Gil, N., Santos, J.L. and Giráldez, R. (1982b) Genotype-dependent effect of B-chromosomes on chiasma frequency in *Eyprepocnemis plorans* (Acrididae: Orthoptera). *Genetica*, **59**, 223–7.

Henriques-Gil, N., Arana, P. and Santos, J.L. (1983) Spontaneous translocations between B chromosomes and the normal complement in the grasshopper *Eyprepocnemis plorans*. *Chromosoma*, **88**, 145–8.

Henriques-Gil, N., Santos, J.L. and Arana, P. (1984) Evolution of a complex polymorphism in the grasshopper *Eyprepocnemis plorans*. *Chromosoma*, **89**, 290–3.

Herrera, J.A., López-León M.D., Cabrero, J. *et al.* (1996) Evidence for B chromosome drive suppression in the grasshopper *Eyprepocnemis plorans*. *Heredity* (in press).

Holmes, D.S. and Bougourd, S.M. (1989) B-chromosome selection in *Allium schoenoprasum* I. Natural populations. *Heredity*, **63**, 83–7.

Holmes, D.S. and Bougourd, S.M. (1991) B-chromosome selection in *Allium schoenoprasum* II. Experimental populations. *Heredity*, **67**, 117–22.

Jamilena, M., Ruiz-Rejón, C. and Ruiz-Rejón, M. (1994) A molecular analysis of the origin of the *Crepis capillaris* B chromosome. *Journal of Cell Science*, **107**, 703–8.

Jiménez, M.M., Romera, F., Gallego, A. and Puertas, M.J. (1995) Genetic control of the rate of transmission of rye B chromosomes. II. 0B × 2B crosses. *Heredity*, **74**, 518–23.

John, U.P., Leach, C.R. and Timmis, J.N. (1991) A sequence specific to B chromosomes of *Brachycome dichromosomatica*. *Genome*, **34**, 739–44.

Jones, R.N. (1985) Are B chromosomes selfish?, in *The Evolution of Genome Size* (ed T. Cavalier-Smith), Wiley, London, pp. 397–425.

Jones, R.N. (1991) B-chromosome drive. *American Naturalist*, **137**, 430–42.

Jones, R.N. and Puertas, M.J. (1993) The B-chromosomes of rye (*Secale cereale* L.), in *Frontiers in Plant Science Research* (eds K.K. Dhir and T.S. Sareen), Bhagwati Enterprises, Delhi (India), pp. 81–112.

López-Fernández, C., Mezzanotte, R. and Gosálvez, J. (1992) Autosomal, sex and B chromosomes in *Eyprepocnemis plorans* (Orthoptera) viewed with restriction endonuclease *in situ* digestion. *Heredity*, **68**, 365–72.

López-León, M.D., Cabrero, J. and Camacho, J.P.M. (1991) A nucleolus organizer region in a B chromosome inactivated by DNA methylation. *Chromosoma*, **100**, 134–8.

López-León, M.D., Cabrero, J., Camacho, J.P.M. *et al.* (1992a) A widespread B chromosome polymorphism maintained without apparent drive. *Evolution*, **46**, 529–39.

López-León, M.D., Pardo, M.C., Cabrero, J. and Camacho, J.P.M. (1992b) Random mating and absence of sexual selection for B chromosomes in two natural populations of the grasshopper *Eyprepocnemis plorans*. *Heredity*, **69**, 558–61.

López-León, M.D., Cabrero, J., Pardo, M.C. *et al.* (1993) Generating high variability of B chromosomes in the grasshopper *Eyprepocnemis plorans*. *Heredity*, **71**, 352–62.

López-León, M.D., Neves, N., Schwarzacher, T. *et al.* (1994) Possible origin of a B chromosome deduced from its DNA composition using double FISH technique. *Chromosome Research*, **2**, 87–92.

López-León, M.D., Cabrero, J. and Camacho, J.P.M. (1995a) Changes in DNA methylation during development in the B chromosome NOR of the grasshopper *Eyprepocnemis plorans*. *Heredity*, **74**, 296–302.

López-León, M.D., Vázquez, P., Hewitt, G.M. and Camacho, J.P.M. (1995b) Cloning and sequence analysis of an extremely homogeneous tandemly repeated DNA in the grasshopper *Eyprepocnemis plorans*. Heredity (in press).

López-León, M.D., Cabrero, J. and Camacho, J.P.M. (1995c) Changes in NOR activity pattern in the presence of supernumerary heterochromatin in the grasshopper *Eyprepocnemis plorans*. *Genome*, **38**, 68–74.

Nur, U. and Brett, B.L.H. (1985) Genotypes suppressing meiotic drive of a B chromosome in the mealy bug *Pseudococcus obscurus*. *Genetics*, **110**, 73–92.

Nur, U. and Brett, B.L.H. (1987) Control of meiotic drive of B chromosomes in the mealy bug *Pseudococcus affinis (obscurus)*. *Genetics*, **115**, 499–510.

Nur, U. and Brett, B.L.H. (1988) Genotypes affecting the condensation and transmission of heterochromatic B chromosomes in the mealy bug *Pseudococcus affinis*. *Chromosoma*, **96**, 205–12.

Östergren, G. (1945) Parasitic nature of extra fragment chromosomes. *Botaniska Notiser*, **2**, 157–63.

Parker, J.S. and Wilby, A.S. (1989) Extreme chromosomal heterogeneity in a small-island population of *Rumex acetosa*. *Heredity*, **62**, 133–40.

Plowman, A.B. and Bougourd, S.M. (1994) Selectively advantageous effects of B chromosomes on germination behaviour in *Allium schoenoprasum* L. *Heredity*, **72**, 587–93.

Randolph, L.F. (1928) Types of supernumerary chromosomes in maize. *Anatomical Record*, **41**, 102.

Romera, F., Jiménez, M.M. and Puertas, M.J. (1991) Genetic control of the rate of transmission of rye B chromosomes. I. Effects in 2B × 0B crosses. *Heredity*, **66**, 61–5.

Sandery, M.J., Forster, J.W., Blunden, R. and Jones, R.N. (1990) Identification of a family of repeated sequences on the rye B chromosome. *Genome*, **33**, 908–13.

Shaw, M.W. (1983) Movement of a cline for supernumerary chromosomes, in *Kew Chromosome Conference II* (eds P.E. Brandham and M.D. Bennett), George, Allen and Unwin, London, pp. 217–23.

Shaw, M.W. (1984) The population genetics of the B-chromosome polymorphism of *Myrmeleotettix maculatus* (Orthoptera: Acrididae). *Biol. J. Linn. Soc.*, **23**, 77–100.

Shaw, M.W. and Hewitt, G.M. (1985) The genetic control of meiotic drive acting on the B chromosome of *Myrmeleotettix maculatus* (Orthoptera: Acrididae). *Heredity*, **54**, 259–68.

Shaw, M.W. and Hewitt, G.M. (1990) B chromosomes, selfish DNA and theoretical models: where next?, in *Oxford Surveys in Evolutionary Biology*, Vol. 7D, (eds D. Futuyma and J. Antonovics), pp. 197–223.

Shaw, M.W., Hewitt, G.M. and Anderson, D.A. (1985) Polymorphism in the rates of meiotic drive acting on the chromosome of *Myrmeleotettix maculatus*. *Heredity*, **55**, 61–8.

Webb, G.C. and Nehaus, P. (1979) Chromosome organization in the Australian plague locust *Chortoicetes terminifera*. II. Banding variants of B chromosome. *Chromosoma*, **70**, 205–12.

Werren, J.H. and Beukeboom, L.W. (1993) Population genetics of a parasitic chromosome: theoretical analysis of PSR in subdivided populations. *American Naturalist*, **142**, 224–41.

White, M.J.D. (1973) *Animal Cytology and Evolution*. 3rd ed. Cambridge University Press, London.

19

Chromosome painting in wheat

J.M. Vega, S. Abbo, M. Feldman and A.A. Levy

19.1 INTRODUCTION

Chromosome painting is a technique that allows the specific visualization of an entire chromosome in metaphase spreads and in interphase nuclei by *in situ* hybridization with a mixture of sequences generated from that particular chromosome (Pinkel *et al.*, 1988; Carter, 1994). In the first chromosome painting experiments the source of composite probes were recombinant libraries from specific human chromosomes isolated by flow cytometry (Pinkel *et al.*, 1988; Lichter *et al.*, 1988). Painting probes are now generated more efficiently from microdissected chromosomes by a general PCR amplification of the chromosomal DNA with a degenerated primer (Telenius *et al.*, 1992a, b; Guan *et al.*, 1994).

Chromosome painting has become a powerful tool that significantly extends the limits of conventional analysis in human cytogenetics. The method allows an unequivocal identification of numerical aberrations, chromosomal rearrangements, such as complex translocations, duplications and deletions, and the rapid determination of the composition of newly generated marker chromosomes (Carter, 1994). Painting probes from aberrant chromosomes hybridized onto metaphase spreads from normal subjects reveal not only the chromosomal origin of the aberrant chromosome but also the regions and breakpoints involved (Telenius *et al.*, 1992b; Carter *et al.*, 1992), hence providing a precise diagnosis. Furthermore, paints used for human chromosomes in comparative studies have disclosed high degree of chromosomal homology in primates (Jauch *et al.*, 1992) and the presence of syntenic groups in distantly related mammals

(Scherthan *et al.*, 1994a).

Following the successful application of chromosome painting in human cytogenetics, the production of chromosome paints in other organisms soon followed. Paints have been recently obtained using flow-sorted pig and mouse chromosomes (Langford *et al.*, 1992; Breneman *et al.*, 1993). Chromosome painting has been extended to plants with the development of whole chromosome-specific probes from microdissected chromosomes of common wheat (Vega *et al.*, 1994). The potential of painting particular chromosomes might be specially useful in plant breeding, allowing for a rapid and direct characterization of small alien chromosome segments that are introgressed into the genome of cultivated plants. Painting probes might also be used to estimate the conservation of chromosome segments among distantly related species. Moreover, since these probes detect chromosomal domains within interphase nuclei, they may be applied to the study of chromosome distribution and nuclear organization in different stages of the cell cycle. This analysis is particularly promising for common wheat, which, being a polyploid plant, contains closely related chromosomes that do not pair at meiosis. Monitoring their relative positions during presynaptic stages by specific painting probes may shed light on the mechanism that determine the exclusive bivalent pairing in polyploid wheat.

In this chapter, we summarize recent advances in the isolation of plant chromosomes, describe the development of chromosome painting in wheat and discuss future applications of this technique.

19.2 ISOLATION OF PLANT CHROMOSOMES

The first step in achieving saturation of DNA markers for a whole chromosome is the isolation of that particular chromosome by flow sorting or microdissection.

Chromosomes can be sorted on the basis of their DNA content and base pair ratio from cell cultures synchronized at mitotic metaphase. This method has been successfully used to construct chromosome-specific libraries from each human chromosome (Van-Dilla *et al.*, 1986). However, for most plant species, attempts to sort chromosomes have been hampered by the lack of sufficient size differences, resulting in the presence of different chromosomes at the same peak of a given karyotype. In a library generated from chromosomes sorted from one peak of the wheat flow karyotype, half of the clones mapped to

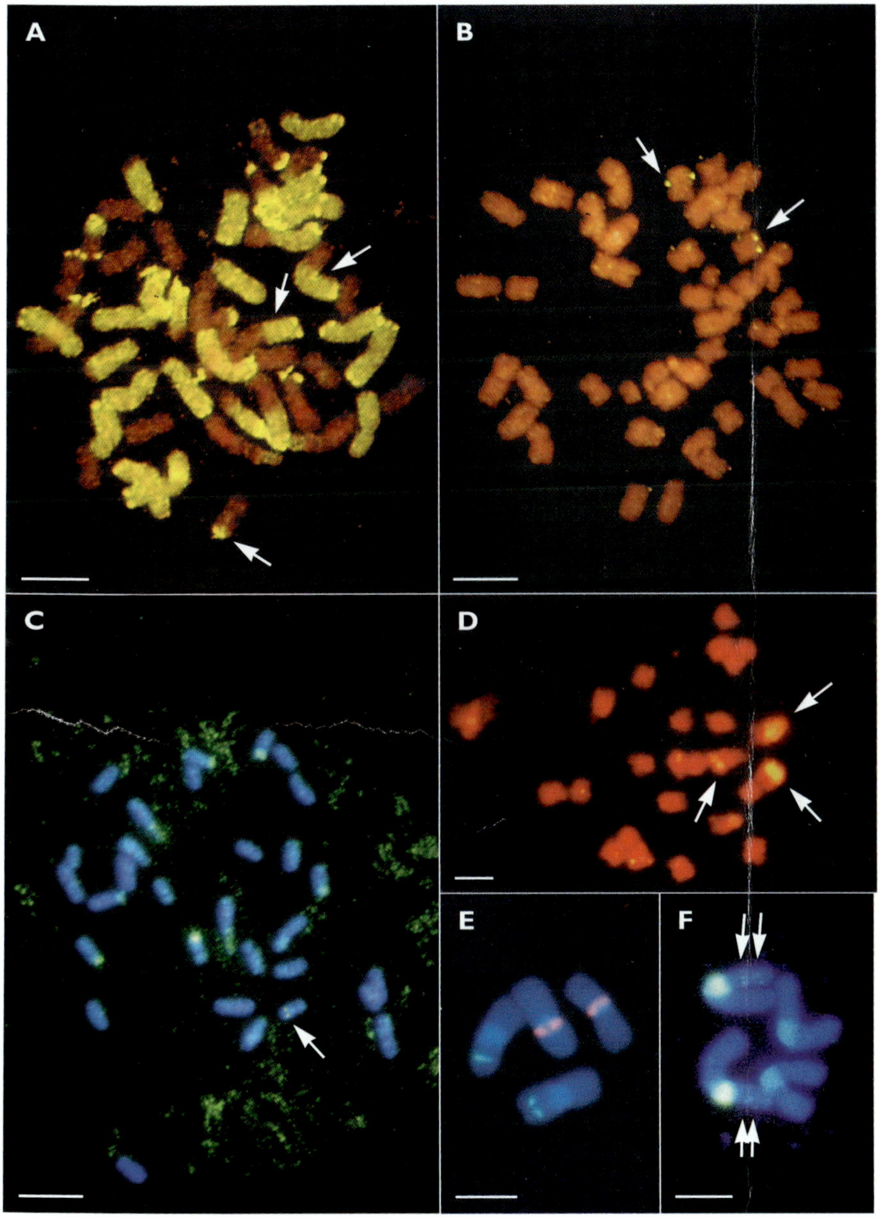

Plate 8 **Fig. 20.1** a–f (see p.340)

chromosome 4A while the rest were dispersed over most of the chromosome complement (Wang *et al.*, 1992). Up to now, high purity fractions have been obtained for chromosomes of field bean (Lucretti *et al.*, 1993; Macas *et al.*, 1993) and for tomato chromosome 2 (Arumuganathan *et al.*, 1994); the former have been used for the physical localization of genes and the latter for the construction of a chromosome-enriched library.

Microdissection allows the isolation of chromosomes that are distinguishable by their length or morphology on metaphase spreads, using glass needles or laser beams. This method was first applied to plant chromosomes for the dissection of supernumerary B-chromosomes from rye meiocytes where they are easily identified by their size and behaviour (Sandery *et al.*, 1991). As in the case of flow sorting, the reconstructed karyotypes of field bean allow single chromosomes to be distinguished and therefore offer excellent material for micromanipulation (Schubert *et al.*, 1993). In humans, microdissection has become especially powerful for the isolation of particular chromosomal regions that are readily identifiable by their G-banding pattern (Senger *et al.*, 1990; Meltzer *et al.*, 1992). At present, the identification of individual plant chromosomes or chromosomal regions by banding techniques can not be useful in microdissection because current banding techniques in plants require acid treatments that damage the DNA. This drawback has been partially overcome through the use of aneuploid lines which carry chromosomes that are distinguishable either by their morphology at mitotic metaphase, e.g. barley and wheat telosomes (Schondelmaier *et al.*, 1993; Albani *et al.*, 1993) or by their behavior at meiotic metaphase, e.g. wild-beet univalents (Jung *et al.*, 1992) and wheat isochromosomes (Vega *et al.*, 1994). In cases where libraries have been prepared from dissected chromosomes (Jung *et al.*, 1992; Schondelmaier *et al.*, 1993), all single- or low-copy clones tested originated from the target chromosome. This clearly indicates that microdissection achieves a higher degree of purity than flow-sorting in the isolation of specific chromosomes.

19.2.1 Microdissection of wheat 5BL chromosome arm

Common wheat, *Triticum aestivum* L., is an allohexaploid ($2n=6x=42$) carrying three closely related genomes (A, B and D). Based on studies of dosage compensation among the chromosomes of the three genomes, the 42 chromosomes have been placed into seven groups

(Sears, 1952). Each group is composed of three pairs of homoeologues, i.e. partially homologous chromosomes. The compensation among homoeologues has facilitated the development of a wide array of viable aneuploid stocks (Sears, 1954) which makes common wheat an easy target for the isolation of particular chromosomes by microdissection. Monosomic lines may be used for the isolation of whole chromosomes since the unpaired chromosome can be identified as a univalent at first meiotic metaphase. Moreover, specific chromosome arms can be isolated from both mitotic and meiotic spreads using telocentric lines, or from meiotic preparations of lines carrying isochromosomes. Cytological preparations of meiocytes offer two main advantages over somatic spreads for the isolation of chromosomes. First, meiocytes are highly synchronized at first meiotic metaphase. Second, the target chromosome, when present in a single dose, behaves as an univalent that is both recognizable and accessible in the majority of the meiocytes, lying at the periphery of the equatorial plate. Moreover, since the univalent is spatially well separated from the bivalents there is a minimal risk for DNA contamination from non-target chromosomes during micromanipulation.

We have recently microdissected isochromosomes for the long arm of chromosome 5 of the wheat B genome (5BL) using a monoisosomic 5BL line of the common wheat cultivar Chinese Spring (Vega *et al.*, 1994). According to our procedure, meiotic spreads are prepared from fresh anthers fixed for a short time to avoid depurination of the DNA. The two homologous arms of the isochromosome pair at first meiotic metaphase to form a ring univalent that is well separated from the bivalents and therefore, accessible for dissection. Microdissections are performed on an inverted microscope with glass microneedles controlled by a micromanipulator. The tip of the needle (~1.0 μm diameter) is used to remove the chromosome (~5.0 μm length) from the surrounding nucleoplasm. Thereafter, the tip carrying the dissected chromosome is broken off into a microfuge tube where the PCR is carried out. To obtain detectable levels of amplification five to ten chromosomes are collected into a single tube.

19.3 MANIPULATION OF DNA FROM ISOLATED CHROMOSOMES

The first attempts to characterize isolated chromosomes involved digestion of the chromosomal DNA with restriction enzymes and subsequent cloning. By this method, however, a relatively low cloning efficiency was obtained (Röhme *et al.*, 1984; Sandery *et al.*, 1991). Two main approaches were developed to amplify by PCR the limited amounts of isolated DNA available, so that later manipulations (e.g., cloning, labelling) can be performed more efficiently. In the first approach, isolated DNA is digested with restriction enzymes and ligated to a vector or to specific adaptors which provide the primer binding sites for PCR amplification. Vector-mediated PCR has been used to generate chromosome-specific libraries in wild-beet and barley (Jung *et al.*, 1992; Schondelmaier *et al.*, 1993), and adaptor-mediated PCR has been used to amplify microdissected chromosome arms of wheat (Albani *et al.*, 1993). A major drawback of these techniques is the multiplicity of handling steps that increase the likelihood of contamination from exogenous DNA. Moreover, since only some of the products of the restriction endonuclease digestion are small enough to be successfully amplified by PCR, certain regions of the microdissected DNA are likely to be absent after ligation-mediated PCR (Albani *et al.*, 1993).

The second approach is based on direct priming along the isolated chromosomes. General amplification of human DNA was achieved using primers homologous to interspersed repetitive *Alu* sequences (Nelson *et al.*, 1989). However, *in situ* hybridization onto metaphase spreads with *Alu*-PCR products revealed that these repeats were not uniformly distributed within the human chromosomes (Baldini and Ward, 1991; Suijkerbuijk *et al.*, 1992). A more general amplification scheme has been developed by a procedure termed degenerate oligonucleotide-primed PCR (DOP-PCR) using a primer based on a partially degenerate sequence (Telenius *et al.*, 1992a). DOP-PCR is becoming the method of choice for the generation of whole chromosome painting probes from isolated human chromosomes since its products provide an even and uniform signal intensity along the target chromosomes (Telenius *et al.*, 1992b; Guan *et al.*, 1994). Moreover, protocols based on DOP-PCR have been developed to produce band specific painting probes from a single microdissected human chromosome region (Guan *et al.*, 1993). Recently, DOP-PCR has been applied with minor modifications to isolated chromosomes from several plant species (Arumuganathan *et al.*, 1994; Pich *et al.*, 1994; Vega *et al.*, 1994). Since DOP-PCR is apparently not dependent on the presence of certain repetitive sequences, it represents a species-independent efficient technique for uniform random amplification of DNA.

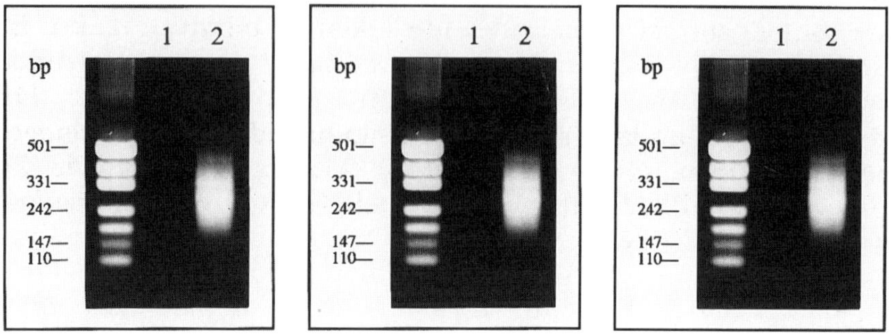

Fig. 19.1 DOP-PCR products from microdissected chromosomes. PCR products were size fractionated on a 1.8 % agarose gel and stained with ethidium bromide. Lane 1, PCR reaction with no added DNA; lane 2, PCR products from 6 microdissected 5BL isochromosomes.

19.3.1 Amplification of the wheat 5BL chromosomal DNA

In our studies (Vega *et al.*, 1994), we have amplified the dissected 5BL isochromosomes of wheat by DOP-PCR using the universal primer 5'-CCGACTCGAGNNNNNNATGTGG-3' in a two-step PCR as described by Telenius *et al.* (1992a). In the first step, eight PCR cycles are performed at low annealing temperature. This allows frequent priming along the template DNA. During the second step, 25 amplification cycles are performed at a higher annealing temperature. Under these conditions, priming is more stringent and the sequences generated in the first step are preferentially amplified.

Ideally, DOP-PCR results in a smear of DNA fragments that are visible by agarose gel electrophoresis. In our experimental conditions, we do not detect amplification of the dissected chromosomes in a first DOP-PCR reaction. Hence, to increase the amount of product, an aliquot from the first reaction mixture is subjected to a 30-cycle second-round PCR under the same conditions of high-stringency cycles as in the first round. The gel in Fig. 19.1 (lane 2) shows that the second round of amplification of the dissected isochromosomes results in a continuous smear of fragments ranging from 150 to 500 bp.

Similar complexity was found in DOP-PCR amplifications of microdissected human chromosomes (Guan *et al.*, 1994) and in flow-sorted plant chromosomes (Arumuganathan *et al.*, 1994; Pich *et al.*, 1994). Because the DOP-PCR primer can amplify DNA virtually from any source, special precautions are taken to eliminate DNA from exogenous sources during the preparation of the meiotic spreads, microdissection and PCR assays. To test the level of DNA contamination, a blank reaction with no added DNA is included in each experiment (Fig. 19.1, lane 1). Southern blots of the PCR reactions are probed with wheat genomic DNA to confirm the origin of the amplified DNA.

19.4 CHROMOSOME PAINTING IN WHEAT

Products of the second-round PCR were directly labelled with fluorochromes and used as a complex probe mixture for fluorescence *in situ* hybridization (FISH) onto meiotic chromosome spreads of wheat (Vega *et al.*, 1994). Since these spreads were prepared from lines carrying 5BL as a distinctive marker (5BL iso- and telo-chromosomes), the specificity of the amplified DNA could be directly verified. Figure 19.2 shows a FISH profile where the 5BL probe was labelled with rhodamine and the sites of hybridization are therefore visualized as red signals. Chromosomes and nuclei were counterstained with the DNA-specific dye DAPI, and appear blue. When the hybridization was visualized using a filter that transmits only rhodamine fluorescence, all chromosomes were stained, with the 5BL marker chromosome being stained most heavily (Fig. 19.2a). This indicates that some of the probe's sequences are shared by all the chromosome complement, but most sequences are 5BL-specific. In addition to the 5BL marker chromosome, two bivalents were intensely labelled in some of the cells (Fig. 19.2a). When the meiocytes were visualized in double-exposure photographs for both rhodamine and DAPI fluorescences, a bright red signal was specifically observed along the entire length of the 5BL telo- and iso-chromosomes (Fig. 19.2b and c). The faint red signals that appeared on the rest of the chromosomes when visualized only with a rhodamine filter (Fig. 19.2a) were masked by the blue DAPI fluorescence, except for two bivalents that were detected in some of the cells (Fig. 19.2b). These are the very same bivalents that were intensely stained with the rhodamine fluorescence alone. It is most likely that these bivalents are 5A and 5D, the 5B

Fig. 19.2 ***Colour Plate 5*** Hybridization pattern of a painting probe for chromosome arm 5BL on first meiotic metaphase and interphase cells of aneuploid wheat lines for 5BL. The 5BL probe was labelled with rhodamine and sites of probe hybridization are visualized as red signals. Chromosomes and nuclei were counterstained with DAPI and appear blue. **a.** Single exposure using a filter that transmits only rhodamine fluorescence. **b** to **d.** Double exposure visualizing rhodamine and DAPI fluorescences. **a** and **b.** Monotelosomic 5BL showing the telochromosome separated from the metaphase plate. The two painted bivalents are assumed to be the homoeologues 5A and 5D. **c.** Monoisosomic 5BL with the isochromosome as a ring univalent at the upper left corner. The two univalents on the right side resulted from asynapsis. **d.** Interphase of a binucleated tapetal cell of monoisosomic 5BL line. (Figure 18.2b reproduced from Vega *et al.* 1994, with permission from the National Academy of Sciences USA).

homoeologues, which presumably share extensive sequence homology with 5BL (Gale *et al.*, 1995). The differential labelling of the 5BL chromosomes was not caused by their separation from the equatorial plate and, hence, a greater accessibility to the probe. No such labelling was found in non-related chromosomes that were apart from the bulk of the chromosome complement as result of asynapsis (Fig. 19.2c). Additional evidence for the probe's specificity was the presence of discrete domains of hybridization in interphase nuclei of the tapetum (Fig. 19.2d).

The observed FISH pattern indicates that PCR products are enriched in sequences specific to 5BL and its homoeologues. However, since labelling of the homoeologous chromosomes appeared somewhat weaker than that of 5BL and was not manifested in all the cells, this suggests that both chromosome-specific (5B) and group-specific (5A, 5B, 5D) sequences are present in the probe. Whereas group-specific sequences are most probably remnants from a common ancestral chromosome, chromosome-specific sequences have evolved during the speciation events that gave rise to the A, B and D genomes, and, presumably, have continued to change at the polyploid level. We hypothesize that chromosome-specific sequences belong to the families of repeat sequences that turned over at a relatively high rate during evolution to create major structural differences between homoeologous chromosomes in related cereal species (Flavell 1982).

19.5 CONCLUSIONS AND FUTURE PROSPECTS

The general amplification of DNA from microdissected chromosomes by DOP-PCR allows the generation of a mixture of sequences that can be used in FISH for the production of chromosome paints in common wheat. Painting was obtained not only for the target chromosome but most likely for its homoeologues too. This needs to be further confirmed in genetic stocks where the homoeologues can be easily identified. It might be possible to enhance the signal ratio between the target chromosome and that of the homoeologues, or the rest of the chromosome complement, by preannealing unlabelled wheat genomic DNA to the probe prior to adding it to the slides. This method, known as chromosome *in situ* suppression (CISS), preferentially depletes the high-copy repeated sequences in the probe and thereby inhibits the hybridization of sequences that bind to multiple chromosomes (Landegent *et al.*, 1987; Lichter *et al.*, 1988).

The DOP-PCR products are currently being cloned as chromosome-arm enriched libraries for use as sources of new markers. Although we have no direct evidence concerning the types of sequences responsible for the 5BL painting we have not observed distinct bands in the smears of the PCR products, which could have suggested preferential amplification of a subset of DNA sequences from the microdissected chromosomal DNA. The fact that intense and specific fluorescent staining of a particular chromosome can be achieved is by itself an indication for the presence of chromosome-specific sequences. In the case of wheat, this observation is meaningful since these sequences have evolved during the divergence of homoeologous chromosomes. Some of these sequences may be recognized by the genetic mechanism that suppresses pairing of homoeologues at meiosis. Therefore, the characterization of chromosome-specific sequences and the determination of their chromosomal location may give relevant information about the type of sequences that play a key role in speciation.

Chromosome paints should be useful for karyotypic studies in common wheat to identify chromosome rearrangements that are unidentifiable by standard banding analysis. Furthermore, paints from chromosomes of related species may facilitate the identification of small chromosomal segments from those species that were introgressed into the genome of cultivated wheat either by spontaneous natural hybridization or in breeding programs. The use of chromosome painting in comparative studies should also be a valuable tool to elucidate karyotype rearrangements during the evolution of the Gramineae.

Since the painting probe revealed a territorial organization of the target chromosome at interphase, this procedure should facilitate studies on the arrangement of particular chromosomes during the cell cycle. In yeast, alignment of homologous chromosomes at premeiotic interphase has been demonstrated by the use of specific probes for whole chromosomes and defined chromosomal regions (Weiner and Kleckner, 1994; Scherthan *et al.*, 1994b). To date, the genomic *in situ* hybridization method has allowed the identification of alien chromosomes in wheat by using total genomic DNA from the alien species as a probe (Schwarzacher *et al.*, 1992). However, judging from the compartmentalization of two distinct genomes in wide hybrids (Leitch *et al.*, 1991) the behaviour of alien chromosomes may well differ from that of the native chromosomes. Further development of probes that paint homologous, homoeologous and non-related wheat chromosomes, within their own genome, should allow the simultaneous analysis of their arrangement at interphase. Such probes will enhance the prospect of studying interphase nuclear architecture, and of understanding certain aspects of chromosome behaviour such as recognition and search for homology prior to meiotic pairing.

ACKNOWLEDGEMENTS

We thank Y. Avivi for editing the manuscript. Part of this work was supported by the Leo and Julia Forchheimer Foundation.

REFERENCES

Albani, D., Côté, M.J., Armstrong, K.C., *et al.* (1993) PCR amplification of microdissected wheat chromosome arms in a simple 'single tube' reaction. *The Plant Journal*, **4**, 899—903.

Arumuganathan, K., Martin, G.B., Telenius, H., *et al.* (1994) Chromosome 2-specific DNA clones from flow-sorted chromosomes of tomato. *Molecular and General Genetics*, **242**, 551—8.

Baldini, A. and Ward, D.C. (1991) *In situ* hybridisation of human chromosomes with Alu-PCR products: A simultaneous karyotype for gene mapping studies. *Genomics*, **9**, 770—4.

Breneman, J.W., Ramsey, M.J., Lee, D.A., *et al.* (1993) The development of chromosome-specific composite DNA probes for the mouse and their application to chromosome painting. *Chromosoma*, **102**, 591—8.

Carter, N.P. (1994) Cytogenetic analysis by chromosome painting. *Cytometry*, **18**, 2—10.

Carter, N.P., Ferguson-Smith, M.A., Perryman, M.T., *et al.* (1992) Reverse chromosome painting: A method for the rapid analysis of aberrant chromosomes in clinical cytogenetics. *Journal of Medical Genetics*, **29**, 299—307.

Flavell, R.B. (1982) Sequence amplification, deletion and rearrangement: major sources of variation during species divergence, in *Genome evolution*, eds G.A. Dover and R.B. Flavell, Academic Press, London, pp. 301—23.

Gale, M.D., Atkinson, M.D., Chinoy, C.N., *et al.* (1995) Genetic maps of hexaploid wheat, in *Proceedings of the 8th International Wheat Genetics Symposium*, eds Z.S. Li and Z.Y. Xin, Scientech Press, Beijing, pp. 29—40.

Guan, X.-Y., Meltzer, P.S., and Trent, J.M. (1994) Rapid generation of whole chromosome painting probes (WCPs) by chromosome microdissection. *Genomics*, **22**, 101—7.

Guan, X.-Y., Trent, J.M., and Meltzer, P.S. (1993) Generation of band-specific painting probes from a single microdissected chromosome. *Human Molecular Genetics*, **2**, 1117—21.

Jauch, A., Wienberg, J., Stanyon, R., *et al.* (1992) Reconstruction of genomic rearrangements in great apes and gibbons by chromosome painting. *Proceedings of the National Academy of Science, USA*, **89**, 8611—5.

Jung, C., Claussen, U., Horsthemke, B., *et al.* (1992) A DNA library from an individual *Beta patellaris* chromosome conferring nematode resistance obtained by microdissection of meiotic metaphase chromosomes. *Plant Molecular Biology*, **20**, 503—11.

Landegent, J.E., Jansen-in-de-Wal, N., Dirks, R.W., *et al.* (1987) Use of whole cosmid cloned genomic sequences for chromosomal localization by non-radioactive *in situ* hybridization. *Human Genetics*, **77**, 366—70.

Langford, C.F., Telenius, H., Carter, N.P., *et al.* (1992) Chromosome painting using chromosome-specific probes from flow-sorted pig chromosomes. *Cytogenetics and Cell Genetics*, **61**, 221—3.

Leitch, A.R., Schwarzacher, T., Mosgöller, W., *et al.* (1991) Parental genomes are separated throughout the cell cycle in a plant hybrid. *Chromosoma*, **101**, 206—13.

Lichter, P., Cremer, T., Borden, J., *et al.* (1988) Delineation of individual human chromosomes in metaphase and interphase cells by in situ suppression hybridization using recombinant DNA libraries. *Human Genetics*, **80**, 224—34.

Lucretti, S., Dolezel, J., Schubert, I., *et al.* (1993) Flow karyotyping and sorting of *Vicia faba* chromosomes. *Theoretical and Applied Genetics*, **85**, 665—72.

Macas, J., Dolezel, J., Lucretti, S., *et al.* (1993) Localisation of seed protein genes on flow-sorted field bean chromosomes. *Chromosome Research*, **1**, 107—15.

Meltzer, P.S., Guan, X.Y., Burgess, A., *et al.* (1992) Rapid generation of region specific probes by chromosome microdissection and their application. *Nature Genetics*, **1**, 24—8.

Nelson, D.L., Ledbetter, S.A., Corbo, L., *et al.* (1989) Alu polymerase chain reaction: A method for rapid isolation of human-specific sequences from complex DNA sources. *Proceedings of the National Academy of Science USA*, **86**, 6686—90.

Pich, U., Houben, A., Fuchs, J., *et al.* (1994) Utility of DNA anplified by degenerate oligonucleotide-primed PCR (DOP-PCR) from the total genome and defined chromosomal regions of field bean. *Molecular and General Genetics*, **243**, 173—7.

Pinkel, D., Landegent, J., Collins, C., *et al.* (1988) Fluorescence *in situ* hybridization with human chromosome-specific libraries: Detection of trisomy 21 and translocations of chromosome 4. *Proceedings of the National Academy of Science USA*, **85**, 9138—42.

Röhme, D., Fox, H., Herrmann, B., *et al.* (1984) Molecular clones of the mouse *t* complex derived from microdissected metaphase chromosomes. *Cell*, **36**, 783—8.

Sandery, M.J., Forster, J.W., Macadam, S.R., *et al.* (1991) Isolation of a sequence common to A- and B-chromosomes of rye (*Secale cereale*) by microcloning. *Plant Molecular Biology Reporter*, **9**, 21—30.

Scherthan, H., Gremer, T., Arnason, U., *et al.* (1994a) Comparative chromosome painting discloses homologous segments in distantly related mammals. *Nature Genetics*, **6**, 342—7.

Scherthan, H., Bahler, J., and Kohli, J. (1994b) Dynamics of chromosome organization and pairing during meiotic prophase in fission yeast. *Journal of Cell Biology*, **127**, 273—85.

Schondelmaier, J., Martin, R., Jahoor, A., *et al.* (1993) Microdissection and microcloning of the barley (*Hordeum vulgare* L.) chromosome 1HS. *Theoretical and Applied Genetics*, **86**, 629—36.

Schubert, I., Dolezel, J., Houben, A., *et al.* (1993) Refined examination of plant metaphase chromosome structure at different levels made feasible by new isolation methods. *Chromosoma*, **102**, 96—101.

Schwarzacher, T., Anamthawat-Jónsson, K., Harrison, G.E., *et al.* (1992) Genomic *in situ* hybridization to identify alien chromosomes and chromosome segment in wheat. *Theoretical and Applied Genetics*, **84**, 778—86.

Sears, E.R. (1952) Homoeologous chromosomes in *Triticum aestivum*. *Genetics*, **37**, 624.

Sears, E.R. (1954) The aneuploids of common wheat. *Missouri Agricultural Experimental Station Research Bulletin*, **572**, 1—58.

Senger, G., Lüdecke, H.-J., Horsthemke, B., *et al.* (1990) Microdissection of banded human chromosomes. *Human Genetics*, **84**, 507—11.

Suijkerbuijk, R.F., Matthopoulos, D., Kearney, L., *et al.* (1992) Fluorescent *in situ* identification of human marker chromosomes using flow sorting and *Alu* element-mediated PCR. *Genomics*, **13**, 355—62.

Telenius, H., Carter, N.P., Bebb, C.E., *et al.* (1992a) Degenerate oligonucleotide-primed PCR: general amplification of target DNA by a single degenerated primer. *Genomics*, **13**, 718—25.

Telenius, H., Pelmear, A.H., Tunnacliffe, A., *et al.* (1992b) Cytogenetic analysis by chromosome painting using DOP-PCR amplified flow-sorted chromosomes. *Genes, Chromosomes & Cancer*, **4**, 257—63.

Van-Dilla, M.A., Deaven, L.L., Albright, K.L., *et al.* (1986) Human chromosome-specific DNA libraries: Construction and availability. *Biotechnology*, **4**, 537—52.

Vega, J.M., Abbo, S., Feldman, M., *et al.* (1994) Chromosome painting in plants: *in situ* hybridization with a DNA probe from a specific microdissected chromosome arm of common wheat. *Proceedings of the National Academy of Science USA*, **91**, 12041—5.

Wang, M.L., Leitch, A.R., Schwarzacher, T., *et al.* (1992) Construction of a chromosome-enriched *Hpa*II library from flow-sorted wheat chromosomes. *Nucleic Acids Research*, **20**, 1897—901.

Weiner, B.M. and Kleckner, N. (1994) Chromosome pairing via multiple interstitial interactions before and during meiosis in yeast. *Cell*, **77**, 977—91.

20

New insights into chromosome evolution in plants from molecular cytogenetics

I.J. Leitch, A.S. Parokonny and M.D. Bennett

20.1 INTRODUCTION

To understand the mechanisms involved in chromosome evolution, the genomes of many different organisms have been analysed. These analyses are regarded as fundamental to evolutionary problems because they detect changes in the material directly concerned with heredity. However, the evolutionary and biological significance of how DNA is organised into chromosomes remains, in many cases, an enigma. This is largely due to the huge variation observed in chromosome number, shape and size among, and often within, plant taxa. In angiosperms, for example, the amount of DNA in a 4C nucleus ranges from less than 1 pg in *Arabidopsis thaliana* to over 500 pg in *Fritillaria assyriaca* (Bennett and Smith, 1976; Bennett *et al.*, 1982). There is also a huge range in chromosome number from 2n=4 such as in *Zingeria biebersteiniana* (Tsvelev and Zhukova, 1974) to over 600 in *Sedum suaveolens* (Uhl, 1978) and around 600 in the palm *Voanioala gerardii* (Johnson *et al.*, 1989). Yet, in neither DNA amount nor chromosome number is there any discernible relationship with evolutionary complexity. Numerous evolutionary mechanisms have been described which can account for some of the karyotypic diversity observed. These include chromosome fission and fusion events, chromosome translocations, substitutions, inversions, aneuploidy and polyploidy.

While classical cytogenetic approaches have provided much information on plant chromosome evolution, the more recent development of molecular cytogenetics has opened up new avenues of investigation. In particular, the technique of fluorescent *in situ* hybridization (FISH), which enables the organization of DNA sequences, chromosomes and whole genomes to be visualised, has generated significant new information on the origin, organization and evolution of chromosomes (Heslop-Harrison, 1991, Kenton *et al.*, 1993a, Leitch *et al.*, 1994) and has shed new light on plant biosystematic studies (Bennett, 1996a).

This paper reviews four areas where molecular cytogenetics has contributed significant new information on chromosome evolution in plants. These are polyploidy, insertion of viral DNA sequences into plant genomes, chromosome changes induced by man and finally possible constraints on chromosome evolution and organization.

20.2 POLYPLOIDY

Polyploidy is considered to be a significant force in plant evolution (Soltis *et al.*, 1992). Unlike the situation in animals where polyploidy is rare, occurring in only a few unrelated families, (e.g. insects, fishes, amphibians and reptiles) it has been estimated that maybe as many as 70% of the angiosperms (Stace, 1993) and 95 % of pteridophytes (Grant, 1981) are polyploids. It perhaps assumes even greater significance to man when one considers that many important crops are or suspected to be polyploid, (e.g. bread wheat, maize, potatoes, sunflower, cotton and oil seed rape).

The development of genomic *in situ* hybridization (GISH) has been particularly useful for the analysis of polyploids as it enables the identity and organization of the different parental genomes to be visualised. This field has been moving very rapidly. Probably, the first use of GISH to determine genome relationships in a wild plant species was undertaken at Kew by Bennett *et al.* (1992). They used GISH to identify one of the genome donors and confirm the allopolyploid origin of the bimodal karyotype in the grass *Milium montianum*. Since then there have been numerous reports using GISH to identify the parental genome donors in polyploids or suspected polyploids (e.g. Ørgaard *et al.,* 1996, Kenton *et al.*, 1993b, Leggett and Markhand, 1996, Humphreys *et al.*, 1995, Chen and Armstrong, 1994).

Not only can GISH identify parental genome donors but it can also examine the organization and evolution of the component genomes in polyploids. It is now apparent from numerous studies that considerable structural rearrangements may take place following polyploid formation. For example, GISH identified nine intergenomic translocations in tobacco (*Nicotiana tabacum*; Kenton *et al.*, 1993b, see below), and, in oats, at least five intergenomic translocations in the allotetraploid *Avena maroccana* (Jellen *et al.*, 1994) and as many as 18 in the allohexaploid *A. sativa* (Chen and Armstrong, 1994).

Using a combination of N-banding and GISH to analyse tetraploid wheats, Jiang and Gill (1994) distinguished two different types of intergenomic chromosome translocations (i) 'random translocations' that occurred in different chromosomes in different populations of the same polyploid species and (ii) 'species-specific' translocations involving specific chromosomes and found in every population. To explain the presence of species-specific chromosome translocations Gill (1991) proposed a 'nucleocytoplasmic interaction (NCI) hypothesis' of genome evolution and speciation in polyploid plants. According to this hypothesis a newly formed amphiploid must pass through a bottleneck of sterility resulting from the adverse interaction between the male nuclear genome and the nuclear and cytoplasmic genomes of the female parent. Certain bottleneck (species-specific) chromosomal changes must occur in the nuclear genome to restore fertility and nucleocytoplasmic compatibility. Jiang and Gill (1994) suggested that the intergenomic translocations they observed in the tetraploid wheats *Triticum timopheevii* and *T. turgidum* belonged to the second type because they were species-specific and involved intergenomic chromosomes so could not pre-exist in the progenitor species.

Species-specific intergenomic translocations have also been observed in tobacco (*Nicotiana tabacum*, 2n=4x=48; Parokonny and Kenton, 1996). Tobacco has long been suspected of being an allotetraploid formed by combining two ancestral genomes known as the S and the T genomes. The S genome donor is widely accepted to be *N. sylvestris* or some close relative. The T genome donor is more controversial, although it is thought to come from the section Tomentosae, it may be an introgressive hybrid between *N. tomentosiformis* and *N. otophora* (Kenton *et al.*, 1993b).

The presence of intergenomic translocations in tobacco was first visually demonstrated by Kenton *et al.* (1993b) using GISH. They analysed three different *N. tabacum* genotypes and found that each line had up to nine different homozygous translocations between the S and T genomes (Plate 1a). By physically mapping two highly repetitive DNA sequences, HRS60.1 (Kenton *et al.*, 1993b) and GRS (Gazdovà *et al.*, 1995), the identity of the chromosomes involved in three of the recombinants were determined (Parokonny and Kenton, 1996). This analysis identified at least one recombinant common to all genotypes examined and thus probably represents a species-specific chromosome translocation that occurred following polyploid formation to restore nucleo-cytoplasmic compatibility (cf. Gill, 1991). While it is likely that some of the other recombinants also represent species-specific translocations, insufficient physical markers are currently available to be certain that these recombinants have involved the same chromosomes in every cultivar tested. Further use of sequential C-banding and ISH/GISH analysis may provide the physical markers that are essential for the precise identification of chromosomes involved in these translocations.

These molecular cytogenetic studies show that considerable reorganization of genomes may occur following polyploid formation. The continued physical mapping of DNA sequences and whole genomes by FISH will likely further uncover other sequence and chromosome reorganizational events that take place following polyploid formation.

20.3 INSERTION OF VIRAL SEQUENCES INTO PLANT GENOMES

Viral DNA is known to integrate into the host nuclear genome in animal and bacterial systems (e.g. Berg and Howe, 1989). However, although DNA viruses are known to replicate in the plant nucleus there had been no direct evidence that they were capable of integrating into the nuclear genome. This situation was recently changed by the discovery that the nuclear genome of tobacco, *Nicotiana tabacum,* carries about 25 direct repeats of a geminivirus-related DNA (GRD) sequence in a unique locus (Bejarano *et al.*, 1996). DNA sequence analysis revealed that each repeat had sequences most closely resembling those for the DNA replication origin plus the adjacent *AL1* gene (encoding the viral replication protein) of the New World geminiviruses.

Using GRD as a probe for FISH followed sequentially by total genomic DNA from *N. sylvestris* mapped the GRD sequence to a single locus on the long arm of a small pair of T genome chromosomes which did not contain any translocated segments or ribosomal DNA sequences (Kenton *et al.,* 1995 and Plate 1b). As yet insufficient physical markers are available to determine the precise chromosome containing GRD in tobacco. However, the visualisation of GRD on a tobacco chromosome confirms that GRD is endogenously integrated into *Nicotiana* DNA and is thus the first report visualising a viral DNA sequence integrated into a plant genome. The presence of the sequence exclusively in the T genome is consistent with the discovery of GRD in *N. tomentosiformis* and not *N. sylvestris* (Warry and Lichtenstein pers. comm. 1995). Thus, GRD was probably present in the T genome donor of *N. tabacum* prior to the formation of the amphidiploid.

As geminiviruses are not seed transmitted the most likely mechanism for entry into the germline was by integration (by illegitimate recombination) into vegetative tissue which subsequently differentiated into floral meristem. The subsequent maintenance of GRD within *Nicotiana* may have been by neutral genetic drift in the small population from which tobacco evolved or it may have been actively selected (Bejarano *et al.,* 1996).

The frequency of viral integration into plant genomes is unknown but these new results suggests that the physical mapping of viral DNA inserts by ISH will be of particular use in phylogenetic and evolutionary studies. The physical location of GRD in relation to known viral-resistance genes will be of special interest.

20.4 CHROMOSOME CHANGES INDUCED BY MAN

Many plant breeding strategies aim to incorporate useful genes from wild, alien species into the cultivated crops. In so doing the organization and evolution of chromosomes in many cultivated plants has been affected. The location and size of the alien chromatin can be identified using GISH and there are many examples where this has been successfully done for a variety of plants including wheat (*Triticum aestivum*; Heslop-Harrison, 1992; Heslop-Harrison *et al.*, 1991; Schwarzacher *et al.,* 1992; Heslop-Harrison and Schwarzacher, 1993), sugar cane (*Saccharum officinarum*; D'Hont *et al.*, 1995), *Lolium-Festuca* hybrids (Thomas *et al.*, 1994), and tomato-potato hybrids (de Jong *et al.,* 1996; Jacobsen *et al.*, 1995).

The amount of alien chromatin introduced varies, depending on the technique used, from whole genomes to small chromosome segments. The smallest chromosome segments detected by GISH so far are estimated to be less than 100 Mbp. For example, Mukai *et al.* (1993) identified an intercalary wheat-rye chromosome segment of about 40

Mbp of DNA, whereas Parokonny and Kenton (1996) detected a 20 to 30 Mbp segment of *Nicotiana glutinosa* DNA in *N. tabacum* cv. Samsun NN (Fig 20.1).

For the successful and stable transfer of alien DNA into the cultivated crop it is necessary to determine that the alien DNA has been stably integrated and transferred into the sexual progeny. Once again GISH has proved invaluable for monitoring the fate of alien chromatin through subsequent generations especially in examples where similar chromosome morphology of donor and alien species excludes the use of more traditional cytogenetic analysis (e.g. Jacobsen *et al*., 1995; de Jong *et al.,* 1996; Wolters *et al*., 1994; Parokonny *et al*., 1994).

Two studies at Kew have highlighted how useful GISH has been in analysing the genetic composition and fate of chromosomes in man-made hybrids.

20.4.1 GISH provides evidence of somatic translocations in potato dihaploids

Dihaploids, which contain two identical genomes, are widely used for the genetic improvement of a number of crops because they simplify the complex genetics of inheritance. However, their value depends on the implicit assumption that they represent genetically pure gametic samples of their parents. This view has been recently challenged by work on potato dihaploids at Kew in collaboration with Mike Wilkinson at the Scottish Crop Research Institute.

The euploid potato dihaploid PDH55 (2n=24) was made by pollinating the cultivated potato (*Solanum tuberosum*) cv. Pentland crown (2n=4x=48), with the dihaploid inducer *S. phureja* line IVP48. It is assumed that following pollination the *S. phureja* genome is eliminated resulting in a genetically pure dihaploid *S. tuberosum*. However, molecular and biochemical analysis suggested that the dihaploid PDH55 expressed DNA from the dihaploid inducer IVP48 (Wilkinson *et al.,* 1995; Waugh *et al.,* 1992). This was confirmed using GISH which identified large chromosome segments of *S. phureja* IVP48 DNA in the dihaploid PDH55 (Fig. 20.1d). Thus it is apparent that during dihaploid induction the *S. tuberosum* and *S. phureja* genomes do not remain discrete, instead somatic translocations take place, leading to the incorporation of *S. phureja* DNA into the dihaploid potato.

These observations may have important implications for the use of dihaploids in plant breeding. First, the assumption that euploid dihaploids represent pure gametic samples of their parents may not always be true. This will complicate meiotic, genetic and molecular studies of dihaploids. Second, although the frequency of such translocations is unknown at present, if they are common this may offer a novel way to introduce useful genes directly from wild dihaploid-inducing species into the cultivated crop. For potato breeding this could be agronomically important since recent observations show that some *S. phureja* populations are resistant to commercially important diseases (Hawkes, 1990).

20.4.2 Monitoring the fate of recombinant chromosomes in *Nicotiana* asymmetric hybrids

Asymmetric somatic hybridization is used in plant improvement strategies to create wide hybrids containing the whole genome from a non-irradiated species (recipient) and a small fraction of the genome from the irradiated species (donor). The method aims to transfer DNA encoding potentially useful genes from the donor to the recipient while minimising the incorporation of unfavourable DNA.

Parokonny *et al.* (1992) used GISH to identify intergenomic recombinant chromosomes in plants regenerated from asymmetric hybrids between *Nicotiana plumbaginifolia* (recipient) and *N. sylvestris* (donor). The fate of these recombinants through meiosis and into the sexual progeny was monitored using GISH (Parokonny *et al.*, 1994). The intergenomic recombinant chromosomes usually formed univalents although they occasionally paired with respectively homologous segments of *N. sylvestris* and *N. plumbaginifolia*. Recombinant chromosomes were transmitted through either male or female gametes and were detected by GISH in sexual progeny obtained on selfing or backcrossing the regenerant plants. The detection of desired genes in the translocated segments using sequential ISH with probes for the gene of interest and total genomic DNA from one of the parents will further improve the analysis of such hybrids.

Fig. 20.1 ***Colour Plate 8*** **A.** Chromosomes of *Nicotiana tabacum* L. line 095-55 probed with total genomic DNA from *N. sylvestris* (probable S genome donor; yellow fluorescence) and counterstained with propidium iodide (red fluorescence). Some of the intergenomic translocations between the S and T genome chromosomes are arrowed. **B.** Chromosomes of *Nicotiana tabacum* L. cv. Samsun probed with the geminivirus-related DNA (GRD) sequence (Bejarano *et al.*, 1995). Two pairs of hybridization sites (arrows; yellow fluorescence) are seen on the long arm of a small pair of submetacentric chromosomes. **C.** *Nicotiana tabacum* L. cv. Samsun NN chromosomes probed with total genomic DNA from *N. glutinosa* (green fluorescence). A segment of *N. glutinosa* DNA estimated to be 20 to 30 Mbp is arrowed. Chromosomes are counterstained with DAPI (blue fluorescence). **D.** Chromosomes of *Solanum tuberosum* dihaploid line PDH55 probed with total genomic DNA from the dihaploid inducer *S. phureja* line IVP48 (yellow fluorescence) and counterstained with propidium iodide (red fluorescence). Chromosome segments containing *S. phureja* DNA are arrowed. **E.** Chromosomes of *Zingeria biebersteiniana* following simultaneous hybridization with 5S (green fluorescence) and 18S-5.8S-28S (red fluorescence) ribosomal DNA probes, and counterstained with DAPI (blue fluorescence). **F.** *Z. biebersteiniana* chromosomes probed with the 18S-5.8S-28S ribosomal DNA probe revealing additional weak hybridization sites (arrowed). Scale bar=10 μm

20.5 POSSIBLE CONSTRAINTS ON CHROMOSOME EVOLUTION

From the above studies it is clear that plant genomes are capable of undergoing considerable changes and rearrangements. These observations raise the question 'What, if any, are the constraints which act on the genome to facilitate or prevent particular evolutionary changes from taking place?'

One way to tackle this problem is to look for patterns in the organization of DNA sequences along the chromosomes. A comparison of the genetic maps for several plant genomes are already revealing exciting similarities in the order of DNA sequences — a phenomenon termed synteny. For example, there is striking conservation in the gene order between wheat, maize, sorghum, sugar cane, foxtail millet and rice even though these grasses differ widely in genome size, chromosome number and taxonomic distance. These observations have led to the proposal of an 'ancestral' order of DNA sequences in the grasses (Moore *et al.*, 1995).

An alternative approach, taken at Kew, is to examine an extremely simple system by physically mapping the order of DNA sequences in

Zingeria biebersteiniana, a grass with 2n=4. For example, Bennett *et al.* (1995) analysed the organization of two highly repeated ribosomal DNA (rDNA) sequences (Fig. 20.1e and f) and the results have highlighted two interesting features of genome organization and evolution.

Firstly, the occurrence of three 18S-28S rDNA sites on the short arm of chromosome 1 is highly unusual and unexpected (Plate 1f). There are no known reports of more than two 18S-28S rDNA sites on a single chromosome arm (Schubert and Künzel, 1990). However, this could be explained if the chromosome arm evolved by the tandem fusion of several ancestral chromosomes which also contained 18S-28S rDNA sites. Tandem fusion has been suggested as an important karyotypic mechanism in evolving low chromosome numbers of several other plant genera (e.g. Charlesworth, 1985; Ikeda, 1987). Evolution by this mechanism would mean that the two chromosome types of *Z. biebersteiniana* are actually composed of more than two ancestral chromosome types.

Secondly, the near-centromeric location of the major 18S-28S rDNA site, the nucleolus organising region (NOR), is unusual for grasses but is known in both the composite *Haplopappus gracilis* (also 2n=4, Ikeda, 1987) and the ant *Myrmecia pilosula* (2n=2; Imai and Taylor, 1989). Comparisons of a wide range of taxonomically diverse organisms show various non-random features in genome organization, together forming what Lima-de-Faria (1954) has termed the chromosome field. In some cases, the reasons for such common recurrent non-random features of gross genome organization may be physical or mechanical rather than genetic. Such effects may place constraints on which of the possible end products of chromosome fusion can function in practice. Thus, the location of the major NOR in eukaryotes with only one or two large chromosomes may be a case in point. These observations suggest some importance of a near-centromeric position in such organisms which share low chromosome numbers. At present limited data preclude firm conclusions, but future work should address this question.

Bennett (1984) proposed that genomes of related and unrelated species might be usefully compared as natural karyotypes drawn as circles to see whether related DNA sequences occurred in physically similar locations, independent of genome size and chromosome number. It was recently shown that the circularised genomes of tetraploid *Zea mays*, whose segments display a hypothetical ancestral genome for grasses (Moore *et al.*, 1995) are very similar to the predicted natural karyotypes for these two genomes (Bennett, 1996b).

Molecular cytogenetics may play an important part in exploring the role of nuclear genome organization in function and phylogeny.

The continued mapping of genes not only in *Zingeria biebersteiniana* but also in other organisms sharing very low chromosome numbers, may highlight further features and constraints on genome organization and evolution. In addition, determining the order of genes in these species will contribute to the testing of the hypothesis as to whether there is indeed one ancestral basic order of genes common, not only to grasses, but perhaps to all angiosperms.

REFERENCES

Bejarano, E.R., Khashoggi, A., Witty, M. and Lichtenstein, C. (1996) Integration of multiple repeats of geminiviral DNA into the nuclear genome of tobacco during evolution. *Proceedings of the National Academy of Sciences* **93**, 759—64.

Bennett, M.D. (1984) The genome, the natural karyptype, and biosystematics, in *Plant Biosystematics*, (ed W.F. Grant), Academic Press, Canada, pp. 41—66.

Bennett, M.D. (1996a) The development and use of genomic *in situ* hybridization (GISH) as a new tool in plant biosystematics, in *Kew Chromosome Conference IV*, (eds P.E. Brandham and M.D. Bennett), Royal Botanic Gardens, Kew UK, pp. 167—83.

Bennett, M.D. (1996b) The nucleotype, the natural karyotype, and biosystematics, in *Symposia of the Society of Experimental Biology, Volume 50*, (ed J.S. Heslop-Harrison), Company of Biologists, Cambridge, (in press).

Bennett, M.D. and Smith, J.B. (1976) Nuclear DNA amounts in angiosperms. *Philosophical Transactions of the Royal Society of London, B,* **274**, 227—74.

Bennett, M.D., Smith, J.B. and Heslop-Harrison, J.S. (1982) Nuclear DNA amounts in angiosperms. *Proceedings of the Royal Society of London, B,* **216**, 179—99.

Bennett, S.T., Kenton, A.Y. and Bennett, M.D. (1992) Genomic in situ hybridization reveals the allopolypoid nature of *Milium montianum* (Gramineae). *Chromosoma* **101**, 420—4.

Bennett, S.T., Leitch, I.J. and Bennett, M.D. (1995) Chromosome identification and mapping in the grass *Zingeria biebersteiniana* (2n=4) using fluorochromes. *Chromosome Research*, **3**, 101—8.

Berg, D.E., Howe, M.M. (eds) (1989) *Mobile DNA*. American Society for Microbiology, Washington, D.C.

Charlesworth, B. (1985) Recombination, genome size and chromosome number, in *The evolution of genome size*, (ed T. Cavalier-Smith), Wiley, New York.

Chen, Q. and Armstrong, K. (1994) Genomic *in situ* hybridization in *Avena sativa. Genome,* **37,** 607—12.

D'Hont, A., Rao, P.S., Feldmann, P. *et al.* (1995) Identification and characterisation of sugarcane intergeneric hybrids, *Saccharum officinarum* × *Erianthus arundinaceus*, with molecular markers and DNA *in situ* hybridization. *Theoretical and Applied Genetics,* **91**, 320—6.

de Jong, J.H., Garriga Caldere, F., Ramanna, M.S. *et al.* (1996) Molecular cytogenetic studies of somatic potato (+) tomato hybrids and their derivatives, in *Kew Chromosome Conference IV*, (eds P.E. Brandham and M.D. Bennett), Royal Botanic Gardens, Kew UK, pp. 57—62.

Gazdovà, B., Siroky, J., Fajkus, J. *et al.* (1995) Characterization of a new family of tobacco highly repetitive DNA, GRS, specific for *the Nicotiana tomentosiformis* genomic component. *Chromosome Research,* **3**, 235—54.

Gill, B.S. (1991) Nucleocytoplasmic interaction (NCI) hypothesis of genome evolution and speciation in polyploid plants, in *Proceedings of the Kihara Memorial International Symposium on Cytoplasmic Engineering in Wheat*, (eds T. Sasakuma and T. Kinoshita), Yokohama, Japan, pp. 48—53.

Grant, V. (1981) *Plant speciation,* Columbia University Press, New York.

Hawkes, J.G. (1990) *The potato: evolution, biodiversity and genetic resources*. Belhaven, London, pp. 208—9.

Heslop-Harrison, J.S. (1991) The molecular cytogenetics of plants. *Journal of Cell Science,* **100**, 15—21.

Heslop-Harrison, J.S. (1992) Molecular cytogenetics, cytology and genomic comparisons in the Triticeae. *Hereditas,* **116**, 93—9.

Heslop-Harrison, J.S. and Schwarzacher, T. (1993) Molecular cytogenetics — biology and applications in plant breeding. *Chromosomes Today,* **11**, 191—8.

Heslop-Harrison, J.S., Leitch, A.R., Schwarzacher, T. and Anamthawat-Jonsson, K. (1991) Detection and characterization of 1B/1R translocations in hexaploid wheat. *Heredity,* **65**, 385—92.

Humphreys, M.W., Thomas, H.M., Morgan, W.G. *et al.* (1995) Discriminating the ancestral progenitors of hexaploid *Festuca arundinacea* using genomic *in situ* hybridization. *Heredity,* **75,** 171—4.

Ikeda, H. (1987) Cytogenetic studies on the chromosome complements of *Haplopappus gracilis* (2n=4) and *H. ravenii* (2n=8). *Journal of Science, Hiroshima University, Series B, Division* 2, **21**, 67—104.

Imai, H.T. and Taylor, R.W. (1989) Chromosomal polymorphisms involving telomere fusion, centromeric inactivation and centromere shift in the ant *Myrmecia (pilosula)* n=1. *Chromosoma,* **98**, 456—60.

Jacobsen, E., de Jong, J.H., Kamstra, S.A., van den Berg, P.M.M.M. and Ramanna, M.S. (1995) Genomic *in situ* hybridization (GISH) and RFLP analysis for the identification of alien chromosomes in the backcross progeny of potato (+) tomato fusion hybrids. *Heredity,* **74**, 240—57.

Jellen, E.N., Gill, B.S., Cox, T.S. (1994) Genomic in situ hybridization differentiates between A/D- and C-genome chromatin and detects intergenomic translocations in polyploid oat species (genus *Avena*). *Genome,* **37**, 613—8.

Jiang, J. and Gill, B.S. (1994) Different species-specific chromosome translocations in *Triticum timopheevii* and *T. turgidum* support the diphyletic origin of polyploid wheats. *Chromosome Research,* **2**, 59—64.

Johnson, M.A.T, Kenton, A.Y., Bennett, M.D. and Brandham, P.E. (1989) *Voanioala gerardii* has the highest known chromosome number in the monocotyledons. *Genome,* **32,** 328—33.

Kenton, A., Khashoggi, A., Parokonny, A. *et al.* (1995) Chromosomal location of endogenous geminivirus-related DNA sequences in *Nicotiana tabacum* L. *Chromosome Research,* **3**, 346—50.

Kenton, A., Parokonny, A.S., Bennett, S.T. and Bennett, M.D. (1993a) Does genome organization influence specieation? A reappraisal of karyotype studies in evolutionary biology, in *Evolutionary patterns and processes*, (eds D.R. Lees and D. Edwards), Academic Press, London, pp. 189—206.

Kenton, A., Parokonny, A.S., Gleba, Y.Y. and Bennett, M.D. (1993b) Characterization of the *Nicotiana tabacum* L. genome by molecular cytogenetics. *Molecular and General Genetics,* **240**, 159—69.

Leggett, J.M. and Markhand, G.S. (1996) The genomic structure of *Avena* revealed by GISH, in *Kew Chromosome Conference IV,* (eds P.E. Brandham and M.D. Bennett), Royal Botanic Gardens, Kew UK, 133—9.

Leitch, A.R., Schwarzacher, T. and Leitch, I.J. (1994) The use of fluorochromes in the cytogenetics of the small-grained cereals (Triticeae). *Histochemical Journal,* **26**, 471—9.

Lima-de-Faria, A. (1954) Chromosome gradient and chromosome field in *Agapanthus. Chromosoma,* **6,** 330—70.

Moore, G., Foote, T., Helentjaris, T. *et al.* (1995) Was there a single ancestral cereal chromosome? *Trends in Genetics*, **11**, 81—2

Mukai, Y., Friebe, B., Hatchett, J.H. *et al.* (1993) Molecular cytogenetic analysis of radiation-induced wheat-rye terminal and intercalary chromosomal translocations and the detection of rye chromatin specifying resistance to Hessian fly. *Chromosoma,* **102**, 88—95.

Ørgaard, M., Jacobsen, N. and Heslop-Harrison, J.S. (1996) Molecular cytogenetics in the genus *Crocus* L, in *Kew Chromosome Conference IV*, (eds P.E. Brandham and M.D. Bennett), Royal Botanic Gardens, Kew UK, 291—9.

Parokonny, A.S. and Kenton, A.Y. (1996) Comparative physical mapping and evolution of the *Nicotiana tabacum* L. karyotype, in *Kew Chromosome Conference IV*, (eds P.E. Brandham and M.D. Bennett), Royal Botanic Gardens, Kew UK, pp. 301—20.

Parokonny, A.S., Kenton, A., Gleba, Y.Y. and Bennett, M.D. (1992) Genome reorganization in *Nicotiana* asymmetric somatic hybrids analyzed by *in situ* hybridization. *Plant Journal,* **2**, 863—74.

Parokonny, A.S., Kenton, A., Gleba, Y.Y. and Bennett, M.D. (1994) The fate of recombinant chromosomes and genome interaction in *Nicotiana* asymmetric somatic hybrids and their sexual progeny. *Theoretical and Applied Genetics,* **89**, 488—97.

Schubert, I. and Künzel, G. (1990) Position-dependent NOR activity in barley. *Chromosoma,* **99**, 352—9.

Schwarzacher, T., Anamthawat-Jonsson, K., Harrison, G.E. *et al.* (1992) Genomic *in situ* hybridization to identify alien chromosomes and chromosome segments in wheat. *Theoretical and Applied Genetics*, **84**, 778—86.

Soltis, P.S., Doyle, J.J. and Soltis, D.E. (1992) Molecular data and polyploid evolution in plants, in *Molecular systematics of plants*, (eds D.E.Soltis, P.S.Soltis and J.J.Doyle), Chapman & Hall, London. pp 177—201.

Stace, C.A. (1993) Rare events in polyploid evolution, in *Evolutionary patterns and processes* (eds D.R. Lees and D. Edwards), Academic Press, London, pp. 157—69.

Thomas, H.M., Morgan, W.G., Meredith, M.R. *et al.* (1994) Identification of parental and recombined chromosomes in hybrid derivaties of *Lolium multiflorum* × *Festuca pratensis* by genomic in situ hybridization. *Theoretical and Applied Genetics,* **88**, 909—13.

Tsvelev, N.N. and Zhukova, P.G. (1974) On the minimal main chromosome number in the family Poaceae. *Botanicheskyi Žurnal SSSR*, **59**, 265—9.

Uhl, C.H. (1978) Chromosomes of Mexican *Sedum*, II. Section *Pachysedum*. *Rhodora* **80,** 491—512.

Waugh, R., Baird, E. and Powell, W. (1992) The use of RAPD markers for the detection of gene introgression in potato. *Plant Cell Reporter*, **11**, 466—9.

Wilkinson, M.J., Bennett, S.T., Clulow, S.A. *et al.* (1995) Evidence for somatic translocation during potato dihaploid induction. *Heredity,* **74**, 146—51.

Wolters, A.M.A., Schoenmakers, H.C.M., Kamstra, S.A. *et al.* (1994) Mitotic and meiotic irregularities in somatic hybrids of *Lycopersicon esculentum* and *Solanum tuberosum*. *Genome,* **37**, 726—35.

21

A saltation model of karyotypic evolution in the Octodontoidea (Mammalia, Rodentia)

M.H. Gallardo

21.1 INTRODUCTION

The suborder Caviomorpha was erected to group all South American and Antillean hystricognath rodents, although hystricognathy also is observed in some African counterparts (Woods, 1982). The abrupt appearance of miocenic caviomorph fossils represents an unsolved phylogenetic problem with two competing hypotheses attempting to explain the source area of the Caviomorpha. One hypothesis proposes a mid-Cenozoic immigration of rodents to South America from the early Eocene Franimorphs of North America, whereas the alternative hypothesis proposes an African origin (Wyss *et al.*, 1993). The fossil fauna in the central Chilean Andes indicate a boundary between 31.5—37.5 myr for the group's occurrence in South America, and favour the African origin over the South-American radiation. (Wyss *et al.*, 1993). A recent 12S rRNA gene phylogeny of selected genera of hystricognaths also supports the monophyletic origin of the Octodontoidea, containing the families Capromyidae, Ctenomyidae, Echimyidae, Myocastoridae and Octodontidae (Nedbal *et al.*, 1994).

Due to the lack of a well corroborated phylogeny for the group, karyotypic evolution of the South American Hystricognaths has been interpreted using the taxonomic classification as an indicator of phylogenetic signal. Chromosomal pathways have been postulated to occur gradually through tandem fusions from an hypothetical ancestor with 2n=98, having mainly acrocentric chromosomes, to the low numbers observed in some ctenomyids and octodontids (George and Weir, 1972, 1974). Nevertheless, C-band analysis has indicated that such an heterogeneous distribution within and among taxa cannot be explained by the fusion hypothesis (Gallardo, 1991, 1992). A bidirectional and gradual trend of karyotypic evolution was advanced for those taxa and, independently, for *Proechimys* (Reig *et al.*, 1980).

These conclusions have been derived from current taxonomic classifications. However, the Linnean system may be biased in assuming monophyly (Lanyon, 1994). The uncertainty of monophyly produces less parsimonious trees by artificially inflating the number of independent origins of a trait. Moreover, it tends to erroneously favour adaptation by assuming that equivalence in age reflects equivalence of taxonomic rank, and therefore sister-taxa relationships (Miles and Dunham, 1993). Since conclusions about extant taxa should not be divorced from an historical context, potential weaknesses are corrected here by using biochemically-derived phylogenetic trees. The extreme range of chromosomal variation in the paucispecific Octodontidae (Reig, 1989) is explained by proposing tetraploidy as a causal mechanism of chromosome change since it better explains the inconsistencies inherent in the gradualistic model. If this unorthodox proposition is proven to be true, a re-examination of most cytogenetically-derived relationships in mammals will be needed, the primary role of chromosomal speciation will be asserted, and instantaneous speciation would define the cytogenetic geometry of mammalian divergence patterns.

21.2 KARYOTYPIC CHARACTERISTICS OF FAMILIES WITHIN OCTODONTOIDEA

21.2.1 Abrocomidae

The abrocomids have been placed within the Echimydae, the Octodontidae and the Chinchilloidea (Glanz and Anderson, 1990). The karyotype of *Abrocoma bennetti* (2n=64, FN=110) includes twenty four pairs of biarmed chromosomes and seven pairs of acrocentrics. The X chromosome is a large submetacentric. Minimal amounts of weak-staining heterochromatin characterise this karyotype (Gallardo, 1992).

21.2.2 Capromyidae

The karyotypic data for this group are scarce and are summarised by George and Weir (1974). *Capromys pilorides* has 2n=40, and *Geocapromys brownii* 2n=88.

21.2.3 Ctenomyidae

This highly speciose group is characterised by large karyotypic differences, with diploid numbers ranging from 10 to 70 (Reig, 1989). This extensive karyotypic variation represents a paradigmatic instance of explosive chromosomal repatternings. In general, each species has a distinct karyotype (Cook, 1990; Reig *et al.*, 1990). Different cytotypes have been reported in *C. boliviensis*, *C. latro* and *C. talarum* (Cook, 1990; Reig *et al.*, 1992). Chromosomal repatternings can result from Robertsonian processes, pericentric inversions and tandem fusions. Heterochromatin transformations (Gallardo, 1991), and full or partial additions/deletions of euchromatic and heterochromatic short arms also have been reported (Reig *et al.*, 1992).

21.2.4 Echimyidae

This is a highly diverse and insufficiently understood taxon (Patton and Reig, 1989). Most species belong to the single, highly polytypic genus *Proechimys* (Reig, 1989). A cladistic analysis of allozymic data indicates that *Proechimys* is genetically close to *Hoplomys*, and *Makalata* to *Isothrix*, whereas *Mesomys* associates equally well with either of these two clades (Patton and Reig, 1989). A high degree of intrageneric chromosome diversity (2n=14—65) is found within *Proechimys* (Reig, 1989). Chromosomal evolution was initially proposed to have been bidirectional, starting from the intermediate numbers 2n=40—50 (Reig *et al.*, 1980). G-band homology indicates the inheritance of intact chromosomes or as separated arms. Pericentric inversions involving autosomal and sex chromosomes are also indicated, whereas chromosomal changes from lower to higher numbers are now postulated for the genus (Aguilera *et al.*, 1995).

21.2.5 Myocastoridae

In spite of the difficulty in establishing cytogenetic relationships among the three capromyid genera (*Capromys*, *Geocapromys* and *Myocastor*), the cytogenetic justification to maintain them as a natural groups was overly simplistic (George and Weir, 1974). The distinctiveness of *Myocastor* as a member of the monotypic Myocastoridae is supported by allozymic (Woods, 1982) and sequencing data (Nedbal *et al.*, 1994). *Myocastor coypus* has a totally biarmed karyotype with a diploid number of 42 (FN=80; González and Brum-Zorrilla, 1995). Small, centromeric C-bands are observed in six autosomes and in the sex chromosomes. Although C-band heteromorphism is present, no conclusion regarding heterochromatin addition or transformation was provided.

21.2.6 Octodontidae

Contrasting with its extensive adaptive radiation, there is a remarkable lack of intrafamilial diversity. The monotypic genus *Tympanoctomys* has the highest diploid number of any mammal (2n=102; Contreras *et al.*, 1990), and the family has the largest range in the number of chromosomes (38—102) recorded in mammals. Octodontid rodents comprise four monotypic genera: *Octodontomys, Octomys, Spalacopus* and *Tympanoctomys*. The polytypic genus *Octodon* (2n=58—78) is formed by four species, whereas three species are recognised in *Aconaemys* (2n=54—58; Gallardo and Reise, 1992). A superimposition of tandem fusions and non-Robertsonian mechanisms involving euchromatic segments accounts for the extremely high variation in diploid and fundamental numbers. Heterochromatic short arms, although present, do not explain the general biarmed karyotypic condition observed across the genera. Any hypothesis dealing with the widespread existence of totally heterochromatic arms in the unparalleled karyotype of *Tympanoctomys* must rest on a series of putative congeneric species, not even recorded as fossils. Apparently, evolution of heterochromatin has proceeded through whole-arm additions and euchromatin transformations, and has been independent of the bidirectional changes in diploid number. Based on heterochromatin content and the communality criterion, extremely low and high chromosome numbers represent derived conditions in the family (Gallardo, 1991, but also see Spotorno *et al.*, 1995).

21.3 ALLOZYMIC VERSUS TAXONOMIC APPROACHES

Adaptive scenarios derived from comparative studies may be explained by factors other than natural selection. By including a phylogenetic approach in evolutionary studies, hypotheses regarding the origin and maintenance of complex organismal traits have been refined (Miles and Dunham, 1993). Mapping characters onto a phylogeny may also assist in determining the ecological conditions associated with specific evolutionary transformations, especially where character states represent putative structures associated with rapid differentiation of a clade. Thus, when information is retrieved from phylogenies, key factors ascribed to causal historical roles in cladogenetic events are refined and homoplasies derived from adaptive responses may be differentiated from homologies. When coincidental patterns are found in both taxonomically- and phylogenetically-based systematic relationships, inferential arguments increase their validity as evolutionary hypothesis in a Popperian sense (Popper, 1972).

Allozymic data of proteins encoded by 19 presumptive gene loci obtained from 228 specimens representing the Abrocomidae, Ctenomyidae, Myocastoridae and all currently recognised genera of the Octodontidae are used (unpublished data). This auxiliary data are used as a guideline because interpretations regarding karyotypic evolution may be subject to retesting as available phylogenetic hypotheses are refined. Genetic relationships, summarised by a majority-rule consensus tree, are based on 100 bootstrap iterations of the Nei's distance matrix. The relationships rest on the molecular clock hypothesis, assuming that genetic distance reflects the degree of relatedness and that the tree topology reflects an accurate evolutionary sequence of events.

This is not intended to be a comprehensive literature survey. The reader is referred to George and Weir (1974) and to Reig (1989) for thorough treatments.

21.4 DISCUSSION

Chromosomal evolution in the Hystricomorpha has been assumed to occur through Robertsonian processes affecting an hypothetical 2n=98, FN=136 ancestor having a full complement of acrocentric chromosomes (George and Weir, 1974). This rampant gradualistic formulation assumes that overall karyotypic similarity reflects phylogenetic proximity, and that the karyotype of specialised species also must be derived.

Based on the taxonomic rank of the abrocomids, and considering the presence of acrocentrics observed in *Abrocoma bennetti* and *Octodon lunatus*, high-numbered karyotypes and acrocentric chromosomes have been considered ancestral to low-numbered karyotypes and to metacentric chromosomes (Contreras *et al.*, 1990, 1994; Spotorno *et al.*, 1995). The 102-chromosome karyotype of the desert specialist *Tympanoctomys barrerae*, having a fully metacentric karyotype with a preponderance of totally heterochromatic arms, is perplexing.

Based on the assumption that morphological specialisation implies karyotypic derivation, the derived nature of this karyotype has been favoured (Contreras *et al.*, 1990; Gallardo, 1991). Nevertheless, to fit karyotypic data with the assumed primitive 1—1 pattern of spikes in the phallic morphology of the Abrocomidae and the Octodontidae, the karyotype of *Tympanoctomys* is considered 'primitive, or better, an early derived genus of octodontids' (Contreras *et al.*, 1993). The claim that large amounts of heterochromatin should be expected in the ancestral genome of *T. barrerae,* as predicted by the fusion hypothesis (Contreras *et al.*, 1990), represents *post hoc* reasoning since no prediction concerning heterochromatin in the ancestor or its descendant lineages was advanced by George and Weir (1974). Apparently, the high chromosomal number, considered ancestral under the fusion hypothesis, together with the widespread existence of heterochromatin (considered to represent a derived trait in the hystricomorph rodents; Gallardo, 1991, 1992), contributed to the ambiguity concerning the ancestral-versus-derived condition of the *Tympanoctomys* karyotype.

Based on C-bands, and using the commonality criterion, a reformulation of the fusion hypothesis proposed that the 2n=56—58 karyotypes of the octodontids *Octomys*, *Spalacopus*, *Aconaemys* and *Octodon degus* represent ancestral conditions (Gallardo, 1992). This inference was disregarded as valid for its difficulty in explaining the chromosomal formula of *Tympanoctomys*, *Abrocoma* and *O. lunatus* (Contreras *et al.*, 1993). However, common ancestry, but not parallel acquisition, is supported by the basal position of *Spalacopus* and by the monophyletic derivation of the *Octodon-Aconaemys* clade (Fig. 21.1).

21.4.1 Karyotypic evolution, allozymic data and historical perspectives

A cladistic analysis of cranial-mandibular morphology of the Octodontidae led Verzi (1994) to conclude that *Octodontomys* is a ctenomyine and not an octodontine. Allozymic data does not support the alleged polyphyly of the Octodontidae, rather its monophyly. Like in the abrocomids, high-crowned molars apparently represent an adaptive response to herbivory, disassociated from the genetic cohesiveness that must define group membership.

The fact that *Abrocoma* appears morphologically derived, but not greatly different at the biochemical-genetic level when compared to the octodontids, challenges its current taxonomic rank. Thus, neither *Abrocoma's* chromosome number and morphology, nor its heterochromatin content are reliable phylogenetic signals of outgroup membership. The tree topology (Fig. 21.1) also brings into question the family rank for Ctenomyidae. By forming sister-group rather than outgroup relationships with the Octodontidae-*Abrocoma* clade, its status as a subfamily (Ctenomyinae) is implied by the data.

The allozymic tree corroborates the methodology's resolving power since topological stability increases with genetic distances between pairs of comparisons, distances which are substantial in the present case (unpublished data). Lack of resolution is constrained by neither taxa representation nor by number of specimens analysed because generic distinction is defined by fixed allelic differences, independent of sample size. Thus the inability to discriminate between *Octodon* and *Aconaemys* (while perfectly segregating the geographic samples of *Spalacopus cyanus*), is a reflection of their recent common ancestry.

Extensive and disparate chromosomal rearrangements would be needed to explain the associations depicted in several clades of Fig. 21.1. First, the ancestral condition of the 58-chromosome karyotype of *Spalacopus* is reasonably supported. This does not imply that the subterranean way of life is ancestral to the group; instead, it represents a derived evolutionary trend also affecting *Aconaemys*. The 64-chromosome karyotype of *Abrocoma* may well represent another divergent karyotypic trend linking it to the radiation of the desert specialists (*Octodontomys*, *Octomys* and *Tympanoctomys*). The sister group to this 58—64-chromosome branch is formed by two clades. The first contains the desert specialists (2n=38, 56 and 102), whereas the second is formed by *Octodon* (2n=58) and *Aconaemys* (2n=54—58). The marked allozymic difference of *O. lunatus* (2n=78) relative to the genetically-similar *Aconaemys*—*Octodon* clade is evident by its

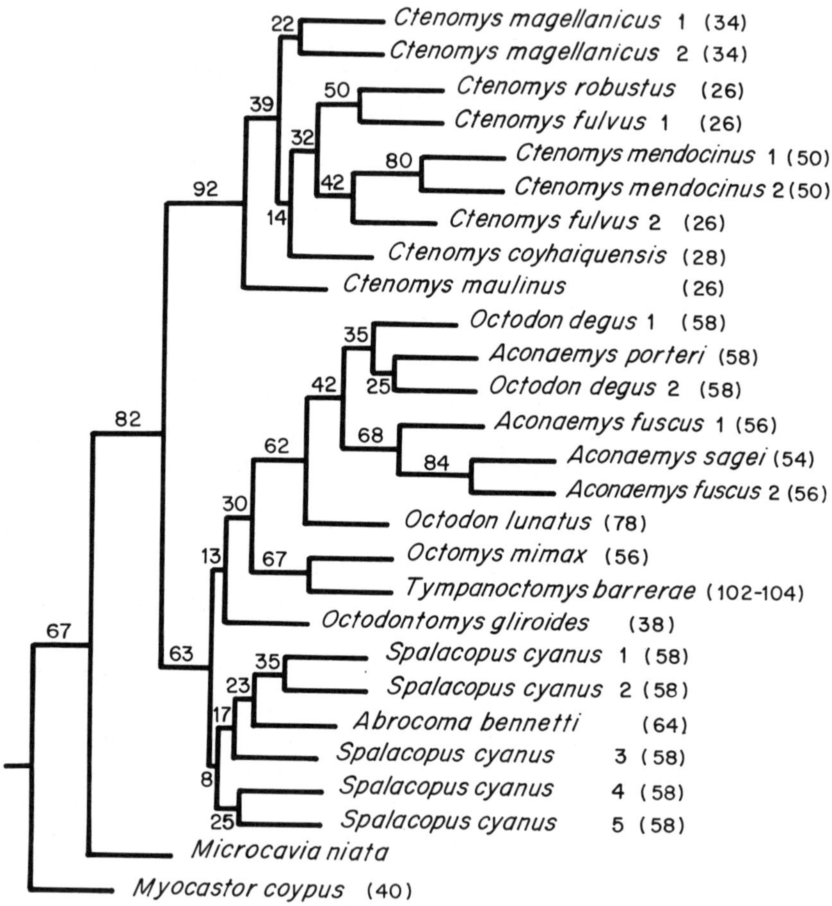

Fig. 21.1 Majority-rule consensus tree summarising the genetic relationships of 19 taxa of Octodontoidea. Nodal values are a measure of repeatability based on 100 bootstrap iterations. Diploid numbers (in parentheses) have been added to each taxon, except for *Microcavia niata*.

outgroup placement. Under a gradualistic model, this implies that extensive Robertsonian processes took place within the *Octodon* lineage. The most parsimonious sequence of events suggests that the ancestral 58-chromosome karyotype was inherited by the *Octodon—Aconaemys* lineage (Gallardo, 1991), while multiple fissions may have created the karyotype of *O. lunatus.* If recently formed chromosome races, genetically indistinguishable from one another are found in a lineage, then chromosomal changes could only have been established as the first stage of differentiation within that lineage (King, 1993). Thus, chromosomal speciation seems to have operated in the genetically cohesive *Aconaemys* cluster. A similar interpretation has been used to explain the identical allozymic and karyotypic pattern in *Proechimys* (Benado *et al.*, 1979).

Outgroup comparisons and allozymic data support the derived nature of asymmetrical sperm morphology in *Ctenomys* (Feito and Gallardo, 1982). Species with asymmetrical sperms (*C. maulinus*, *C. magellanicus* and *C. coyhaiquensis*; 2n=26—34) are ancestral to northern *C. fulvus,* differing in sperm morphology while sharing the same overall karyotypic similarity (Fig. 21.1). Increasing genetic divergence is correlated with higher karyotypic divergence (Cook and Yates, 1994) and with sperm morphology, regardless of chromosome number (*C. maulinus* and *C. fulvus*; Gallardo and Palma, 1992). Higher distance values between chromosomally-distinct populations of *C. boliviensis* suggest specific differentiation (Cook, 1990). The challenge in establishing the ancestral chromosome number, and the relationship between chromosome change and speciation in *Ctenomys*, is to distinguish between changes that have preceded, are causal of, or subsequent to, the speciation event (John and Miklos, 1988; King, 1993). Thus, if chromosomal differences are demonstrated to become established before genic or morphological divergence become fixed in the genus, then it follows that chromosome change is the primary and causative factor in speciation (King, 1993). Because genetic distance increases with karyotypic divergence and with differences in sperm morphology in *Ctenomys*, it is difficult to assert at this point whether geographic isolation accompanied or imposed reproductive isolation.

The basal branch position depicted by low-numbered *C. maulinus* (Fig. 21.1) and by Bolivian *C. opimus* (2n=26; Cook, 1990) suggests their relatively distant divergence from the clade that includes the other forms. The phylogenetic divergence of *C. opimus* is also supported by its highly repetitive DNA sequence (RPCS; Rossi *et al.*, 1995). This sequence, moderately amplified in some octodontids but highly amplified in other *Ctenomys* species, also suggests the ancestrality of *C. opimus* having low chromosome numbers. Nevertheless, it is too early to determine if these coincidental associations truly reflect phylogenetic trends (Gallardo, 1991).

Since extensive chromosomal repatternings and genetic distances separate the Octodontidae, the shared presence of acrocentrics between *Abrocoma* and *Octodontomys* can hardly be conceived as an homology. Likewise, the type and degree of chromosomal reshufflings observed in *Ctenomys* argue against homologous similarity with the acrocentrics of the Octodontidae s.s., because it would require only Robertsonian events when in fact pericentric inversions and heterochromatin transformations have also contributed to the karyotypic diversification of that genus (Gallardo, 1991). Although acrocentrics may be ancestral to a particular lineage, similar chromosome morphologies may result from a panoply of chromosomal repatternings affecting euchromatic segments of independent taxa.

Tympanoctomys barrerae and the other desert specialists formed a well-defined lineage where the most extreme karyotypic repatternings took place (Fig. 21.1). A decrease in chromosome number, involving complex chromosome rearrangements, characterised the lineage leading to *Octodontomys*; however, the opposite was true for the derived karyotype of *Tympanoctomys*. Inconsistencies between allozymic, karyotypic and morphological data indicate that the pathways leading to the low- and high-numbered karyotypes of the brush-tailed octodontids (that is, the three desert specialists, and *Octodon*) need to be reinterpreted, while the importance of chromosomal repatternings as a major cladogenetic event is maintained. The ubiquitous presence of metacentrics in the karyotypes of the desert specialists, and the high heterochromatin content in karyotypically-derived *Tympanoctomys,* contradict tandem events (Contreras *et al.*, 1990). Moreover, tandem events are primarily observed in lineages with acrocentric karyotypes, and usually involve euchromatic chromosome regions, not heterochromatic arms (King, 1993, p. 74).

While the highest chromosomal number and the largest span for chromosomal variation in a mammal family are observed in the Octodontidae, extreme gaps between chromosome numbers and intriguing chromosomal repatternings also characterise this family. Gaps, if taken as evidence of diversity, imply that the extant octodontid groups are living remnants of a past diversification that encompassed the most extreme chromosomal disparities observed today. No widespread progressions in chromosome number or in heterochromatin content as observed in polytypic *Proechimys* or *Ctenomys* can be proposed here. This point is raised by Reig (1989) in connection with the perplexing reality of the 'hollow-curve' of taxic diversity:

> 'for any taxonomic major group, there are only a few of their subordinated taxa that include most of the species (speciose taxa), whereas the remaining species are distributed into a smaller number of paucispecific (< 10 species) and a large array of monotypic taxa of the same subordinated rank' (Reig, 1989, p. 251).

21.4.2 The saltational model of karyotypic repatterning by tetraploidy in the Octodontidae

Chromosomal variation among the genera of the Octodontidae (including *Ctenomys* as a monotypic subfamily; Reig, 1989), fits a hollow-curve distribution, with *Ctenomys* containing by far the most species. Since no fossil evidence exists for *Octomys*, *Tympanoctomys*, *Spalacopus* or *Octodontomys* (Reig, 1989), any model will have to deal with 'missing karyotypic links' to explain the hollow-curve distribution of the Octodontidae. According to the gradualistic view, this peculiar distribution is explained by low rates of chromosomal evolution in the depauperate Octodontinae of Reig (1989). Nevertheless, an average of 21 ± 3 chromosomes separate each octodontid taxa. Aside from the ancestral-versus-derived controversy, an extensive sequence of putative chromosomal forms would be required to explain this striking gap in chromosome numbers. The necessary assumption of these intermediate forms constitutes an *ad hoc* hypothesis of gradualism. Moreover, the extreme gaps in chromosome number contradict the assumption of slow rates of chromosomal change if one adheres to the ruling gradual paradigm. Heterochromatization of the genomic arm number of *Tympanoctomys* is more likely the result of an instantaneous event (i.e. tetraploidy) than the outcome of gradually-occurring, coincidental and unparsimonious rearrangements.

The saltational explanation for the magnitude of the numerical gap in chromosome number in the Octodontidae would assume tetraploidization within the clade containing *Octomys mimax* and *Tympanoctomys barrerae*. Although this is an admittedly unorthodox

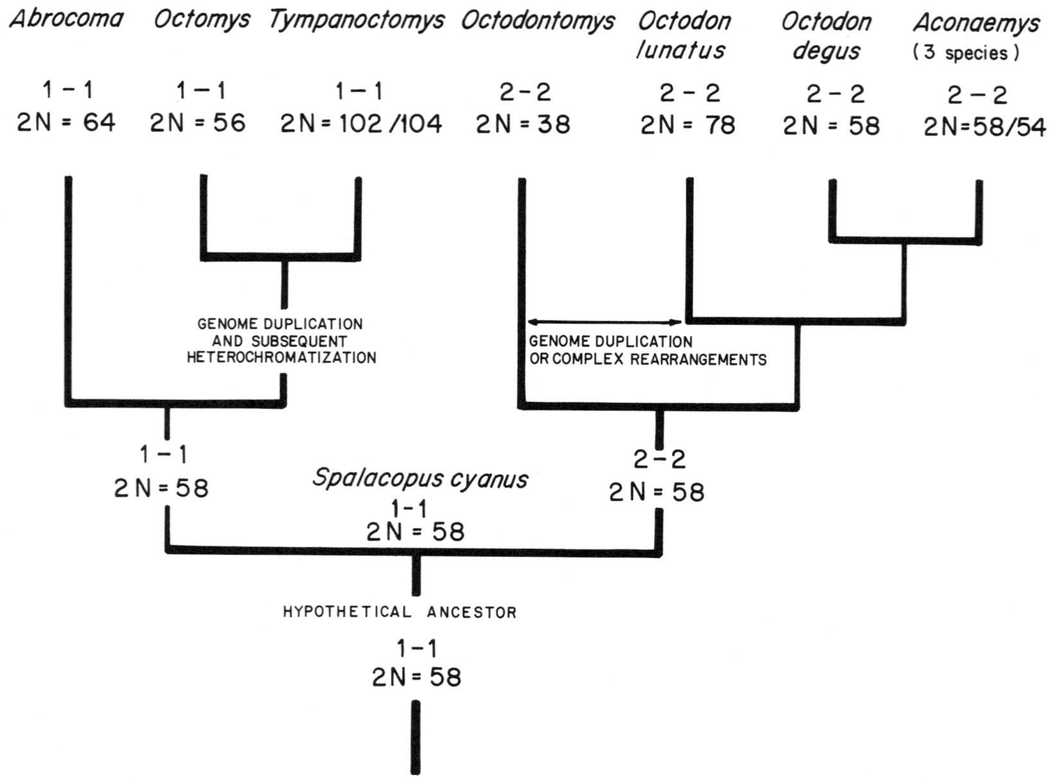

Fig. 21.2 Proposed phylogeny of the Octodontidae (containing *Abrocoma*) based on allozymic data, and assuming tetraploidization within two related lineages. The patterns of spike morphology (1—1, 2—2) and the diploid numbers (2n) are also shown.

model for mammals, it seems to be the only cytogenetic alternative that explains the kind and degree of chromosomal divergence and the distribution of heterochromatin in *Tympanoctomys.* The model does not invoke hypothetical chromosome forms, and more parsimoniously explains the complex phylogenetic pattern of reproductive structures within the Octodontidae (Fig. 21.2). Moreover, it proposes a simpler and a more parsimonious explanation of the heterochromatin distribution in the unparalleled karyotype of *Tympanoctomys barrerae*. Heterochromatin would not be acquired by saltatory replication occurring in hypothetical ancestors, but would reflect genomic compensation through imprinting and chromosome inactivation (Lyon, 1992). Finally, it is consistent with genetic predictions of similarity and coincidental geographic distribution between putative parental and derived species. With this evidence in mind, we can reason that *Octomys* and *Tympanoctomys* are the phylogenetic representatives of an underlying autotetraploidy that took place after the *Octomys*-like ancestor diverged from the 58-chromosome ancestor of the Octodontidae (Fig. 21.2).

Regardless of their importance in the copulatory-induced estrous of octodontids (Spotorno, 1979), the interlineage pathways of phallic spikes are extremely complex and clouded by homoplasies and evolutionary reversals (Contreras *et al.*, 1993). These difficulties stem from assuming the validity of the Abrocomidae as an outgroup, and by trying to fit data into the predominant gradualistic model. A reinterpretation of phallic morphology data supports homology, thereby eliminating the assumption of reversals. Thus, the ancestral 1—1 spike pattern present in *Octodon lunatus*, is also shared by its putative parental species, *Octodontomys*. Likewise, the derived 2—2 pattern observed in parental *Octomys* is shared by the tetraploid-descendant *Tympanoctomys* (Fig. 21.2).

As is also predicted by the saltational model, high genetic similarity must characterise the taxa involved in polyploidization. In fact, these four species form a cohesive clade where the *Tympanoctomys*—*Octomys* association is supported by a bootstrap value of 67%, while the outgroup placement of *Octodon lunatus* is supported a bootstrap value of 62% (Fig. 21.1). When the octodontids (as defined by genetic data) are rooted by *Abrocoma,* the association *Octomys*—*Tympanoctomys* persists, and a substantially higher bootstrap value (96%) supports their close affinity within the phylogenetic tree.

Although the possibility of tetraploidization for the origin of the *Tympanoctomys* karyotype has been suggested, it was disregarded because the bone-marrow C-banded chromosomes did not appear in clustered pairs (Contreras *et al.*, 1990, their Fig. 3). Since this assertion was not based on a meiotic analysis, it does not represent a valid objection to tetraploidization.

Although not discussed in their work, an additional small submetacentric pair was observed in bone-marrow cells of male *Tympanoctomys* (Contreras *et al.*, 1990). Accordingly, the chromosome number of the species is 2n=102 (FN=200) in females, and 2n=104 (FN=204) in males. This indicates either a sex-related somatic mosaicism in *Tympanoctomys*, or reflects meiotic drive as reported in allied *Proechimys iheringi* (Yonenaga-Yasuda *et al.*, 1985). Differences between expected and observed diploid numbers by tetraploidization could reflect underlying physiological mechanisms of genomic adjustment (i.e. extra sex chromosomes, etc.).

Full-arm heterochromatization of the metacentric karyotype of *Tympanoctomys* may represent a selective mode for maintaining genic balance in the whole genome. The epigenetic phenomena of imprinting and X-chromosome inactivation are the common mechanisms to co-ordinate gene regulation (Lyon, 1993). Imprinting may be manifested by differential gene expression, by differential inactivation, or by elimination of chromosomes derived from either parent (Lyon, 1993). Somatic elimination of one or more chromosomes has been described in Australasian (Hayman, 1990), and South American marsupials (Gallardo and Patterson, 1987). Differences in chromosome number may also represent meiotically-derived phenomena, like the polymorphism involving the euchromatic, early replicating supernumerary chromosomes of *Proechimys iheringi* (Yonenaga-Yasuda *et al.*, 1985). In eutherians there are also clear methylation differences between the active and the inactive X chromosomes, interpreted as a means of stabilising inactivation by a feedback mechanism involving a maintenance methylase (Lyon, 1992). The challenging questions posed by *Tympanoctomys* may help to discern whether imprinting and X chromosome inactivation have a common evolutionary origin in eutherians, and whether tetraploidy is a common phenomenon among rodents or mammals.

The causal mechanism of inter-individual chromosome differences between the desert specialists, although still unexplained, is consistent with meiotic instabilities derived from genomic duplication. Consequently, the saltational model is not contradicted by any available data. It is simpler and more assumption-free than the gradual model. It better explains cladistically-analysed morphological data, and it is consistent with similarities in both genetic and geographic distributions.

Because no banding data exist for *Octodon lunatus*, and because its genetic distance to *Octodontomys* is rather high, its derivation from *Octodontomys* through tetraploidy is less apparent. If tetraploidy also links these species, then complex chromosomal rearrangements took place in the differentiation of *Octodon lunatus* from the 54—58-chromosome *Octodon—Aconaemys* clade (Fig. 21.2).

Several predictions can be made from the saltational model. One is that double, or close to double, amounts of nuclear DNA would be expected in *Tympanoctomys* relative to *Octomys*. Moreover, because a single X chromosome remains active for every two sets of autosomes, two X chromosomes, and two Barr bodies would be expected to remain active in female somatic cells of *Tympanoctomys*. Because genome inactivation is extensive in *Tympanoctomys*, double doses of gene products would be difficult, although not impossible, to find.

The primacy of chromosomal repatterning as a factor that triggers speciation is emphasised in the saltational model. Another corollary is that reproductive isolation would be the consequence of a cytological event. Finally, the similar morphology of the putative desert-specialists (*Octomys* and *Tympanoctomys*) would not be a convergent adaptation to xeric environments, but the result of the same, amplified genome. Direct cytogenetic research and mtDNA analyses are in progress to test the predictions already advanced.

ACKNOWLEDGEMENTS

This work is dedicated to the memory of Dr. Osvaldo A. Reig, a leading figure in South American mammalogy. I thank Nelida Köhler and Cristian Araneda for laboratory and computer assistance, respectively. The meticulous editorial assistance of S. Fraver is greatly appreciated. The collecting permits granted by Corporación Nacional Forestal and Servicio Agrícola y Ganadero are also appreciated. This research was supported by grant FNC 92/0178 and by grant A-13219/1-000070 from Fundación Antorchas.

REFERENCES

Aguilera, M., Reig, O. A. and Perez-Zapata, A. (1995) G- and C-banding karyotypes of spiny rats (*Proechimys*) of Venezuela. *Revista Chilena de Historia Natural*, **68**, 185—96.

Benado, M., Aguilera, M., Reig, O. A. *et al* (1979) Biochemical genetics of chromosome forms of Venezuelan spiny rats of the *Proechimys guairae* and *Proechimys trinitatus* superspecies. *Genetica,* **50**, 89—97.

Contreras, L. C., Torres-Mura, J. C. and Spotorno, A. E. (1990) The largest known chromosome number for a mammal, in a South American desert rodent. *Experientia,* **46**, 506-08.

Contreras, L. C., Torres-Mura, J. C., Spotorno *et al.* (1993) Morphological variation of the glans penis of South American octodontid and abrocomid rodents. *Journal of Mammalogy,* **74**, 926-35.

Contreras, L. C., Torres-Mura, J. C., Spotorno *et al* (1994) Chromosomes of *Octomys mimax* and *Octodontomys gliroides* and relationships of octodontid rodents. *Journal of Mammalogy,* **75**, 768—74.

Cook, J. A. (1990) Rapid chromosomal evolution and the systematics of the Bolivian tuco-tucos (Rodentia, Ctenomyidae). Unpublished Ph.D. dissertation, The University of New Mexico.

Cook, J. A. and Yates, T. L. (1994) Systematic relationships of the Bolivian tuco-tucos, genus *Ctenomys* (Rodentia, Ctenomyidae). *Journal of Mammalogy,* **75**, 583—99.

Feito, R. and Gallardo, M. H. (1982) Sperm morphology of the Chilean species of *Ctenomys*. *Journal of Mammalogy,* **63**, 658—61.

Gallardo, M. H. (1991) Karyotypic evolution in *Ctenomys* (Rodentia, Ctenomyidae). *Journal of Mammalogy,* **72**, 11—21.

Gallardo, M. H. (1992) Karyotypic evolution in Octodontid rodents based on C-band analysis. *Journal of Mammalogy,* **73**, 89-98.

Gallardo, M. H. and Patterson, B. D. (1987) An additional 14—chromosome karyotype and sex-chromosome mosaicism in South American marsupials, in *Studies in Neotropical mammalogy: essays in honor of Phylip Hershkovitz* (eds B. D. Patterson and R. M.Timm), Fieldiana, Zoology, new series, **39**, 111—15.

Gallardo, M. H. and Palma, R. E. (1992) Intra- and interspecific genetic variability in *Ctenomys* (Rodentia: Ctenomyidae). *Biochemical Systematics and Ecology,* **20**, 523-34.

Gallardo, M. H. and Reise, D. (1992) Systematics of *Aconaemys fuscus* (Rodentia, Octodontidae). *Journal of Mammalogy,* **73**, 779-88.

Glanz, W. E. and Anderson, S. (1990) Notes on Bolivian mammals. 7. A new species of *Abrocoma* (Rodentia) and relationships of the Abrocomidae. *American Museum Novitates,* **2991**, 1—32.

George, W. and Weir, B. J. (1972) The chromosomes of some Octodontids with special reference to *Octodontomys* (Rodentia; Hystricomorpha). *Chromosoma (Berl.),* **37**, 53—62.

George, W. and Weir, B. J. (1974) Hystricomorph chromosomes, in *The Biology of Hystricomorph Rodents*. (eds I. W. Rowlands and B. J. Weir), Academic Press, London, pp. 79—108.

González, S. and Brum-Zorrilla, N. (1995) Karyological studies of the South American rodent *Myocastor coypus* Molina 1782 (Rodentia: Myocastoridae). *Revista Chilena de Historia Natural,* **68**, 215—26.

Hayman, D. L. (1990) Marsupial cytogenetics. *Australian Journal of Zoology*, **37**, 331—49.

John, B. and Miklos, G. L. G. (1988) *The eukaryote genome in development and evolution,* Allen and Unwin, London.

King, M. (1993) *Species evolution: the role of chromosome change,* Cambridge University Press, Cambridge.

Lanyon, S. M. (1994) Polyphyly of the blackbird genus *Agelaius* and the importance of assumptions of monophyly in comparative studies. *Evolution,* **48**, 679—93.

Lyon, M. F. (1992) Some milestones in the history of X—chromosome inactivation. *Annual Review of Genetics*, **26**, 17—28.

Lyon, M. F. (1993) Epigenetic inheritance in mammals. *Trends in Genetics*, **9**, 123—28.

Miles, D. B. and Dunham, A. E. (1993) Historical perspectives in ecology and evolutionary biology: the use of phylogenetic comparative analyses. *Annual Review of Ecology and Systematics*, **24**, 587—619.

Nedbal, M. A., Allard, M. C. and Honeycutt, R. L. (1994) Molecular systematics of hystricognath rodents: evidence from the mitochondrial 12S RNA gene. *Molecular Phylogenetics and Evolution*, **3**, 206—20.

Patton, J. L. and Reig, O. A. (1989) Genetic differentiation among echimyid rodents with emphasis on spiny rats, genus *Proechimys*. *Advances in Neotropical Mammalogy*, **1989**, 75—96.

Popper, K. R. (1972) *Objective Knowledge: an Evolutionary Approach,* Oxford University Press, Oxford.

Reig, O. A. (1989) Karyotypic repatternings as one triggering factor in cases of explosive speciation, in *Evolutionary Biology of Transient Unstable Populations* (ed A. Fondevila), Springer Verlag, Berlin, pp.246—89.

Reig, O. A., Aguilera, M. Barros, M. A. *et al* (1980) Chromosomal speciation in a rassenkreis of Venezuelan spiny rats (genus *Proechimys*, Rodentia, Echimyidae). *Genetica* **52/53**, 291—312.

Reig, O. A., Bush, C. Ortells, M. O. *et al* (1990) An overview of evolution, systematics, population biology, cytogenetics, molecular biology and speciation in *Ctenomys,* in *Evolution of Subterranean Mammals at the Organismal and Molecular Levels* (eds E. Nevo and O. A. Reig), Alan R. Liss, New York, pp. 71—96.

Reig, O. A. Massarini, A. Y., Ortells, M. O. *et al* (1992) New karyotypes and C-banding patterns of the subterranean rodents of the genus *Ctenomys* (Caviomorpha, Octodontidae) from Argentina. *Mammalia*, **56**, 603—23.

Rossi, M. S., Pesce, C. G., Kornblihtt, A. R. *et al* (1995) Origin and evolution of major satellite DNA from South American rodents of the genus *Ctenomys*. *Revista Chilena de Historia Natural*, **68**, 171—83.

Spotorno, A. E. (1979) Contrastación de la macrosistemética de roeodres caviomorfos por análisis de la morfología reproductiva masculina. *Archivos de Biología y Medicina Experimentales*, **12**, 97—106.

Spotorno, A. E., Walker, L. Y., Contreras, L. C. *et al* (1995) Chromosome divergence of *Octodon lunatus* and *Abrocoma bennetti* and the origins of the Octodontoidea (Rodentia, Histricognathi). *Revista Chilena de Historia Natural*, **68**, 227—39.

Verzi, D. H. (1994) Origen y evolución de los Ctenomyinae (Rodentia, Octodontidae): un análisis de anatomía cráneo-dentaria. Unpublished Ph.D. dissertation, Universidad Nacional de La Plata, Argentina.

Woods, C. A. (1982) The history and classification of South American hystricognath rodents: reflections on the far away and long ago, in *Mammalian Biology in South America* (eds M.A. Mares and H.H. Genoways), Special Publications Series, Pymatuning Laboratory of Ecology, Univ. Pittsburgh, **6**, 1-539, pp. 377—92.

Wyss, A. R., Flynn, J. J., Norell, M. A. *et al* (1993) South America's earliest rodent and recognition of a new interval of mammalian evolution. *Nature,* **365**, 434—37.

Yonenaga-Yasuda, Y., De Sousa, M. J., Kasahara, S. *et al* (1985) Supernumerary system in *Proechimys iheringi iheringi* (Rodentia, Echimyidae), from the state of São Paulo, Brazil. *Caryologia*, **38**, 179—94.

EPILOGUE

Peter Moens

22

The who is who of chromosomes

P.B. Moens

The title, 'The Who is Who of Chromosomes' is taken from the many 'Who is Who' books published in the United States, such as the 'Who is Who of Science' or Medicine or Government which serve to identify and to recognize individuals by listing their education, professional accomplishments, family and their hobbies. The September supplement of *Chromosome Research* (vol **3**, suppl 1) is to me the 1995 edition of the 'Who is Who of Chromosomes'. Incidentally, it is also the 'Who is Who' of the 1995 temporal guardians of these chromosomes. This supplement publicizes the personalities of chromosomes, what they look like, how they have evolved, what they have accomplished, what are their foibles and which people they have recruited to publicize their eminence. Thus the 12th International Chromosome Conference is a fertile ground to learn all about the different kinds of chromosomes that are seeking distinction in 1995.

Once upon a time, long, long ago, chromosomes where just that: chromo-somes, coloured bodies. They were seen with primitive microscopes in tissue sections which were cut with a razor blade by hand. The observations were then summarized, as Harry Scherthan showed on Wednesday, in elegant pen or pencil drawings (p.217—48). If the fingers that made the drawings are digits, then the drawings are the original digitally-enhanced images of chromosomes. The results are very much akin to the more recent technology of image enhancement with cooled CCD camera and by a somewhat different form of digital embellishment. With these techniques the 1995 images of chromosomes have become exquisite but the line between fact and fancy is sometimes treacherously narrow.

The primitive chromosomes had weak staining regions, the centromere and sometimes the secondary constriction. Since their positions often varied for the different chromosomes, they could therefore be used to identify individual chromosomes. Such were the good old days. But nostalgia for those simpler times has no place in science and with courage we have to face the proliferation of banding techniques, chromosome paints and the ever greater complexities of chromosome structure and function at this, the 12th International Chromosome Conference and recorded here in Chromosomes Today, volume 12.

Take, for example, the never-ending telomere story. These extensive repeats at the ends of chromosomes are supposed to provide a simple mechanism that counteracts the inadequacies of DNA polymerase in completing its job at the ends of chromosomes. Apparently not so. I understand that in spite of the telomeric repeats, my telomeres are getting shorter with age even as I am talking to you. Only those cells where the enzyme telomerase is turned on can restore and extend telomeres. Unfortunately that seems to happen only in my reproductive cells. What, I ask myself, is the good of that? It is like wearing a hairpiece. Such rescue methods are not likely to improve my Darwinian fitness at this time.

The remarkable characteristic of the telomere repeats is that they are fairly similar from *Tetrahymena* to man. At least that was so until Biessmann and colleagues (p.104—21) perplexed us by showing that *Drosophila* chromosomes construct their ends with transposons. These transposons are so totally different from the canonical telomere sequences that if one were to naïvely construct a parsimonious phylogenetic tree based on telomere sequences and on a constant rate of nucleotide substitution of *Tetrahymena*, yeast, plants, *Drosophila* and man, it would likely show that life is older than the universe itself. And if not life in general, certainly that of *Drosophila*. Such a contradiction also surfaces in astronomy where it appears that by using the Hubble constant, some very old stars exceed the estimated age of the Universe. Not to mention the 4th book of the Douglas Adams' *Hitchhiker* trilogy, in which Marvin, the whiny, rusty robot, finally falls apart when he is many times the age of the Universe. Fortunately the evidence of transposed *Drosophila* telomeres point to possible faulty assumptions about the homology of sequences at the ends of chromosomes in different phylogenetic groupings. This can be an instructive warning as we seek homologies between mitotic and meiotic chromosome events.

At breakfast Harold Biessmann told us in confidence, and I pass on to you in confidence, a possible evolutionary explanation for the exceptional telomeres in the Diptera. Organisms with many chromosomes, such as the *Tetrahymena* macronucleus, yeast and man, have a lot of ends that need to be looked after. Groups of organisms with few chromosomes presumably can get by with occasional transposition events and they could afford to loose the more complex genetic machinery for conventional chromosome end repair. The loss would presumably be a secondary event since some of the more primitive arthropods such as the crustaceans, appear to have very high chromosome numbers. It sounded like a sensible proposition. But then Herbert Macgregor had to spoil this attractive generalization with the fact that while amphibians have the canonical telomere sequence, the newt does not. Here we go again.

The fortuitous useful function of transposons in the repair of *Drosophila* chromosome ends may suggest that transposons are maintained in the population because they have a selective advantage. A related case seems to exist in *E. coli.* Apparently, there are a number of cryptic genes that can provide enzymes to metabolize unusual substrates. These genes are normally silenced by the presence of a transposon in the promoter region. If, under challenged conditions, the transposon happens to leave the promoter, the result may be an adaptive mutation and the lucky *E. coli* clone can resume growth. Such fortuitous contributions to the well being of the organism may ensure the persistence of strains that carry transposons.

It is appropriate that since the 12th International Chromosome Conference is being held in Spain, a great deal of attention should be paid to FISH. For the uninitiated, FISH is the acronym for fluorescent *in situ* hybridization. At the 1994 Chromosome Meeting in the Royal Botanical Gardens at Kew, a plenary speaker of this meeting, Harry Scherthan, presented a list of approximately ten different ISHes that are now in use. To that we add our sincere WISH not to have any more. It is old news and by now a platitude, to proclaim that FISH technology has revolutionized chromosome research. It has obviously completely replaced and immeasurably improved the cumbersome radioactive isotope technology. At one of the most minimal levels, it is now possible to detect the location of some genes along the chromosome using a cDNA probe as short as 600 base pairs. This is a tricky proposition if one considers the enormous sizes of the intervening introns in the genomic DNA. Yet Dr. Henry Heng seems to do this routinely and accurately. Henry, I should say, is an inventor of sorts. In a 1992 PNAS article (Heng, H.H.Q., Squires, J. and Tsui, L.C. (1992) *Proc. Natl. Acad. Sci.*, 9505—13), he demonstrated the use of extended chromatin fibres in the ordering of cosmids with multicolour FISH. A tribute to the value of this technology are the publications of other laboratories who have more recently pioneered the same technique.

More conventionally-sized probes, derived from 40 kb cosmids or even longer YACs, can be used to define the position of genes relative to mammalian chromosome bands, analogous to the way ISH was started by Gall and Pardue on *Drosophila* polytene chromosomes. These types of mapping have become routine and there are numerous posters that report such mapping results at this meeting.

Missing are presentation on how the positions of genes relate to scaffold-associated regions of the DNA, the SARs, and matrix-associated regions of the DNA, the MARs. There are no reports of how those stretches of chromatin are integrated with the scaffolds or the matrix. For many years, it has been forcefully postulated that the relationship between nuclear matrix and chromatin is essential for gene regulation. One would expect that the existence of such a physical relationship would have been verified by now with the isolation of MARs from the matrix, or by a combination of immunofluorescence of the matrix and FISH of specific chromatin segments. However, I noticed that on the contrary, in the chromosome model of Peter Cook (p.147—60), neither SARs nor MARs seem to play any role. What I would like to see is the type of experiments that have already been completed for meiotic chromosomes. In that case, the proteins of the chromosome cores and their genes have been identified. The chromatin loops are less mystical than their somatic counterparts, they can be painted individually and the DNA segments that are associated with chromosome cores have been isolated, cloned and sequenced.

Intricate issues of chromosome evolution have also proven to be accessible by FISH of entire chromosomes and by GISH, genomic *in situ* hybridization. This involves the production of complex probes that specifically recognize one or more chromosomes. With the initiative and leadership of Mike Bennett and Pat Heslop-Harrison (p.333—46), the detection of entire genomes and their ancestral origins has become of general use and of immense practical agricultural value in the transferal of valuable genetic material to economically-important crops. They furthermore applied the technique to the study of the selective loss of chromosomes when two foreign genomes are brought together.

With due respect to the immense amount of progress made in the field of somatic cytogenetics, it is always disappointing that the cytogenetics of the reproductive cells take a second place. To my mind, reproduction and the regulation of reproduction are, in fact, more urgent issues than the proliferation of somatic cells. I have a paranoid opinion that the administrators of grants are of an age when control of somatic cell proliferations appears more important to them than reproduction and that the distribution of grant money reflects that attitude. I am sure that this is incorrect and my view is unfounded. Much of the molecular information gleaned from somatic cytogenetics has found its application in to the meiotic process. The neglect of meiosis may in part be due to practical reasons. For example, because the differentiation of reproductive cells is a one-way and mostly a dead-end street, no mass cultures can be grown continuously to test various genetic and cytogenetic parameters. One of the exceptions is the yeast, *Saccharomyces cerevisiae*, where large batches of cells can be sent synchronously into meiosis. It is therefore no surprise that much of the progress in meiosis research has come from yeast. Even the meiotic chromosome visualization has become reasonably good, thanks to the work by Byers, Dresser and Loidl.

The results are already spilling over into distant relatives of yeasts: the plant and animals. The yeast recombination experts demand a double-strand break and repair mechanism for genetic recombination at meiosis. Simple reasoning therefore predicts that the main players of that scenario, RecA and Rad 51-type proteins, should be located at the site of recombination, possibly the famous, and occasionally infamous, recombination nodule. Doug Bishop and Nancy Kleckner have shown that antibodies to the Rad 51 protein are, in fact, associated with synaptonemal complexes which can be interpreted as supportive evidence for a recombination complex at the SC. Now homologues of RecA-like proteins have been reported by Ogawa in lily meiotic chromosomes and they are under investigation by Heyting and Anderson in the tomato and by Ashley in the mouse.

Finally some philosophical considerations. The presently popular constructionist philosophy explores the idea that all that we do see or think is a property of the mind, embedded in its social cultural background. The method is anti-science in that it claims that reality is no more than what we make it to be. Everything around us is constructions of our mind. Thus Madame Curie did not discover radium, she constructed it. Although the philosophy has no clear idea of the scientific method, it must be admitted that we do, in fact, work in a societal context and on topics that society has prioritized in direct or indirect ways. In the 1960's and 70's, sociobiology, altruism, kin selection and parent-offspring conflict were popular topics. Now in the 90's feminism and sexual discrimination are in vogue. It is therefore with amusement that I read about the rejection of the paternal genome in a number of species, the advantages of parthenogenesis, the cost of meiosis and the parasitic invasion of paternal B chromosomes. It may be my imagination but I thought I sensed some acrimony when Dr. Barlow (p.201—14) did not wish to be lectured on the Haig hypothesis on the conflict between maternal and paternal alleles in the embryo and she was quick to give evidence to discredit the hypothesis. Such confrontations fire the scientific enterprise. I hope that we will be able to continue our search for answers and I urge everyone not to let any authority define our knowledge.

Index

Abnormal anaphase resolution, (*aar*) 93
Abrocomidae 348, 351, 352, 359
5-AC see 5-Azacytidine
ACS 161—63
Allele-specific methylation 205—10
Allozymes 349—51, 353—6, **358**
Alu-PCR 58—60
Alu interspersed repeat 66
Amphiplasty 184
Anaphase 272
Anaphase I 271—3
Anaphase II 272
Aneugenic effects 123, 128
Arcyptera fusca 39
ARS, Autonomously replicating sequences 161—62, 164, **166**, 170, 172
ARS consensus sequence *see* ACS
5-Azacytidine 37, 186—9, 192—3
AZF (Azoospermia Factor) 51

B chromosomes 301, 374
  meiotic behaviour 251—67
  effects 302—5, 308—11
  evolution 301—2, 308—13
Bands
  C- 23, 28, 32—3, 39, 65, 67, **286—8**, 290—2, 336, 348, 350, 352, 360
  C-like 25, *26—7*, 28, 31, 33—4, 37
  Cryptic 33, 43
  G- 23, 31, 33—4, 65—79
  G-like 25, *26, 27*, 33—4
  general properties *67*
  N- 335
  negative 33
  R- 31, 33—4, 42—3, 65—79, 134
  R-like 25, *26*, 34
  T- 65—79
Base composition 67, 77—8
Bent DNA 164
Breast carcinomas 133, 135, 136—8

Cadherin 154
Cancer *see* gliomas; breast carcinomas
Capping 104, 108
Capromyidae 347, 349
Cell hybrids 56—62
  human—hamster 122—8
Cellular adhesion 152
Centromere 223—7, **232**—4, 270—2, 274, 277—9
Centromere Linking Proteins (CLiPs) 273
Centromeric heterochromatin 88
CENP 88
CGH 133—40
Chiasmata 283—95
Chorthippus parallelus *26*, 41
Chromatid adhesion, *see* pasc
Chromatin 147—51, 155—6
  fibre 32, 34—5
  organisation 190
  packaging 88, 90, 92, 95—7
Chromomeres 66, 78, 148, 150—1, 153, **155**, 156
Chromosomes
  1B (wheat) 186, 188
  1R (*Secale*) 186—9
  2L (*Drosophila*) 112—6, 118
  5BL (wheat) 321—7
  6B (wheat) 186, 188
  10 (human) 57—8, *59*
  A (*Drosophila*) 113—4, 116
  B (*Drosophila*) 114—4
  B *see* B chromosomes
  definition 4, 369—70

Chromosomes (*cont.*)
human 65—79
in human sperm *see* human sperm
mammalian 65—79, 347
theory of heredity 6
Y *see* Y chromosome
X *see* X chromosome
Chromosome alignment 217, 218, 222—4, 227—34
Chromosome behaviour
amphitelic orientation 249, **259**, **261**, *264*, 267
anaphase I laggards 249—50, 253, *256—7*, 260, 263—5
effects of heat 252, 262—5
orientation velocity 252, 253—60, 263, 266
pairing 217—34
polarization 224—5, 229—30, 233
reorientation 249, 252—7, 260, 263, 266
Chromosome evolution in plants 333—42
Chromosome *in situ* suppression (CISS) 327
Chromosome painting 133, 140, 319—20, 323, 325—8
Chromosome structure 147—8
breakage 95
condensation 87—90, 93, 94, 96—8, 147, **153**, **155**, 156
overcondensation 94
rearrangements 56—7, 139—40
region 133
structural aberrations 122—3, 128
Clastogenic effects 122—3, 128
CLM *see* Cytogenetic ladder maps
Comparative genomic hybridization *see* CGH
Cot 1 DNA 74, 133—4
CpG island 65, 68—77, 205—7, 209
Crossover 283—9, 291—5
Ctenomyidae 347, 349, 351, 353
Cytogenetic markers 285—6, **291**—2, 295
Cytogenetic ladder maps 292, 293
Cytogenetics 4—12
contents 11
molecular 333—4, 336, 342
Cytoskeleton 228—9

DAPI **226**, 229, 325, 32**6**, **340**
DAZ (deleted in azoospermia) 51
Deletion **62**
Dense fibrillar component 150
Digital image analysis 34, **35**
dmin 132—6
DNA methylation 182, 185—9, 193, 202, 303, 314
DNase I 25
hypersensitivity 40—1, 43
DOP-PCR 323—5, 327
Double minute chromosomes *see* dmin
Drosophila melanogaster 87—98, 104—18

Echimyidae 347, 349
EGFR gene 134, 136, 140
Electron microscopy 32, 34, 41
Elongation mechanism 104—5, **106**, 111—3, 117
ERBB2 gene 136, 138, 139
Exonuclease III **35**, 41—3
Eyprepocnemis plorans 252—67, 301—14

Fibrilar centres 150—1
FISH *see in situ* hybridization
Flavours 66, 77
Foci 150, 156

γ-Rays 122—8
Gene
amplification 132, 133, 136, 140 *see* hsr; dmin
density 66, *67*, **68**, 71—2, 74, 77
expression 182—8, 193—4
meiotic reprogramming 189
preferential inactivation 183—5, 187—8
Genetic maps 283—6, 289, **291**—5
Genome
interaction 182—7, 190—4
nuclear compartmentalization 190
Genome *in situ* hybridization *see* GISH
GISH 334—9, 372
Gliomas 133—6, **135**, 140
Granular component 150
GRD (Geminivirus-related DNA) probe 336—7, **340**

H19 gene *202*, 207, 209
Hamster DNA probes 125—7
Heat shock proteins 184
Heterochromatin 65, 67, 74, **75**, **76**, 78
condensation 92, 95
constitutive 29, 31, 33—4, 36—7, 39, 42—3
intercalary 78
organization 95
HeT-A element 105—18
structure 107
transposition mechanism 108
Heteracris littoralis 252, 255, *257*, 260, 266
Homoeologous chromosomes 322, 326—8
Homogeneously staining regions *see* hsr
Hordeum 183—4, 191, 193
Housekeeping genes 65, 70
HP-1 protein 96, 98
HSP *see* Heat shock proteins
hsr 132—40, **138**
Human DNA probes 125—7
Human sperm chromosomes 122—8

IGF2R/MPR300 locus 201—10
Imprinted genes in mammalians *202*
Imprinting 187—9
gametic 201—10
in situ hybridization 23—4, 28, **35**, 41, 56—7, 69, 371—2
genomic **188**, 190—2, *see* GISH
inter-CpG islands 71—4
in meiotic chromosomes 217, 220—1, 223, 225—8
in radiation-induced micronuclei 125, 127—8
repetitive DNA 303
telomeric DNA 114, **116**
tumour DNA 133, 136—8, 140
in wheat 319, 323—8
X chromosome 59, 60, **62**
Y linked loci 52, **53**
in situ nick-translation 24, 36, 39, 40
in situ restriction endonuclease digestion 24, *26*—*7*, 28—9, 31—2, 36—37
Inner Centromere proteins (INCENPs) 272
Insulin-like growth factor type 2 receptor *see* IGF2R
Inversion **62**

Karyotypic evolution 347—8, 350—3
Kinetochore 249, 265—7, **271**, 273

l(1) mutants 91, 92—5, 97
l(2) mutants 113—7
Lamins 146
Leptotene 218, 221, 224—7, 231—3
LINE element 105, **107**
Linkage maps *see* genetic maps
Locusta migratoria 252, 255, *257*, 260—6

mei-S332 gene 273—7
MEI-S332 protein 274—9
Metaphase 272
Metaphase I 271—3
Metaphase II 272, 278—9
Microdissection 320—2, 325
Micronucleus test 122—8
Minichromosome 150
Mitosis 147—53, **155**, 156—7
mit(1) mutant 89, **91**, 92—3
mu2 locus 108, 117
Mutagenicity 123
mutator-2 see mu2
mus-101 mutant 89, **91**, 92
Myocastoridae 347, 350—1

NE *see* nuclear envelop
Nicotiana 335—**40**
NOR 185—9, 192—3, 302—4, 314
Nuclear architecture 182, 190, 193
Nuclear envelop 224—9, 231—3
Nucleoskeleton 148, **149**, 156
Nucleosome 147, 149, 153, **155**, 156

Octodontidae 347—359
Octodontoidea 347—8, **354**
Omocestus burri 252, 255, *257*, 260, **262**, 266
Ontogenic map of rye rRNA 188—9
Oocytes 124, 127
Open chromatin 69, 70
ord gene 273—7
ORD protein 274—9

Pachytene 218, 221, 224, 226, 229, **232**
PAR 50
Parallel sister chromatids (pasc) **91**, 95—6, 98
PCR 319, 322—7
PDGFRA gene 136
PEV 89, 90, 95—7
Physical maps 65
Polymerases 148—51
Polyploid cells **91**, 93—5
Poliploidy 333—6
Position effect variegation, *see* PEV
PP1 mutants 95—6, 98
Prometaphase 249, 252—5, 259—67, 274
Protein phosphatase 1, *see* PP1
Pseudoautosomal region *see* PAR

Radiation-induced aberrations 122—8
Rana perezi 41
rDNA 162—75, 185—9, 192—4
Recombinant chromosomes 336, 339
Recombination 217—8, 224, 231, 233, 283—94, 373
  nodules 222
Replication
  timing 67
  fork barrier 169—70, **174**
  initiation 161—64, 167—8, **175**, 176
  intermediates 162, 165, 16**6**, 169, **171**—73
  origin recognition complex 162
  origins 161—70, 175
  termination 161, 168, 170, 173, 175
Retroposons *see* HeT-A; LINE; TART
RFLP-markers 283—5, 289, 292, 295
Ribosomal DNA *see* rDNA
RNA polymerase 156—7

SC see synaptonemal complex
Scaffold 87, 147—8, 372
SCPL 56—62
Secale 184, 191
Selfish DNA 301, 308, 311
Sequence Tagged Sites *see* STS
Sex-determining locus *see* SRY
Sister-chromatid asynchrony 202, 208—10
Sister-chromatid cohesion 270—4, 276, 278, 279

Sister-chromatids *see* parallel sister-chromatids
Solanum 338—**40**
Solenoid 147
Southern analysis 133, 140
Splicing apparatus 150
SRY 50
STS 57, 60
Su(var) mutants 96
Subchromosomal painting libraries *see* SCPL
Substitution lines **287**, 292
Subterminal regions 112, 116
Synapsis 217—34
Synaptonemal complex 217—8, 233

TART element 105—18
  structure 107
  transposition mechanism 108
Tenebrio molitor *26*, 29
Tenebrio obscurus *26*, 31
Telomeres 104—18, 370—1
  DNA components 108, 113, 116
  meiotic movements 217, 223, 234
  polymorphism 113
  silencing 114
  see elongation mechanism; capping; subterminal regions
Temperature sensitive lethals, *see* *mus-101*; *mit(1)*
Terminal deficiencies 108, 115, 117
Testis specific protein Y-encoded *see* TSPY
Tetrahymena 105, **106**
Tetraploidy 348, 357—61
T^{hp} mouse mutants 204
Tissue-specific genes, 68, 70
Transcription
  factories 147, 150, **155**, 156
  initiation 162—4
  termination 170—73
Translocation 57, 60, **62**
  random 335
  species-specific 335—6
Transposons *see* HeT-A; LINE; TART
Triticale 184—93
Triticum 184—5, 286—295, 319—28
Triturus marmoratus 39
TSPY 51—2, **53**, 54
Tumour DNA 133—4
Two-dimensional agarose gel electrophoresis 164, **166**

Univalent behaviour 249—66

Viral sequences in plants 334, 336—7

Wheat *see* *Triticum*

X Chromosome
  human 60, *61*
  Drosophila 108, **109—10**
  meiotic behaviour 251—66
  octodontoidea 347, 360—1
  see terminal deficiencies; subterminal regions
X-ray microanalysis 28

Y chromosome 50
Y linked genes 50—4
Y RNA Recognition Motif *see* YRRM
YRRM 51—2, **53**, 54

Zygotene 218—9, 221—2, 224—33